Water Science and Technology Library

Volume 124

The aim of the *Water Science and Technology Library* is to provide a forum for dissemination of the state-of-the-art of topics of current interest in the area of water science and technology. This is accomplished through publication of reference books and monographs, authored or edited. Occasionally also proceedings volumes are accepted for publication in the series. *Water Science and Technology Library* encompasses a wide range of topics dealing with science as well as socio-economic aspects of water, environment, and ecology. Both the water quantity and quality issues are relevant and are embraced by *Water Science and Technology Library*. The emphasis may be on either the scientific content, or techniques of solution, or both. There is increasing emphasis these days on processes and *Water Science and Technology Library* is committed to promoting this emphasis by publishing books emphasizing scientific discussions of physical, chemical, and/or biological aspects of water resources. Likewise, current or emerging solution techniques receive high priority. Interdisciplinary coverage is encouraged. Case studies contributing to our knowledge of water science and technology are also embraced by the series. Innovative ideas and novel techniques are of particular interest.

Comments or suggestions for future volumes are welcomed.

Vijay P. Singh, Department of Biological and Agricultural Engineering & Zachry Department of Civil and Environment Engineering, Texas A&M University, USA
Email: vsingh@tamu.edu

All contributions to an edited volume should undergo standard peer review to ensure high scientific quality, while monographs should also be reviewed by at least two experts in the field.

Manuscripts that have undergone successful review should then be prepared according to the Publisher's guidelines manuscripts: https://www.springer.com/gp/authors-editors/book-authors-editors/book-manuscript-guidelines

Svenja Fischer · Andreas H. Schumann

Type-Based Flood Statistics

An Interlink Between Stochastic and Deterministic Flood Hydrology

Springer

Svenja Fischer
SPATE Research Unit
Ruhr-Universität Bochum
Bochum, Nordrhein-Westfalen, Germany

Andreas H. Schumann
SPATE Research Unit
Ruhr-Universität Bochum
Bochum, Nordrhein-Westfalen, Germany

ISSN 0921-092X ISSN 1872-4663 (electronic)
Water Science and Technology Library
ISBN 978-3-031-32710-0 ISBN 978-3-031-32711-7 (eBook)
https://doi.org/10.1007/978-3-031-32711-7

This Springer imprint is published by the registered company Springer Nature Switzerland AG
The registered company address is: Gewerbestrasse 11, 6330 Cham, Switzerland

The authors dedicate this book to their families, without their support and patience such a project would not have been possible.

Foreword

On the one hand, both traditional and current Flood Frequency Analysis (FFA) is largely a statistical endeavour combining advances in the probabilistic theory of extremes with modern statistical methods such as copulas, L-moments, maximum likelihood estimation, etc. On the other hand, hydrologists have always understood that the mechanisms which give rise to floods have a rich physical basis, so limiting one's analysis to a purely statistical approach while ignoring the physical basis of floods can only provide a partial solution. This book is an effort to change that paradigm, by combining both the physical and well-established statistical basis of FFA into a unified approach. Imagine trying to determine the risk of a spaceship failure without accounting for the myriad of interacting failure modes. That is analogous to determining the probability of a flood without considering the multiple types of floods which could occur. NASA felt that such formal mathematical risk analysis gave unrealistically pessimistic results, thus the original Shuttles were designed without risk analysis, which led to poor designs and consequent Shuttle tragedies. Jones (2018) attributes those Shuttle tragedies to the Apollo era decisions to abandon mathematical risk analysis. This book provides an extension to the traditional probabilistic assessment of floods which treats the series of annual maximum river discharges as all arising from the same probability distribution, with equal likelihood of occurrence in any given year, as if all causes of floods are equally likely and each year is a separate unrelated experiment. Such flood series are said to be independent and identically distributed (i.i.d.) which leads the hydrologist to the wealth of literature and purely statistical approaches to the frequency analysis of i.i.d. processes. In fact, an entire mathematical field known as extreme value theory lends itself to such analyses and provides the hydrologist with numerous distributions and other methods with which to model such processes. Such traditional approaches to FFA are succinctly summarized in Chaps. 2 and 3 for both univariate and multivariate settings; yet the thrust of this book is quite different. Analogous to the multiple failure modes associated with a spaceship, river floods at a single location can arise from a number of different processes. Convective storms can produce very intense flash floods during the summer months, and the same watershed can experience snowmelt-driven floods during the spring, a frontal storm during nearly any season and possibly floods caused by ice jams during the winter months. There are usually several physical processes that give rise to the annual maximum flood, and resulting flood frequency curves often exhibit sudden breaks in curvature so that the i.i.d. assumption is no longer reasonable, nor is it a strictly statistical approach. This book provides a comprehensive treatment of alternative approaches to FFA which accommodate BOTH the physical and statistical mechanisms which give rise to floods. Such approaches, by necessity, are more time-consuming and detailed than a purely statistical treatment, requiring the collection of additional sources of data and assembling a greater physical basis and set of models describing the mechanisms, which together give rise to a more realistic representation of the spatial and temporal dimensions of flood series and ultimately to a better informed and more realistic FFA. This book introduces the reader, step by step, to a generalized type-based approach to FFA, which is an amalgamation of nearly all of our current regional statistical approaches to FFA with careful attention to the temporal and spatial dimensions of the varied physical mechanisms which together form dependent and non-identically distributed partial duration series. The overall methodologies outlined in this book represent the future of FFA, and integrating both flood types with proven statistical methods offers guidance and methods which can only lead to improved estimates of the frequency and magnitude of future floods.

Boston, USA
March 2023

Richard M. Vogel
Professor Emeritus Tufts University

Reference

Jones HW (2018) NASA's understanding of risk in apollo and shuttle, AIAA 2018–5235, Published Online:15 Sept 2018 https://doi.org/10.2514/6.2018-5235.

Acknowledgements

The authors would like to thank the German Research Foundation (Deutsche Forschungsgemeinschaft, DFG) for funding the research group SPATE (Space-Time Dynamics of Extreme Floods, funding code FOR 2416), within the framework of which this research was carried out. Great thanks also go to the State Offices of Saxony (LfULG) and Bavaria (LfU) as well as the Global Runoff Data Center of the BFG and the Copernicus Climate Change Service for providing the runoff, precipitation and temperature data used in this book. The authors would like to thank Philipp Bühler for his support in the creation of the corresponding R package.

Contents

About the Authors

Svenja Fischer studied mathematics at the Ruhr-University in Bochum, Germany, and holds a Ph.D. in statistics from the Technical University of Dortmund. Since 2021, she is an assistant professor for stochastic hydrology at the Ruhr-University Bochum. Her scientific activities focus on flood research with special emphasis on flood frequency analyses. She has published more than 30 papers, was vice-president of the IAHS Commission for statistical hydrology (ICSH) from 2019 to 2023, and received the Tison award from IAHS in 2021 for her research paper on flood types.

Andreas H. Schumann studied Hydrology and water management at the Technical University in Dresden, Germany. After his Ph.D. in 1981, he worked several years in different water management agencies. In 1989, he went back to academia and became a Senior-Lecturer at the Ruhr-University in Bochum, Germany. In the year 2001, he became a Full Professor for Hydrology and Water Management at this university. Many of his scientific activities were dedicated to flood research. He has published more than 200 papers and is the editor of the book "Flood Risk Assessment and Management", published by Springer in 2011. After retiring in 2019, he is researching as a Senior Professor at the Ruhr-University. He initiated the Research Unit "Space-time Dynamics of Extreme Floods" (SPATE) which was funded by the German Research Foundation from 2017 to 2023.

Acronyms

AIC	Akaike Information Criterion: Criterion to evaluate goodness of fit of distribution
AMS	Annual maximum series: Sample of the maximum discharge in a water year
BIC	Bayesian Information Criterion: Criterion to evaluate goodness of fit of distribution
CART	Classification And Regression Tree: a classification approach based on a tree-like structure
CLD	Clustering of Longitudinal Data
Delete-d jackknife	Jackknife method to evaluate the performance of regionalisation where *d* catchment are treated as ungauged at the same time
DFH	Design Flood Hydrograph
FFA	Flood Frequency Analyses
GEV	Generalized Extreme Value Distribution
GIS	Geographical Information System
GPD	Generalized Pareto Distribution
i.i.d.	Independent and identically distributed
IQR	Interquartile Range: a measure for variance
L-Moment	Linear-moment: a distribution parameter estimation method
MCDM	Multi-criterial decision-making
ML	Maximum-likelihood: a distribution parameter estimation method
MoM	Method of Moments: a distribution parameter estimation method
MQ	Mean long-term discharge
NRMSE	Normalized Root Mean Squared Error: the difference between observed and simulated value is normalized by the observed value to allow for scale-independent comparison
pdf	Probability distribution function

POT	Peak-over-threshold: Approach that considers all (peak) values above a given threshold
$Prec_D$	D-Day precipitation sum: the sum of the precipitation of D-days, where D is chosen flood type-specifically
PWM	Probability weighted moments
QR	Quantile range: parameter space of rise and recession time to derive DFHs
RMSE	Root Mean Squared Error: a comparison measure for the goodness of a model
rv	Random variable
TCEV	Two-component extreme value distribution
TMPS	Type-based Mixture Model of Partial Duration Series
TNRMSE	Trimmed Normalized Root Mean Squared Error: Trimmed version of the NRMSE, where the largest and smallest $\alpha\%$ of the relative differences are removed before calculating the NRMSE. Usually, $\alpha = 5$
TOPSIS	Technique for Order Preference by Similarity to Ideal Solution: an approach for MCDM
TSS	Total Sum of Squares: the total sums of squared distances of each attribute from the clusters centres, a measure to define the optimal number of clusters
USGS	United States Geological Survey

Part I
Introduction and Motivation

Chapter 1
Why Flood-Type-Based Statistics

Abstract Flood type-based statistics offer the possibility to include knowledge on the deterministic geneses of flood events in flood frequency analyses. This introductory chapter briefly summarises, why traditional flood statistics might include inhomogeneous samples and how flood types could overcome this issue. It provides an overview on how this whole research area on type-based statistics has developed and which aspects have to be taken into account when applying flood types and type-based statistics.

In his publication *The Return Period of Flood Flows*, Emil J. Gumbel (Gumbel 1941) stated as one basic assumption for flood statistics

> We suppose that the observations are homogeneous, i.e., subject to a common set of forces.

According to this basic assumption, the theory of hydrological frequency analysis rests on the postulates (1) that the hydrological entity X being analysed is an independent identically distributed random variable (i.i.d. rv) with a distribution F(X); and (2) that its observation record is a random sample from this distribution (Klemes 2000). However, one must acknowledge that flood events may be dominated by different processes, that different mechanisms may prevail at different scales and that the whole i.i.d. concept is hydrologically impeachable. This is not a new finding. In the year 1976, the Bulletin 17 *Guidelines for determining flood flow frequency* was published by the Hydrologic Committee of the U. S. Water Resources Council. At page 7, the problem of different flood types was discussed:

> At some locations, flooding is created by different types of events. For example, flooding in some watersheds is created by snowmelt, rainstorms, or by combinations of both snowmelt and rainstorms. Such a record may not be homogeneous and may require special treatment.

Despite an almost universal recognition that some observed flood samples may not be drawn from a single, homogeneous population, the practical application of this insight took place only slowly. Potter (1958) analysed a sample of 69 gauged watersheds with periods of record ranging from 26 to 48 years and specified two straight lines designated as upper and lower frequency curves as the result of sampling from

S. Fischer and A. H. Schumann, *Type-Based Flood Statistics*, Water Science and Technology Library 124, https://doi.org/10.1007/978-3-031-32711-7_1

two different populations of runoff peak rates. With the introduction of mixture distributions, the combination of different populations of flood events became possible (Singh 1974; Rossi et al. 1984). In these models, the streamflow data alone were analysed to objectively define a mixture of distributions rather than decompose the data on basis of additional climatic information. As a result Klemes (1971)

> thus what we usually find is not, in fact, statistical or stochastic hydrology but merely an illustration of statistical and probabilistic concepts by means of hydrological data. In trying to improve this situation, the main problem is to find the ways in which the physical features of a phenomenon can be introduced into the analysis.

Several researchers started to consider the differences in flood generation processes. Waylen and Woo (1982) looked, e.g., at the differences between rainfall-driven and snowmelt-induced floods to examine the benefits of mixed distributions in hydrologic data. Snowmelt-induced and rainfall-driven annual maxima series were estimated, two Gumbel distributions fitted and combined. A more holistic hydroclimatic approach was followed by Hirschboek (1988). Special emphasis was given in this work to the atmospheric circulation pattern and their impact on the flood generating processes.

Since then, several options to subdivide a sample of observed floods into more or less homogenous groups by causative criteria were applied (Tarasova et al. 2019):

- according to the hydro-climatic processes which result in floods (large-scale circulation patterns and atmospheric state at the time of the event) (Hirschboek 1988; Ashley and Ashley 2008; Grams et al. 2014),
- by hydrological characteristics, defining the reaction of watersheds (spatial and temporal precipitation patterns at the catchment scale, antecedent catchment state characterized by the soil moisture status and snow cover) (Todorovic and Rousselle 1971), Nied et al. (2014), Kampf and Lefsky (2016),
- by hydrograph-based analyses which group flood events by similarities of peak-volume relationships and the shape of hydrographs (Elliott et al. 1982; Fischer et al. 2016).

These different approaches have different requirements for the data basis, lead to different numbers of subsamples, and thus to different sample sizes in these quasi-homogeneous groups. To find an appropriate classification method for the task at hand—in this case flood statistics—there are some practical aspects that should be considered when selecting the criteria for the classification of flood events:

- The number of classes should fit to the number of available flood events. In many parts of Europe, the average number of floods in any year is limited to 2–3 events. Of course, often the runoff shows a higher variability and a higher number of rises and falls. However, a flood event is usually associated with a discharge that results in inundation of land or, at a minimum, a significant exceedance of the mean annual discharge at a river cross section. This results in smaller number of floods that can be separated into single events. The number of such events should be sufficient to specify the probability distribution function (pdf) of the respective class of floods, but the number of classes should also be sufficient to specify the differences between flood types.

- The classes should be defined according to practically relevant characteristics. One main limitation of classical flood statistics is the focus on the flood peak. This is an important characteristic for the assessment of the inundated areas and the comparison of the maximal discharge to the hydraulic capacity, but it is insufficient for all aspects of flood storage and retention which depend on the flood volume and the shape of the hydrograph. These three characteristics (peak, volume, shape of hydrograph) are interdependent in a flood-type-specific way and can be used to differentiate between these types. In Fig. 1.1, an example is given to compare two floods measured at the same gauge with similar peaks but different other characteristics, e.g., by duration, volumes, runoff coefficients, and finally hydrographs.
- The classification should be based on data which are generally available. The validity of flood statistics depends on the length of the available time series. Over the decades, an evolution of meteorological observation systems took place. Some of these new data categories (e.g., weather radar) that could be used to analyse flood types became available for two or three decades only. Differentiation of floods according to their causal atmospheric circulation pattern is scientifically interesting, but is less suitable when one has to differentiate flood series with a length of 100 years into types. It would be a loss of information if analyses are focused on such data only as their limited availability would result in reduced and short flood samples. Instead, in many analyses, effective mechanisms can be applied to assess the hydro-meteorological characteristics of flood events from the past. Examples are long series of daily rainfall and temperature, which could be used to assess the snow cover and snowmelt conditions as well as the initial soil

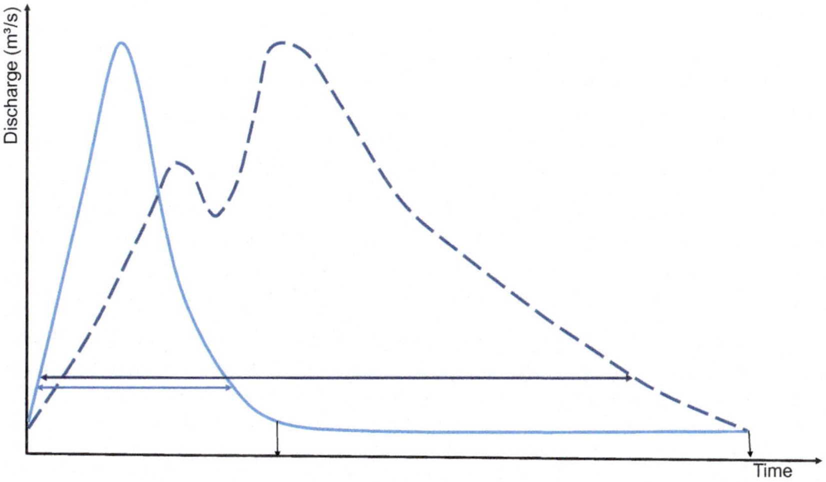

Fig. 1.1 Two flood events with a similar flood peak which differ in all other characteristics

moisture by application of hydrological modelling, or analyses of rain series in the region of interest, which characterise the meteorological conditions that resulted in floods.

These different aspects should be considered before floods are classified. The selection of criteria determines the benefits one can expect from the subdivision of flood series into homogeneous groups. All three aspects were considered in the structure of this book. It is subdivided into six different parts: After the discussion of the motivation and the basics of univariate and multivariate flood statistics (Parts I and II), the flood-type-specific statistic is reviewed with emphasis on the estimation of flood types and the application of statistical mixture models. Additionally, the newly developed Type-based Mixture Model of Partial Duration Series (TMPS) is introduced in Part III. In Part IV, the main aspects of the application of type-based flood statistics are presented. It starts with general data processing to obtain classified flood events and the estimation of flood probabilities under consideration of the different flood types. There, the essential steps for the parametrisation of the model and the determination of the flood probabilities are described. Compared to traditional statistical methods, the implementation of this new model is more complicated but also offers more benefits. Two major advantages of the model are highlighted in Part IV: the ability to account for transient flood conditions as reflected in changes in the frequency of individual flood types over time, and the assignment of type-specific flood hydrographs to design floods based on type-specifically estimated peaks and volumes. All these considerations concentrate at one gauge at a time. Yet, floods are a spatially distributed phenomenon. This is discussed in Part V, where the spatial dependencies between synchronous floods in large river basins are discussed and multivariate statistical assessments of most probable flood scenarios are developed for large rivers and their tributaries. The results of these multivariate assessments give a more holistic view on design flood estimates in large river basins under consideration of tributaries. To specify the variability of flood events in river basins, a stochastic generator is presented that can be used to estimate a wide variety of consistent flood scenarios. As the frequencies of events which belong to a certain flood type depend strongly on the landscape and more generally on the location of a catchment, it can be expected that the regionalisation of flood statistics benefits from the differentiation of flood series into flood types. In Part VI, this evidence is kept. Based on the results of the previous sections, not only are the statistics of flood peaks regionalised, but also the associated type-specific volumes and hydrographs. World-wide, several attempts exists to derive floods from known precipitation series (Alfieri et al. 2014, e.g.). More or less sophisticated hydrological models are applied to transfer rain data into discharges. The methodology for flood type classification presented in this book is based on the relationships between flood peaks and volumes. Thus, it is sufficient to determine the expected flood volume from series of precipitation events by selecting a realistic rainfall-runoff coefficient derived from event analyses to statistically determine the type, peak and hydrograph of the resulting flood. The developed model is described at the end of Part VI, as it is transferable from gauged to ungauged catchments.

In the whole book, the discussion of the practical aspects of the application is given wide scope to illustrate the advantages of the type-based approach, which compensates for the higher computational effort. According to the subtitle of the book, the consideration of deterministic aspects in flood statistics is taken into account both with respect to the complexity of flood generation and the necessary type-specificity of flood statistical statements. Three different flood generators are described in this book: One designed to generate a large number of type-specific peak-volume-hydrograph combinations, one to derive plausible and spatially consistent flood scenarios in large river basins, and a third is based on a more deterministic approach to estimate flood volume from rainfall and runoff coefficient statistics before assigning type-specific peaks and hydrographs. These generators could replace the unrealistic practice to specify one single design flood instead of a variety events which are similar in their peaks but differ in volumes and shapes of hydrographs. Consideration of the diversity of flood events is essential as long as floods are characterized by their peaks and associated statistical features (e.g., return periods).

The flood types considered in this book are derived from a hybrid causative-hydrograph-based flood typology, resulting in five different flood types: three types of rainfall-driven floods, associated with different intensity and duration of rainfall, rain-on-snow floods and snowmelt-induced floods. When these flood types are considered in flood statistics, the information content is increased and design flood estimation can be extended by a type-specific scenario estimation.

For example, consider a cross-section of a river, where a flood retention pond is planned. It has to be designed such that the flood peak of a flood event with a return period of 100 years is significantly reduced. The question is how to evaluate the performance of this reservoir, which has a fixed storage capacity. To answer this question, usually the annual maximum series (AMS) of discharges is analysed statistically, a pdf is fitted and one would obtain the flood peak with a return period of 100 years. In classical flood analyses, this is all one can get from statistics. There is no further information on the volume of this flood or the shape of the hydrograph. If one considers that different flood types are represented in the AMS and therefore uses a mixture model that combines separate type-based pdfs, additional information is gained. The mixture model delivers not only the joint probability of the mixed flood types (here the quantile with the non-exceedance probability of 0.99) but gives the opportunity to compare the non-exceedance probabilities of this quantile separately for the flood types. With regard to the flood retention pond, the rain-driven floods can be separated into three types:

1. flood events with a short duration, steep hydrograph shapes, high peaks and small volumes which result from single heavy rain events,
2. flood events which result from synoptic rain: these events have a longer duration and a higher volume than flood events of the first type,
3. very-long-duration rain-driven floods which consist of a sequence of non-separable single flood waves.

With regard to the limited storage capacity, one can expect the highest performance of the reservoir in the first case. This is where the reservoir will be able to most significantly reduce the peaks of such flash floods. The inflowing flood waves have a smaller volume and could be retained completely in the best case. For the second type of floods, this would be more difficult because the volume of flood water is higher and overflow of the dam is possible before the peak of the inflow is reached. The third option may be less problematic because the intermediate release from the reservoir could ensure a free storage capacity at the time of maximum inflow. Based on the type-specific statistic, one can now compare the probabilities that the flood peak with a return period of 100 years results from one of the three flood types mentioned above, e.g.: Type 1 is 1.5 times more probable than Type 2 and 4.8 times more probable than Type 3. Type 2 is three times more probable than Type 3 (these values result from the example, which is discussed in Chap. 7.2 *Estimation of Design Flood Peaks*).

For more detailed analyses of the flood retention, typical hydrographs for the three flood types can be provided (Chap. 9). Using type-specific statistics, it now becomes possible to evaluate the performance of the planned retention pond in a differentiated and more realistic way. It enables to estimate the probabilities that a more or less significant reduction of the flood peak in the river can be achieved.

By using flood types, one also has the chance to better understand statistical results by attributing them with deterministic knowledge. A straight-forward example are copulas. While copulas are a quite simple statistical tool when aiming to consider more than one random variable at a time, they are used to a limited extent in hydrology. One reason is that for a given return period, copulas do not deliver one combination of, e.g., flood peak and volume but instead they deliver several combinations of peaks and volumes, all having the same joint probability. This is a limitation for practical purposes, where more specific estimations are required. Flood types help to deterministically define such specific estimations. In accordance to event analyses that demonstrate how the relationships between peak and volume differ between flood types, the probabilities of the combinations vary according to the frequency of flood types. Not only do flood types represent in a natural way the different skew levels of the copula, they also distinguish the many pairs of peak and volume with the same return period into different ranges associated with different flood-generating processes. In the example in Fig. 1.2, such a copula of the pseudo-observations (i.e., transformed by the empirical probability to [0, 1]) of pairs of peak and volume is given. Clearly, the different layers defined by the flood types are visible. The two iso-lines represent all the combinations of peak and volume that have a return period of either 20 or 100 years. However, when considering heavy-rainfall floods, only those pairs are relevant where the peak is large and the volume is small. Contrarily, when considering floods caused by long-duration rainfall of several successive rainfall events, the peak can be expected to be small while the flood volume is large. The relevant critical ranges thus can be reduced according to the flood-generating processes.

These are only two example of how flood types can improve the knowledge on floods, extend flood statistics by deterministic knowledge and support decisions in flood protection. Many more like the construction of type-specific hydrographs, the estimation of flood scenarios, the consideration of flood types in spatial statistics and

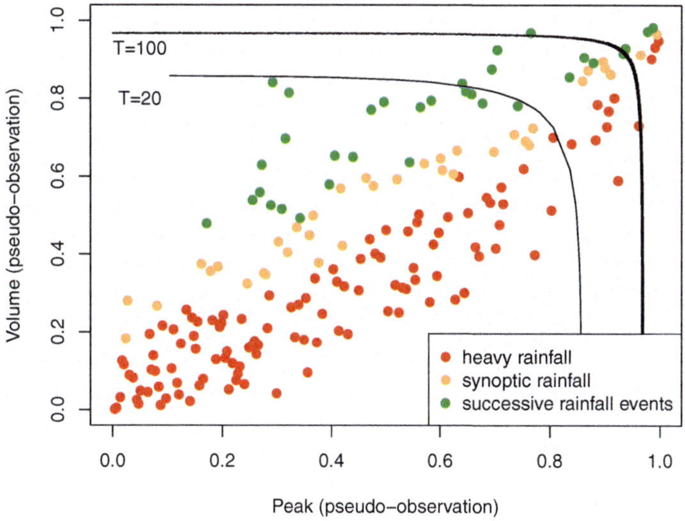

Fig. 1.2 Pseudo-observation of flood peak and corresponding volume for different flood types at one gauge and iso-lines of the joint AND-return periods for $T = 20$ and $T = 100$ years

the improvement of regionalisation by flood types are given in this book. The statistical models as well as the deterministic reasoning are explained and supported by real-life examples. Exemplary programme code for the main steps of the framework are given for an application in R.

This book therefore offers a comprehensive framework on how to use flood types in flood statistics, their practical use and the information they contain.

References

Alfieri L, Salamon P, Bianchi A, Neal J, Bates P, Feyen L (2014) Advances in pan-European flood hazard mapping. Hydrol Process 28:4067–4077

Ashley ST, Ashley WS (2008) The storm morphology of deadly flooding events in the United States. Int J Climatol 28:493–503

Elliott JG, Jarrett RD, Ebling JL (1982) Annual snowmelt and rainfall peak-flow data on selected foothills region streams, South Platte River, Arkansas River, and Colorado River Basins, Colorado. U.S. Geological Survey. Retrieved from https://pubs.er.usgs.gov/publication/ ofr82426

Fischer S, Schumann A, Schulte M (2016) Characterisation of seasonal flood types according to timescales in mixed probability distributions. J Hydrol 539:38–56

Grams CM, Binder H, Pfahl S, Piaget N, Wernli H (2014) Atmospheric processes triggering the central European floods in June 2013. Nat Hazard 14:1691–1702

Gumbel EJ (1941) The return period of flood flows. Ann Math Stat 12(2):163–190

Hirschboek KK (1988) Flood hydroclimatology. In: Baker VR, Kochel RC, Patton PC (eds.): Flood geomorphology: Wiley-Interscience, pp 27–49

Kampf SK, Lefsky MA (2016) Transition of dominant peak flow source from snowmelt to rainfall along the Colorado front range: historical patterns, trends, and lessons from the 2013 colorado front range floods. Water Resour Res 52(1):407–422
Klemes V (1971) Some problems in pure and applied stochastic hydrology. In: Proceedings symposium on statistical hydrology, U.S. Dept. Agri., Tucson, Arizona. Misc. Publ. No. 1275
Klemes V (2000) Tall Tales about Tails of Hydrological Distributions. I. J Hydrol Eng 5(3):227–231
Nied M, Pardowitz T, Nissen K, Ulbrich U, Hundecha Y, Merz B (2014) On the relationship between hydro-meteorological patterns and flood types. J Hydrol 519:3249–3262
Potter WD (1958) Upper and lower frequency curves for peak rates of runoff. Trans AGU 39(1):100
Rossi F, Fiorentino M, Versace P (1984) Two-component extreme value distribution for flood frequency analysis. Water Resour Res 20(7):847–856
Singh KP (1974) A two-distribution method for fitting mixed distributions in hydrology. In: Proceedings of a symposium on statistical hydrology, U.S. Dept. of. Agriculture, Agric. Res. Service. Misc. Publ. No., vol 1275, pp 371–382
Tarasova L, Merz R, Kiss A et al (2019) Causative classification of river flood events. WIREs Water 6:e1353
Todorovic P, Rousselle J (1971) Some problems of flood analysis. Water Resour Res 7(5):1144–1150
Waylen P, Woo MK (1982) Prediction of annual floods generated by mixed processes. Water Resour Res 18:1283–1286

Part II
Basics of Univariate and Multivariate Statistical Flood Models

Flood frequency analysis, i.e., the statistical analysis of flood samples and the estimation of flood probabilities or design flood peaks, is one of the key tasks of hydrology. Classical approaches are based on the fitting of one (or several) distributions functions to a series of flood samples. However, recently also more complex approaches are applied, e.g., based on different types of samples or more complex distributions. Moreover, now often more than one flood characteristic is considered in flood frequency analyses, which requires multivariate statistical approaches. In this part, the basics of flood statistics are introduced, both for the univariate and the multivariate case. These basics will form the benchmark for any newly developed statistical models.

Chapter 2
Principles of Univariate Statistics

Abstract Univariate statistics, i.e., the statistical analyses of one time series at one given location, forms the basis for any flood frequency analyses. In this part, the basic principles of classical univariate flood frequency analyses are presented. The concepts of annual maximum series and peak-over-threshold series are explained, as well as the most common statistical distributions that are essential for flood statistics in each case. To fit these distribution functions to the data, the three most commonly used parameter estimation methods, Maximum-Likelihood, Method of Moments and L-Moments, are described. Finally, methods to select the most appropriate statistical model are presented, ranging from graphical to statistical approaches.

Flood frequency analysis (FFA) is used to obtain probabilities for flood peaks within a series of observations at a gauge site. The application of extreme value statistics for this purpose dates back to works of Gumbel (1941). In this context, the concept of annualities plays a crucial role. Annualities T are expressed in years and describe the return period of floods. The return period is defined as statistical average time period between two exceedances of a flood peak of a given value. The return period is useful if an annual series of maximum values is used. In this case, it is the reciprocal of the exceedance probability. It is a statistical feature to simplify the handling of large non-exceedance probabilities $\mathbb{P}$ and has nothing to do with the real time intervals between exceedances. For example, one can make an assessment for a flood peak with an exceedance probability of 10^{-4}, but the corresponding return period of 10,000 years would not be relevant because flood conditions after (or before) 10,000 years are different under global climate change conditions. Since the concept of return periods is often misleading, it would be better to use the concept of probabilities instead, but the use of return periods has long been established in practice and is important for hydraulic engineering. Return periods are a fundamental part of many guidelines such as the Bulletin 17c of the United States for flood statistics or the DIN standards for reservoir design in Germany. Flood protection systems, dams and other water engineering structures are constructed to be secure for or withstand flood events of certain annualities. For example, dams in Germany are designed to resist flood peaks with an annuality of $T = 1000$ or $T = 10{,}000$ years. Of course, the probability of rarely exceed a length of 100 years of observation. Instead, an

S. Fischer and A. H. Schumann, *Type-Based Flood Statistics*, Water Science and Technology Library 124, https://doi.org/10.1007/978-3-031-32711-7_2

extrapolation of a probability distribution function (pdf) has to be done to obtain results with return periods that are beyond the observed record length. Standard FFA is usually based on either the fitting of a statistical distribution function to either the annual maximum series (AMS) or the peak-over-threshold series (POT). The AMS suits perfectly the concept of return periods as quasi-equidistant values (annual flood peaks) are analysed. However, it limits the information content of the observed time series as only the highest value of a flood is considered in every year. Partial duration or POT series extend this information by considering several flood events per year, their number depending on the discharge threshold which defines a flood. Various approaches and models exist for both approaches. In this section, the basic concepts are introduced that often serve as benchmark when evaluating the performance of new statistical methods.

2.1 The Annual Maximum Series

The annual maximum series consists of the block maxima of observed discharge values at a gauge site. According to this definition it contains one value per year of observation. *Year* in this case often refers to the hydrological year, which consists also of a 12 month period but may differ from the calendar year. It begins usually in a period of low flows to ensure a balanced relationship between annual precipitation and runoff. The 12-month-period of the hydrological year is chosen such that the transfer of water in the snowpack from one year to another is avoided. In the United States, the water years begins at October 1st and ends at September 30th, while in some countries in Europe it begins at November 1st and ends at October 31st. The AMS then takes the maximum discharge of each water year, such that the time series consists of the flood peak of one event per year. This approach is referred to as block maximum, since for pre-defined blocks, in this case one year, the maximum value is considered. For this approach, the single flood peaks usually are assumed to be independent because of the large time span between them and the selection of a low-flow period as the end or beginning of the hydrological year. However, it can occur that flood peaks close to the beginning or end of a hydrological year are selected for the AMS. In this case, often independence criteria are applied to ensure that the flood peaks of two consecutive years do not belong to the same flood event. Such independence criteria frequently make use of a pre-defined span between two consecutive flood peaks to classify both flood peaks as independent. For example, Rahimi et al. (2021) define a time period of five days for catchments with an area smaller than 45,000 km^2, ten days for catchments with an area between 45,000 and 100,000 km^2 and 20 days for larger catchments. The annuality has direct relationship to the probability of the AMS. It can be written as a function of the non-exceedance probability $\mathbb{P}$:

$$T = \frac{1}{1 - \mathbb{P}}. \tag{2.1}$$

This means that a design flood for a given annuality can be easily derived from the quantile of the respective probability $\mathbb{P}$. Similarly, as the exceedance probability $\mathbb{P}'$ is related to the non-exceedance probability by $\mathbb{P}' = 1 - \mathbb{P}$, the return period can be derived by

$$T = \frac{1}{\mathbb{P}'}. \tag{2.2}$$

To obtain these quantiles for standard flood frequency analyses, a statistical distribution function is fit to the AMS. There exist various distribution functions and parameter estimation methods for fitting.

Since the AMS consists of block maxima, a suitable distribution arises naturally from statistical extreme value theory, the Fisher-Tippett-Gnedenko Theorem (Theorem 2.1). The theorem states the following:

Theorem 2.1 (Fisher-Tippett-Gnedenko 1928) *Let $X_1, \ldots, X_n$ be independent and identically distributed (iid) random variables. Suppose there exists a sequence of constants $a_n > 0$ and $b_n \in \mathbb{R}$ such that*

$$\frac{\max(X_1, \ldots, X_n) - b_n}{a_n}$$

has a nondegenerate limit distribution for $n \to \infty$. Then this limit distribution is the extreme value distribution $F_\xi(ax + b)$ with $a > 0$ and $b \in \mathbb{R}$ given by

$$F_\xi(x) = \exp\left(-(1 + \xi x)^{-\frac{1}{\xi}}\right),$$

for $1 + \xi x > 0$ with $\xi \in \mathbb{R}$. If $\xi = 0$, the right-hand side can be interpreted as $\exp(-\exp^{-x})$.

The limit distribution that is obtained for block maxima therefore is the Generalized Extreme Value (GEV) Distribution. The distribution function of the GEV is given by

$$F(x) = \exp\left(-\left(1 + \xi\frac{x - \mu}{\sigma}\right)^{-\frac{1}{\xi}}\right) \tag{2.3}$$

for $1 + \xi(x - \mu)/\sigma > 0$, where $\xi \in \mathbb{R}$ is the shape parameter, $\sigma > 0$ is the scale parameter and $\mu \in \mathbb{R}$ is the location parameter. The special case $\xi = 0$ is called the Gumbel distribution and is given by

$$F(x) = \exp\left(-\exp\left(-\frac{x - \mu}{\sigma}\right)\right). \tag{2.4}$$

The parameter ξ is not only the shape parameter but also called extreme value index and defines the distribution type as well as the tail-behaviour (Fig. 2.1). The

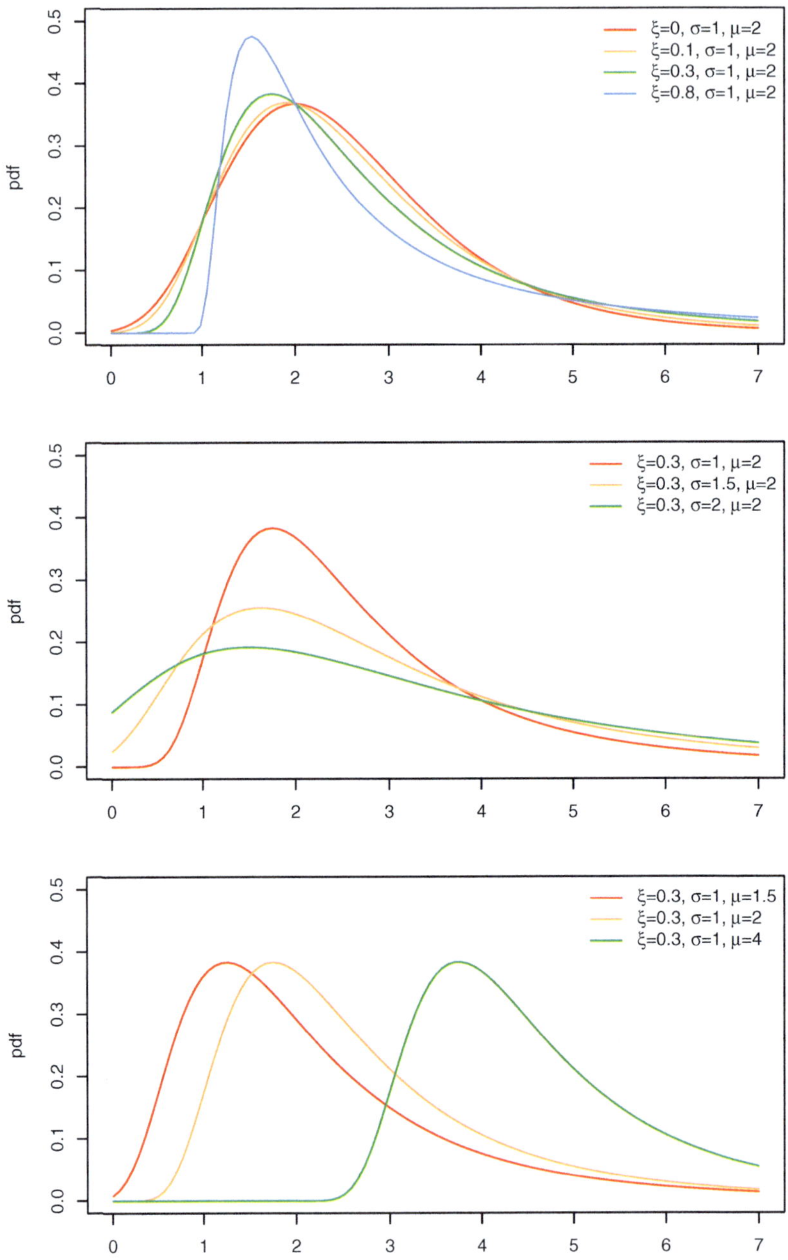

Fig. 2.1 Probability density function (pdf) of the GEV distribution for different parameter combinations

regular case in flood statistics is a positive ξ, indicating an unbounded right part of the distribution and a heavy right tail.

For flood statistic in most European countries, the GEV is the most recommended distribution function and suitable for many discharge series (N.E.R.C. 1975; Hosking et al. 1985; Salinas et al. 2014).

Another distribution function frequently used to describe annual extremes is the PearsonIII distribution. In the US, this distribution is most frequently used to describe the annual minima. However, it is also used (in its logarithmic form) to describe the annual maxima and was recommended in Bulletin 17b by the Water Resources Council of the United States. Similar to the GEV, the PearsonIII distribution has three parameters.

The distribution function of the PearsonIII distribution, which is a shifted Gamma-distribution, is given by

$$F(x) = \begin{cases} \frac{1}{\Gamma(\xi)} \int_0^{\frac{x-\mu}{\sigma}} t^{\xi-1} e^{-t} dt, & \text{for } \sigma > 0 \\ 1 - \frac{1}{\Gamma(\xi)} \int_0^{\frac{x-\mu}{\sigma}} t^{\xi-1} e^{-t} dt, & \text{for } \sigma < 0 \end{cases} \tag{2.5}$$

with parameters $\mu \in \mathbb{R}, \sigma \neq 0, \xi > 0$. When the logarithmic PearsonIII distribution is considered, the data are transformed logarithmically first.

Both distributions, GEV and (logarithmic) PearsonIII, are capable of modelling different tail behaviour and skewness due to their three parameters. This makes these distributions more flexible than common two parametric distributions like the Gumbel distribution. However, the third parameter also introduces additional uncertainty and hence the use of these distributions is always a trade-off between flexibility and uncertainty, especially when having short record lengths. It is recommended to have a data basis of at least $n = 30$, i.e., 30 years of observation in case of the AMS, for applying these approaches.

2.2 The Peak-over-Threshold Approach

While the AMS is widely used due to its simple relation between probability and annuality, it has also some major drawbacks. The greatest drawback certainly is the limitation to only one event per year. Due to this fixed assumption, in years where no flood occurs, small discharge values are included in the AMS. This may lead to an underestimation of the general flood regime. At the same time, in years where more than one major flood event occurs, the restriction to only include one (the largest) event might ignore several other important large flood peaks and thus might lead to an underestimation of the frequency of large floods. To overcome this drawback, the more flexible Peak-over-threshold (POT) approach, also denoted as partial duration series, was developed.

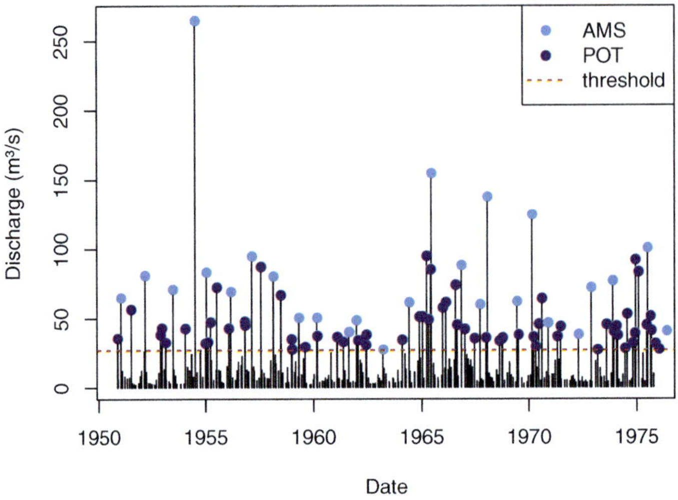

Fig. 2.2 AMS and POT series based on monthly maximum discharges

When referring to POT, samples are described for which all exceedances of a given threshold are considered (Fig. 2.2). The definition of the threshold is not fixed and depends on the application of the approach. Indeed, the choice of the threshold is usually the greatest challenge when applying POT since it has a crucial impact on the results. It is recommended to be chosen such that the model assumptions from extreme value theory are valid but also as low as possible (Lang et al. 1999; Ashkar and Rousselle 1987). There exist many different approaches to define a POT-threshold. Examples are statistical methods such as the mean residual life plot (Coles et al. 2001). However, statistical methods often result in quite high thresholds, which is contradictory to the aim of including more flood events than in the AMS approach. Moreover, expert statistical knowledge is often required to define such statistical thresholds, making them rarely applied in hydrology. Hence, in hydrology often more pragmatic choices are made: a flood should be a singular runoff event (this means it has a beginning and end) which should occur sporadically in a year, i.e., on average no more than two to four times per year. A threshold value for the POT series can therefore be defined such that the partial duration series contains two to four times as many events as the AMS (Lang et al. 1999) or a given number of events per year at least. Alternatively, a flood event can be defined as an event where discharge temporally exceeds a thresholds, which could be estimated in relation to the mean long-term discharge (MQ). This is based on the idea to use statistical moments for defining the threshold. Here, it is therefore recommended to define the threshold as three times the respective MQ.

Similar to the AMS, statistical theory delivers the limit distribution for the POT approach by the Balkema-de Haan-Pickands-Theorem (Balkeema and de Haan 1974; Pickands 1975).

For a given sample $X_1, \dots, X_n$ of a series of i.i.d. random variables and a threshold u, the probability distribution of the exceedances $(X - u)$ conditional on $X > u$ can be shown to follow the Generalized Pareto Distribution (GPD) approximately. The distribution function of the GPD is given by

$$G(x; u) = \begin{cases} 1 - \left(1 + \kappa \left(\frac{x-u}{\beta}\right)\right)^{-\frac{1}{\kappa}}, & \kappa \neq 0 \\ 1 - \exp\left(-\frac{x-u}{\beta}\right), & \kappa = 0 \end{cases}$$

with threshold $u \in \mathbb{R}$, scale-parameter $\beta > 0$ and support $x \geq u$ for $\kappa \geq 0$ and $u \leq x \leq u - \beta/\kappa$ for $\kappa < 0$. Often, κ is called the tail index, referring to the exceedances as tail of the distribution. It mainly defines the shape of the tails of the distribution (Fig. 2.3).

Similar to the GEV distribution, a special case with $\kappa = 0$ of the GPD exists, which is called the shifted exponential distribution

$$G(x) = 1 - \exp\left(-\frac{x-u}{\beta}\right),$$

for $x > u$. This distribution can be preferable in the case of light tails ($\kappa < 0.1$) as shown by (Rosbjerg et al. 1992).

In Sect. 2.1, the concept of annuality was presented: the hydrological practice demands annual return periods respectively design floods for annual return periods. These cannot be calculated directly from the POT approach, since no block maxima are considered. Instead, the distribution of the excesses has to be transformed to obtain the annual distribution function F_a, which can be used to estimate annualities.

For this purpose, the total probability theorem is applied:

$$F_a(x) = \sum_{k=0}^{\infty} \mathbb{P}(l = k)\,(F(x))^k,$$

where $\mathbb{P}(l = k)$ is the probability that the annual number of exceedances of u equals k. The most frequently used discrete distribution for describing the occurrence of rare events is the Poisson distribution with

$$\mathbb{P}_P(l = k) = \frac{\lambda^k}{k!} e^{-\lambda} \tag{2.6}$$

based on the assumption that the underlying process is a Poisson process, but also other discrete distributions like the Binomial distribution are possible. The parameter λ represents both the mean and the variance of the distribution and is estimated by the mean of the annual number of exceedances $\hat{\lambda} = \frac{1}{d}\sum_{i=1}^{n} 1_{[Y_i > u]}$, where d denotes the number of years in the observation period.

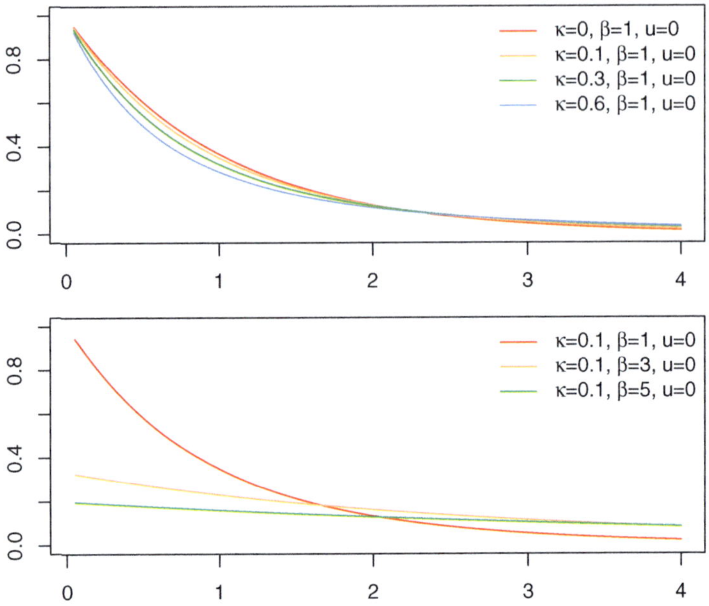

Fig. 2.3 Probability density function (pdf) of the GPD distribution for different parameter combinations

The distribution function F_a then can be expressed as

$$F_a(x) = \sum_{k=1}^{\infty} \frac{\lambda^k}{k!} (G(x;u))^k. \tag{2.7}$$

There can be found direct relationships between POT and AMS. When assuming G to follow a GPD in the previous definition of F_a, the so-called Poisson-Pareto model (Stedinger et al. 1993), one obtains a GEV distribution

$$\begin{aligned} F_a(x) = & \sum_{k=1}^{\infty} \tfrac{\lambda^k}{k!} e^{-\lambda} \left(1 - \left(1 + \kappa \tfrac{x-u}{\beta}\right) - \tfrac{1}{\kappa}\right)^k \\ &= e^{-\lambda} \exp\left(\lambda \left(1 - \left(1 + \kappa \tfrac{x-u}{\beta}\right)^{-\frac{1}{\kappa}}\right)\right) \\ = & \quad \exp\left(-\lambda \left(1 + \kappa \tfrac{x-u}{\beta}\right)^{-\frac{1}{\kappa}}\right). \end{aligned} \tag{2.8}$$

with parameters $\xi = \kappa, \sigma = \beta\lambda^\kappa$ and $\mu = u - \beta(1 - \lambda^\kappa)/\kappa$. Moreover, the average arrival rate can be directly linked to the annual exceedance probability. From Eq. 2.8, one can then directly obtain the quantile function for the partial duration series (Eq. 2.9) to derive design floods for given return periods T:

$$X(T) = u - \beta(1 - \lambda^{\kappa})/\kappa - \frac{\beta\lambda^{\kappa}}{\kappa}\left((-\log(1 - 1/T))^{-\kappa} - 1\right) \qquad (2.9)$$

The main advantage of the POT compared to the AMS lies in the larger number of events that is considered. In fact, it was determined by Madsen et al. (1997) in which cases the POT model is preferable over the AMS approach. By comparison of the asymptotic variances it was shown that the POT series with $\kappa = 0$ is most efficient of all models as long as $\lambda > 1.64$, meaning that in average more than 1.64 flood events per year occur. For the general case of $\kappa > 0$, which mostly appears in flood frequency analysis, (Madsen et al. 1997) recommend the POT model due to the higher efficiency.

2.3 Parameter Estimation

In order to fit a distribution function to the data for any of the introduced concepts, AMS or POT, the parameters of the chosen distribution function have to be estimated. For this purpose, different estimation methods exist with different advantages and drawbacks regarding robustness, efficiency or bias. The three most common parameter estimators in hydrological practice are the Maximum-Likelihood (ML) estimation, the Method of Moments or the L-Moments introduced in the following.

2.3.1 Maximum-Likelihood Estimation

The Maximum Likelihood method makes use of the likelihood which measures the plausibility of a distribution with density f with a given parameter set θ for the data sample. More precisely, for i.i.d. random variables $X_1, \ldots, X_n$ and respective realisations $x_1, \ldots, x_n$, their joint density can be expressed as

$$f_\theta(x_1, \ldots, x_n) = \prod_{i=1}^{n} f_\theta(x_i).$$

This function can be interpreted as function of the parameter set θ for given (observed) values of $x_1, \ldots, x_n$, the so-called likelihood function:

$$L(\theta) = \prod_{i=1}^{n} f_\theta(x_i). \qquad (2.10)$$

The maximum likelihood estimation then is the value of θ for which L takes its maximum:

$$\hat{\theta}_{ML} = \arg \max_{\theta \in \Theta}(\theta).$$

If L is differentiable, the maximum can be obtained by taking the first derivative of L. However, due to complex forms of some densities, the calculation of the maximum of the product can be very difficult. A simplification delivers the log-likelihood function, where the logarithmic likelihood function, which has its maximum at the same location as the standard likelihood function, is applied:

$$l(\theta) = \log\left(\prod_{i=1}^{n} f_\theta\right)(x_i) = \sum_{i=1}^{n} \log(f_\theta(x_i)).$$

The solution of the resulting system of equations is often complex and for many distributions only possible in an iterative way since no analytic expression exists. Therefore, mathematical optimisation methods are usually used to directly determine the maximum of the likelihood function. Other parameter estimates such as the Method of Methods introduced in the next section are then frequently used as start values. The advantage of the maximum likelihood method is primarily its efficiency, i.e., compared to other estimation methods it has the smallest variance in the estimate. However, it is generally less robust than, e.g., the L-moments, can lead to inconsistent estimates if the underlying distribution is assumed incorrectly, and the estimates can be strongly biased for small samples.

2.3.2 Method of Moments

The estimation of distribution parameters by the Method of Moments (MoM) delivers a straightforward relation between the basic statistical characteristics of the data sample like mean, standard deviation or skewness and the distribution parameters. In parameter estimation, the moments of the population are replaced by the corresponding moments of the sample. If k is the number of parameters to be determined, the first k moments are required for parameter estimation. This makes this estimation method preferable for an easy calculation and hence it was in common use when computer systems were not easily available.

The moments of a continuous distribution for $k \in \mathbb{N}^{>0}$ are given by

$$m_k = \int_{-\infty}^{+\infty} x^k \cdot f(x)\, dx$$

$$M_{k,n} = \mathbb{E}\left((X - \mathbb{E}(X))^k\right) = \int_{-\infty}^{\infty} (x - m_1)^k \cdot f(x)\, dx$$

where the second definition is the centred moment, normalised by the mean. The sample equivalent is given by

$$\hat{m}_k = \frac{1}{n}\sum_{i=1}^{n} x_i^k$$

$$\hat{M}_k = \frac{1}{n}\sum_{i=1}^{n}(x_i - \bar{x})^k$$

In some applications, also the unbiased versions of the respective estimators for variance (second moment $\hat{m}_2$) and skewness (third moment $\hat{m}_3$) are used.

With this method, the extreme events in the sample influence the course of the fitted distribution function strongly. Therefore, the extrapolation of the function to large annualities can lead to comparatively large design floods if the series contains very large individual values. Moreover, moment estimators are not always unbiased. Due to the comparatively simple calculation, MoM estimators are suitable as a first approximation of the estimate, for example as starting values in the numerical determination of the Maximum Likelihood method.

In the following, specific estimators for the parameters of the GEV and PearsonIII distributions are given. For both distributions, the following moments are required

$$\bar{x} = \hat{m}_1 = \frac{1}{n} \cdot \sum_{i=1}^{n} x_i \quad \text{(mean)}$$

$$s = \hat{M}_2^{1/2} = \sqrt{\frac{1}{n-1} \cdot \sum_{i=1}^{n} (x_i - \bar{x})^2} \quad \text{(standard deviation)}$$

$$c_v = \frac{s}{\bar{x}} \quad \text{(coefficient of variation)}$$

$$c_S = \frac{n \cdot \sum_{i=1}^{n} (x_i - \bar{x})^3}{(n-1)(n-2) \cdot s^3} \quad \text{(skewness).}$$

With these moments, the estimators for the parameters of the GEV distribution are given by

$$\hat{\sigma} = \bar{x}\left[1 - \frac{2c_v}{c_S}\right]$$

$$\hat{\xi} = \frac{4}{c_S^2} - 1$$

$$\hat{\mu} = \frac{2}{\bar{x} \cdot c_v \cdot c_S}.$$

For the PearsonIII distribution, it holds that (Stedinger et al. 1993)

$$\hat{\xi} = \frac{4}{{c_S}^2}$$

$$\hat{\sigma} = \frac{1}{2}\sqrt{\hat{m}_2} c_S$$

$$\hat{\mu} = \hat{x} - 2\frac{\sqrt{\hat{m}_2}}{c_S}.$$

For the Method of Moments estimation of the parameters of the GPD, no closed form exists but more details are given in Ashkar and Ouarda (1996).

2.3.3 L-Moments

Similar to the Method of Moments, other estimators have been developed that directly relate the moments of a sample to the parameters of a distribution. This makes the estimations easily interpretable and several distribution functions can be fitted with the same basic moments. For example, Greenwood et al. (1979) developed probability weighted moments (PWM). However, a direct relation between, e.g., mean, standard deviation and skewness to the first three moments is not straightforward. This was the reason for Hosking (1990) to extend the PWM by the so-called Linear-moments (L-moments), which are estimated by a linear combination of order statistics. Compared to the ML-estimation and the Method of Moments, this estimation method has the advantage of being robust, i.e., single extraordinary large or small observations do not affect the estimation much. Besides that, L-moments are also comparably efficient to the ML estimators and exist for distributions where the classical moments do not exist.

Let X be a real-valued random variable with cumulative distribution function F and quantile function $x(F)$. The rth L-moment is defined as

$$\lambda_r = \frac{1}{r}\sum_{k=0}^{r-1}(-1)^k\binom{r-1}{k}\mathbb{E}X_{(r-k:r)}, \tag{2.11}$$

where $X_{(i:n)}$ is the ith value of the order statistics of a sample $X_1, \ldots, X_n$ drawn from F and $r = 1, 2, \ldots$. For example, the first two L-moments are given by

$$\lambda_1 = \mathbb{E}X = \int_0^1 x(F)dF,$$
$$\lambda_2 = \frac{1}{2}\mathbb{E}\left(X_{(2:2)} - X_{(1:2)}\right) = \int_0^1 x(F)(2F-1)dF.$$

Hosking (1986) has shown that λ_r exists for $r = 1, 2, \ldots$ if and only if $\mathbb{E}|X|$ exists.

The sample L-moment then is given by

$$\begin{aligned} l_r &= \frac{1}{\binom{n}{r}}\sum_{1\le i_1<\ldots<i_r\le n} r^{-1}\sum_{k=0}^{r-1}(-1)^k\binom{r-1}{k}x_{(i_{r-k}:n)} \\ &= \sum_{k=0}^{r-1}(-1)^{r-1-k}\binom{r-1}{k}\binom{r-1+k}{k}\frac{1}{n}\frac{1}{\binom{n-1}{k}}\sum_{j=k+1}^{n}\binom{j-1}{k}x_{(j:n)}. \end{aligned}$$

It is easy to see that l_r is asymptotically normal under independence and short-range dependence. Additionally, l_r is an unbiased estimator for λ_r. Besides the simple L-moments, often the so-called L-moment ratios are used

$$\tau_r = \lambda_r / \lambda_2, \tag{2.12}$$

respectively their sample estimates, with the special cases of τ_3, the L-skewness, and τ_4, the L-Kurtosis.

With these basic formulas, specific parameter estimators for certain distribution functions can be derived. Here, estimators for the GEV distribution, PearsonIII distribution and GPD are given, the most commonly used distributions in flood statistics.

For the parameters of the GEV distribution, the L-moment estimators are Hosking (1990)

$$\begin{aligned}
\hat{\xi} &= -7.8590 \cdot f - 2.9554 \cdot f^2 \\
\text{with } f &= \frac{2l_2}{l_3+3l_2} - \frac{\ln 2}{\ln 3} \\
\hat{\sigma} &= \frac{-\hat{\xi} \cdot l_2}{\Gamma\left(1-\hat{\xi}\right)\left(1-2^{\hat{\xi}}\right)} \\
\hat{\mu} &= l_1 - \hat{\sigma}\left[\Gamma\left(1-\hat{\xi}\right) - 1\right] / \hat{\xi}.
\end{aligned}$$

For the PearsonIII distribution, two cases of the L-moment ratio $t_3 = \frac{l_3}{l_2}$ representing skewness have to be considered.

For $0 < |t_3| < \frac{1}{3}$, the corresponding shape parameter ξ can be estimated with $z = 3\pi \cdot t_3^2$ as

$$\hat{\xi} = \frac{1 + 0.2906\, z}{z + 0.1882\, z^2 + 0.0442\, z^3}.$$

Otherwise, if $\frac{1}{3} \leq |t_3| < 1$, it is $z = 1 - |t_3|$ and

$$\hat{\xi} = \frac{0.36067z - 0.59567\, z^2 + 0.25361\, z^3}{1 - 2.78861\, z + 2.56096\, z^2 - 0.77045\, z^3}.$$

Independent of the value of the skewness parameter, the remaining parameters of location and scale are estimated by

$$\begin{aligned}
\hat{\sigma} &= \frac{l_2 \cdot \Gamma\left(\hat{\xi}\right) \cdot \pi^{0,5}}{\Gamma\left(\hat{\xi}+0,5\right)} \cdot \operatorname{sgn} |t_3| \\
\hat{\mu} &= l_1 - \hat{\sigma} \cdot \hat{\xi}.
\end{aligned}$$

For the estimation of the parameters of the GPD distribution, also two cases have to be distinguished: if the parameter of location is known or unknown.

In practical applications, the location parameter, i.e., the threshold u, will always assumed to be known. Hence, the following two estimators are given:

$$\begin{aligned}
\hat{\kappa} &= 2 - \frac{l_1 - u}{l_2} \\
\hat{\beta} &= (1 - \hat{\kappa})(l_1 - u).
\end{aligned}$$

2.4 Selection of a Distribution Function

As shown above, there exists a variety of different distribution functions with different properties. Some of them are able to model heavy tails, some are bounded, some are symmetric. None can be rejected from the outset, e.g., for physical reasons, such that initially several distribution functions should certainly be fit to the sample. Subsequently, however, the goodness of fit, i.e., how well the fitted theoretical distribution reproduces the observed distribution of the sample, should be evaluated. There exist various criteria for helping the applicant to choose a distribution function.

The simplest way to evaluate the goodness of fit is to visually inspect the fit. In this case, the probabilities or quantiles obtained by the fitted distribution are compared with the observed, empirical probabilities or quantiles. One possibility for such a comparison offer plotting positions. Plotting positions p_i of an observation x_i are empirical non-exceedance probabilities of the observed data, derived by dividing the rank of each observation by the number of observed data n plus one:

$$p_i = \frac{Rank(x_i)}{n+1}. \tag{2.13}$$

By plotting these empirical probabilities together with the quantiles of the fitted distribution function, the distribution with distribution function closest to the empirical probabilities is considered best fitted. A drawback of plotting positions is their limitation to the sample size: return periods derived for plotting positions will never exceed the number of years of observation. For example, plotting positions for the AMS of a 30-year observation period will have a maximum return period of 31 years, no matter how extreme single flood events may have been. Alternatively, the so-called QQ-plot can be used for visual inspection of the goodness of fit: for this plot, the observed, empirical quantiles are contrasted with those of the fitted distribution of equal probability. If the fitted distribution fits well to the data, the points in the plot lie on a straight line. Deviations in the tail, which are most critical for hydrological purposes, can thus easily be detected (Fig. 2.4).

This example already demonstrates that the greatest challenge that has to be faced in statistical flood hydrology is the extrapolation. For this, the distribution function is used to estimate quantiles for annualities that are beyond the observed record length, e.g., an annuality of 1000 years where only 100 years of discharge data are available. The problem that arises when selecting a distribution is that for the lower quantiles, where many observations are available, most distributions deliver very similar results. However, in the extrapolation range, where only few or none extreme flood events at all can be observed, the distribution functions differ significantly (Fig. 2.5). This can be affiliated to the distributions that are considered. Mostly, two or three parametric distributions are used, which aim to preserve mean and variance of the observed data (Koutsoyiannis 2004). The large uncertainties in the right tail at the same time lead to large uncertainties in the estimation of these quantiles and an incorrectly selected distribution can lead to enormous errors in this range.

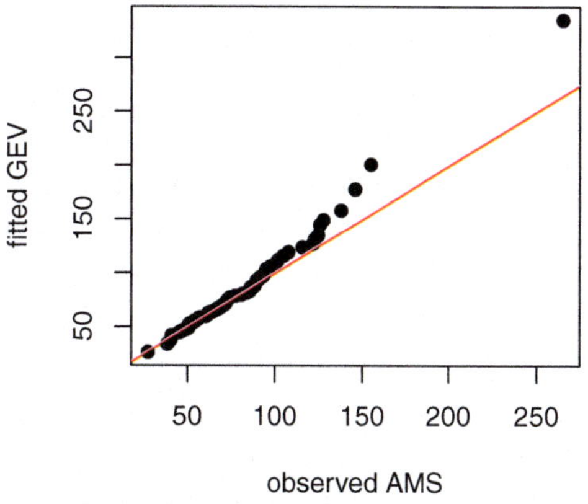

Fig. 2.4 QQ-plot between an observed AMS and the fitted GEV distribution (with L-moments). The deviation of the high quantiles implies that the tail behaviour of the observed AMS is not captured well be the fitted GEV

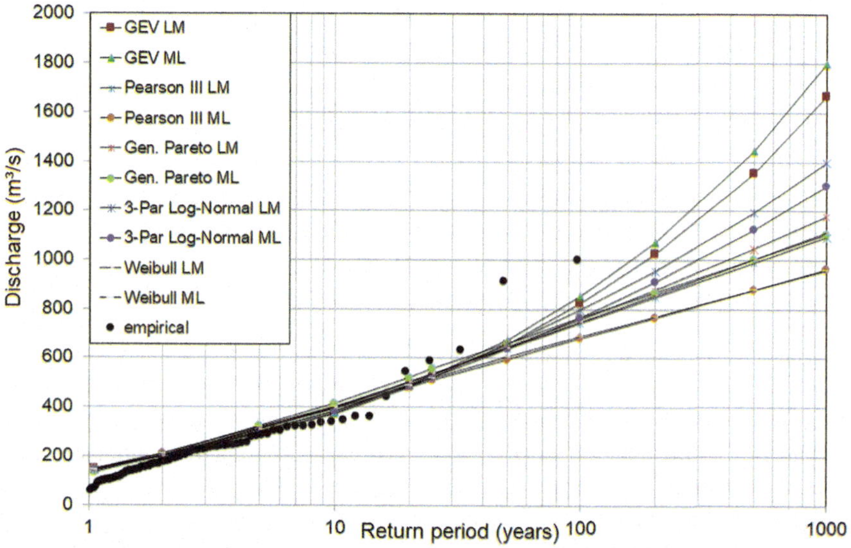

Fig. 2.5 Several distributions fitted with two different parameter estimation methods (Maximum Likelihood (ML) and L-moments (LM)) and comparison to empirical probabilities

Moreover, the parameter estimation method has an impact on the resulting shape of the distribution, too. With the choice of the estimation method, it can for example be adjusted which weight should be given to extreme events and hence the right tail.

Besides, the limited sample sizes of hydrological discharge time series increase the problem. The small number of available events leads to sample uncertainty.

A drawback of all visual inspections is the subjectivity of the assessment of the fit. This subjective decision is also strongly influenced by the mean values of the sample, as these have the highest frequency. The relevant areas of the fit, i.e., the extrapolation range, where the greatest differences between the distributions occur, are usually underlain by only a few events and can therefore only be assessed visually to a limited extent. Therefore, it makes sense to supplement visual assessments with statistical, objective procedures.

Such objective procedures are, for example, statistical tests. Goodness-of-fit tests are based on statistical measures that determine the differences between observed and theoretical fitted distributions. For a given significance level (usually 5%), the corresponding test statistic is evaluated. Most commonly, statistical tests are applied that test the hypothesis of equality between theoretical and observed distribution. A well-known test in this case is the Anderson-Darling test (Anderson and Darling 1952).

The Anderson-Darling test is a statistical test that tests the null hypothesis of equality between theoretical and sample distribution by transforming the ranked observations by the theoretical distribution and comparing the distance to uniform distribution. Its test statistic for a sample $x_1, \ldots, x_n$ is given by

$$A^2 = -n - \sum_{i=1}^{n} \frac{2i-1}{n} \left(\ln(F(x_{(i)})) + \ln(1 - F(x_{(n+1-i)})) \right),$$

with F being the distribution function of the theoretical distribution of interest. Compared to the Kolmogorov-Smirnov test, this test has sufficient power even for estimated parameters of the theoretical distribution. Besides these hypothesis tests, there exist numerous other goodness-of-fit metrics for probability distributions such as Cramer-von-Mises test, L-moment-ratio test or regression-based tests. A detailed overview is for example given by Chowdhury et al. (1991). However, goodness-of-fit tests can only reject distribution functions as being inappropriate, but cannot evaluate the goodness of the non-rejected distribution functions among themselves.

One way to directly compare the goodness of fit of different distribution functions is to use information criteria. These criteria compare the goodness of fit with the complexity of the model, i.e., the number of distribution parameters. Unnecessary complexity of the distribution, which may lead to a slightly better fit but also to significantly higher uncertainty, is therefore penalised. The two most commonly used information criteria are the Akaike Information Criterion (AIC) and the Bayesian Information Criterion (BIC). The AIC makes use of the likelihood function L (see Sect. 2.3). Suppose that F is the theoretical distribution function with k parameters that should be evaluated concerning the goodness of fit for a parameter set θ. Then, the AIC and BIC of it are given by

$$\begin{aligned} AIC &= \quad 2k - 2\ln(\max_{\theta}(L(\theta))) \\ BIC &= k\ln(n) - 2\ln(\max_{\theta}(L(\theta))). \end{aligned}$$

If the AIC criterion is calculated for all distributions of interest, the one is chosen with minimum AIC. The minimum AIC is interpreted as minimum information loss when applying the respective distribution. However, the differences in the information criteria for different distribution functions are sometimes only small. Rafferty et al. (2020), for example, defines a difference in the BIC of less than 2 as weak evidence of a difference, a difference of 2–6 as positive evidence and only from a difference of 10 does he speak of strong evidence. Therefore, it is recommended that all distribution functions that deviate by less than 6 from the smallest BIC are considered to be potentially the best fit. A further selection should then always be made under hydrological aspects and by means of information expansion.

These considerations show that the selection of a distribution in the flood statistical context cannot be distinct and each selection is afflicted with uncertainty.

Flood frequency analyses are one of the key tools in hydrology. They form the basis for almost all design purposes and are topic of many research activities. Basically, one differentiates between annual maximum series and peak-over-threshold series with varying distribution functions and parameter estimation methods. However, usually these approaches are also limited in their meaningfulness as their analyses are mostly focused on the flood peak and one sample only. As an extension to classical flood frequency analyses concepts, this book introduces new statistical models that overcome these drawbacks by considering flood types.

References

Anderson TW, Darling DA (1952) Asymptotic theory of certain “Goodness of Fit” criteria based on stochastic processes. Ann Math Stat 23(2):193–212

Ashkar F, Ouarda TBMJ (1996) On some methods of fitting the generalized pareto distribution. J Hydrol 177(1):117–141

Ashkar F, Rousselle J (1987) Partial duration series modeling under the assumption of a Poissonian flood count. J Hydrol 90:135–144

Balkema AA, de Haan L (1974) Residual life time at great age. Ann Prob 2:792–804

Chowdhury JU, Stedinger JR, Lu L (1991) Goodness-of-fit tests for regional generalized extreme value flood distributions. Water Resour Res 27(7):1765–1776

Coles S, Bawa J, Trenner L, Dorazio P (2001) An introduction to statistical modeling of extreme values, vol. 208. Springer, London, p 208

Fisher RA, Tippett LHC (1928) Limiting forms of the frequency distribution of the largest or smallest member of a sample. Math Proc Cambridge Philos Soc 24(02):180–190

Greenwood JA, Landwehr JM, Matalas NC, Wallis JR (1979) Probability weighted moments: Definition and relation to parameters of several distributions expressable in inverse form. Water Resour Res 15(5):1049–1054

Gumbel EJ (1941) The Return Period of Flood Flows. Ann Math Stat 12(2):163–190

Hosking JRM (1986) The theory of probability weighted moments. IBM Research Report, IBM, p 1986
Hosking JRM (1990) L-moments: Analysis and estimation of distributions using linear combinations of order statistics. J Roy Stat Soc. Ser B (Methodol) 52(1):105–24
Hosking JRM, Wallis JR, Wood EF (1985) Estimation of the Generalized Extreme-Value distribution by the method of probability-weighted moments. Technometrics 27(3):251–261
Koutsoyiannis D (2004) Statistics of extremes and estimation of extreme rainfall: I. Theoretical investigation. Hydrol Sci J 49(4)
Lang M, Ouarda TBMJ, Bobée B (1999) Towards operational guidelines for over-threshold modeling. J Hydrol 225(3–4):103–107. https://doi.org/10.1016/S0022-1694(99)00167-5
Madsen H, Rasmussen PF, Rosbjerg D (1997) Comparison of annual maximum series and partial duration series methods for modeling extreme hydrologic events: 1 At-site modeling. Water Resour Res 33(4):747–757
N.E.R.C. (1975) Flood Studies Report, vol 1. Institute of Hydrology Wallingford, London
Pickands J III (1975) Statistical inference using extreme order statistics. Ann Stat 3(1):119–131
Rafferty AN, Jansen RA, Griffiths TL (2020) Assessing Mathematics Misunderstandings via Bayesian Inverse Planning. Cognit Sci 44:e12900. https://doi.org/10.1111/cogs.12900
Rahimi L, Deidda C, De Michele C (2021) Origin and variability of statistical dependencies between peak, volume, and duration of rainfall-driven flood events. Sci Rep 11:5182. https://doi.org/10.1038/s41598-021-84664-1
Rosbjerg D, Madsen H, Rasmussen PF (1992) Prediction in partial duration series with generalized Pareto-distributed exceedances. Water Resour Res 28(11):3001–3010
Salinas JL, Castellarin A, Viglione A, Kohnová S, Kjeldsen TR (2014) Regional parent flood frequency distributions in Europe - Part 1: Is the GEV model suitable as a pan-European parent? Hydrol Earth Syst Sci 18(11):4381–4389
Stedinger JR, Vogel RM, Foufoula-Georgiou E (1993) Frequency analysis of extreme events. In: Maidment DR (ed) Handbook of hydrology. McGraw-Hill, New York

Chapter 3
Principles of Multivariate Statistics

Abstract Multivariate statistics allow the simultaneous consideration of two or more random variables in a statistical analysis. To take into account the dependency between the random variables, copula models can be applied. In this chapter, the basic concepts of copulas and their extension, Vine-Copulas, are introduced. Different copula models are explained as well as how to derive joint and conditional probabilities. These multivariate statistics form the basis for the spatial statistics and consideration of flood types.

When considering more than one random variable at a time, univariate models as described in Chap. 2 are no longer sufficient. Instead, multivariate distributions and models have to be used. These models do not only take into account the univariate distributions of each variable but also the dependence structure between the random variables. For example, if two neighbouring gauges are considered jointly for flood statistics, it has to be considered that the flood events are most likely dependent since they are caused by similar circumstances, e.g., weather patterns. Hence, large flood events will occur at a similar time and so do periods of low flow.

To measure the dependency between random variables, several coefficients exist. These will be introduced in more detail in the following section. When the dependency between two or more random variables is known, they can be modelled jointly in a multivariate distribution function. The most natural extension of the univariate case would be the use of a multivariate density, e.g., for the bivariate case of two random variables X and Y

$$F(x, y) = \mathbb{P}(a \leq X \leq b, c \leq Y \leq d) = \int_a^b \int_c^d f(x, y)dxdy.$$

This way, e.g., the multivariate normal distribution can be obtained by using

$$f_{XY}(x, y) = \frac{1}{2\pi\sigma_X\sigma_Y\sqrt{1-\rho^2}} \exp\left(-\frac{1}{2(1-\rho^2)}\left(\left(\frac{x-\mu_X}{\sigma_X}\right)^2 - 2\rho\left(\frac{x-\mu_X}{\sigma_X}\right)\left(\frac{y-\mu_Y}{\sigma_Y}\right) + \left(\frac{y-\mu_Y}{\sigma_Y}\right)^2\right)\right),$$

S. Fischer and A. H. Schumann, *Type-Based Flood Statistics*, Water Science and Technology Library 124, https://doi.org/10.1007/978-3-031-32711-7_3

where ρ measures the dependence between both random variables X and Y. This bivariate distribution consists of the univariate normal distributions of X and Y with parameters μ_X and σ_X respectively μ_Y and σ_Y as well as the dependence measure to describe the structure of interrelation between both.

However, even for this rather simple bivariate case, the analytical formula becomes complex. Multivariate distributions of higher order then become even harder to handle. Moreover, this distribution is limited to the case where all random variables belong to same family of distributions, in this case the normal distribution. This is a disadvantage, since in practice often random variables are considered jointly that may have different distributions. For example, if precipitation and discharge are considered jointly, often the PearsonIII distribution is used for the precipitation while the GEV distribution is used for the discharge series. Such a combination would not be possible with the model described above.

Instead, often the multivariate models based on copulas are used. These models allow arbitrary combinations of univariate distributions while being able to model all cases of dependence. Their concept will be introduced in greater detail in this chapter. In hydrology, copulas form the basis for most multivariate statistical modelling.

3.1 Dependence Measures

There exist several measures for dependence that can be applied in very different cases. For two random variables, the three most popular measures are introduced in the following. All are defined on $[-1, 1]$, where $|1|$ indicates direct relationship and 0 implies independence. Negative values indicate anti-proportional dependence, positive values indicate proportional dependence.

The most common measure is the Pearson (product-moment) correlation coefficient. For two random variables X and Y, it is given by

$$\rho_{XY} = \frac{\mathbb{E}(XY) - \mathbb{E}(X)\mathbb{E}Y}{\sqrt{\mathbb{E}(X^2) - \mathbb{E}(X)^2}\sqrt{\mathbb{E}(Y^2) - \mathbb{E}(Y)^2}}$$

and basically divides the covariance by the standard deviations of each random variable. A sample version of it for two samples $x_1, \ldots, x_n$ and $y_1, \ldots, y_n$ is given by

$$\rho_{xy} = \frac{\sum_{i=1}^{n}(x_i - \bar{x})(y_i - \bar{y})}{\sqrt{\sum_{i=1}^{n}(x_i - \bar{x})^2 \sum_{i=1}^{n}(y_i - \bar{y})^2}}.$$

Due to its definition, it can be easily seen that the Pearson correlation coefficient is not robust, i.e., values which are distant from the mean have a large impact on the correlation. Moreover, it is constructed to measure mostly linear dependence.

To overcome the problem of non-robustness, rank-correlation coefficients have been developed. A straightforward transformation of Pearson's correlation is Spearman's correlation coefficient, where the observations are replaced by their ranks. The sample version is defined as

$$\rho_S = \frac{\sum_{i=1}^{n}(rank(x_i) - \bar{rank}(x))(rank(y_i) - \bar{rank}(y))}{\sqrt{\sum_{i=1}^{n}(rank(x_i) - \bar{rank}(x))^2 \sum_{i=1}^{n}(rank(y_i) - \bar{rank}(y))^2}}.$$

This correlation coefficient delivers reliable results also for a non-linear relationship between X and Y.

The third correlation coefficient was developed by M. G. Kendall and is most frequently used in the context of copulas. It is based on ranks, too, but takes into account concordant and discordant pairs. 'Concordant' means that for two pairs (x_i, y_i) and (x_j, y_j), $i < j$, it holds that $x_i > x_j$ and $y_i > y_j$ respectively $x_i < x_j$ and $y_i < y_j$. Similarly, discordance then is present if $x_i > x_j$ and $y_i < y_j$ respectively $x_i < x_j$ and $y_i > y_j$. Kendall's τ is defined as

$$\tau = \frac{\#\text{concordant pairs} - \#\text{discordant pairs}}{\binom{n}{2}}.$$

It has the advantage of being efficient even for small sample sizes.

3.2 Copulas

Copulas are a comparably new tool in the mathematical statistics. They have been used more frequently since the 1990s s for multivariate modelling of random variables with arbitrary margins. Roughly speaking, copulas represent the dependence structure between two or more random variables without taking into account the marginal distribution: they couple multivariate distribution functions to their one-dimensional margins (Nelsen 2006). With this, copulas are an alternative to multivariate distribution functions like the multivariate normal distribution, which are rather complex and limited to the case of the same marginal distribution for all random variables. in this chapter, the focus will be laid on bivariate copulas, i.e., considering two random variables jointly. The copula operates on uniformly distributed random variables $U, V \sim U[0, 1]$.

A copula is a function $C : [0, 1]^2 \rightarrow [0, 1]$ with

1. For every $u, v \in [0, 1]$

$$C(u, 0) = 0 = C(0, v)$$
$$C(u, 1) = u \text{ and } C(1, v) = v.$$

2. For every $u_1, u_2, v_1, v_2 \in [0, 1]$ with $u_1 \le u_2$ and $v_1 \le v_2$,

$$C(u_2, v_2) - C(u_2, v_1) - C(u_1, v_2) + C(u_1, v_1) \ge 0.$$

All copulas are bounded by limit cases, the so-called Frechét-Hoeffding bounds:

$$\max(u + v - 1, 0) \le C(u, v) \le \min(u, v).$$

The upper limit represents perfect correlation between both variables.

A special case is given by the product copula, which represents independence between the random variables

$$\prod(u, v) = u \cdot v.$$

This is in line with the classical definition of independence for random variables.

Analogous to univariate distribution functions, a density can be defined for copulas

$$c(u, v) = \frac{\partial^2 C(u, v)}{\partial u \partial v}.$$

What makes copulas such an important tool in multivariate statistics is the following Theorem of Sklar (1959), who also gave the construct the name "copula".

Theorem 3.1 *Let H be a joint distribution function with margins F and G. Then there exists a copula C, such that for all $x, y \in \mathbb{R}$*

$$H(x, y) = C(F(x), G(y)).$$

Basically, this theorem states that for any two (or even more) random variables with arbitrary distribution functions there can be found a copula to model the joint distribution function. Vice versa, if two random variables are modelled with the copula as stated above, this gives their joint distribution function. This is a strong result and makes copula theory applicable for any case where two or more random variables should be considered jointly. For a detailed discussion and the history of copulas it is referred to Nelsen (2006).

In hydrology, copulas are used to model several characteristics of a flood or realisations of the same flood events at different locations (Salvadori et al. 2007, 2011). A basic requirement is that the two variables belong to the same flood event, i.e., flood peak and flood volume of the same flood event or simultaneous flood peaks of the same flood event at two gauges. If this assumption is not fulfilled, the application of copulas is not meaningful. The definitions and theorem above give a simple way to obtain the joint distribution function: first, the margins F and G of the random variables have to be estimated. This can be done analogously to the univariate case. Second, a copula model has to be selected and the parameters have to be estimated, similarly to a distribution function. Then, the probability-transformed random variables $u = F(x)$ and $v = G(y)$ can be coupled with the copula to obtain the joint distribution. The choice of the copula is crucial. There exist numerous copula models with very different properties. For example, some copulas emphasise upper tail dependence, while other exhibit lower tail dependence or even both. Moreover, copulas can be symmetric or asymmetric. Some of the most important copulas in the field of hydrology are introduced in the following.

There exist two classes of copulas which include almost all important copulas applied in hydrology. They provide easy-to-handle analytical expressions which can be varied according to the desired copula.

First, the class of Archimedean copulas is introduced. Archimedean copulas are defined by a function $\phi : [0, 1] \to [0, \infty]$, which is continuous and strictly monotonously increasing with $\phi(1) = 0$. If ϕ is convex, it is the so-called generator of the Archimedean copula

$$C : [0, 1]^2 \to [0, 1],\ C(u, v) := \phi^{-1}(\phi(u) + \phi(v)),$$

where ϕ^{-1} is the pseudo-inverse function of ϕ. Well-known members of the family of Archimedean copulas are the Gumbel copula with $\phi(t) = (-\ln t)^{\lambda}$, $\lambda \geq 1$, the Clayton copula with $\phi(t) = \frac{1}{\theta}\left(t^{-\theta} - 1\right)$, $\theta > 0$, or the Frank copula with $\phi(t) = -\ln\left(\frac{e^{-\theta t}-1}{e^{-\theta}-1}\right)$, $\theta > 0$.

The second important class are extreme value copulas. Extreme value copulas are multivariate extensions of extreme value distribution for the copula case. A copula C is an extreme value copula if and only if for $u, v \in [0, 1]$ and $k > 0$ it is $C(u^k, v^k) = C^k(u, v)$. Similar to Archimedean copulas, an analytical expression of extreme value copulas exists. If C is an extreme value copula, then

$$C(u, v) = \exp\left(\ln(uv) A\left(\frac{\ln v}{\ln(uv)}\right)\right),$$

where $A : [0, 1] \to [0.5, 1]$ is the convex dependence function with $A(0) = A(1) = 1$ and $\max(t, 1 - t) \leq A(t) \leq 1$. Examples are the product copula ($A(t) = 1$) or the Gumbel copula with $A(t) = \left(t^{\theta} + (1 - t)^{\theta}\right)^{1/\theta}$, $\theta \geq 1$. For more examples of Archimedean and extreme value copulas it is referred to Nelsen (2006).

Depending on the chosen copula model, the tail dependencies may vary significantly (Fig. 3.1). In this case, clearly neither the Clayton nor the Frank copula represent the observed data well, while the Gumbel Copula seems to be an adequate model for the observed flood peak and volume pairs.

Similar to univariate distributions, return periods can be estimated based on the copula. However, there exist different possibilities on how to consider the two variables jointly. On the one hand, one might be interested in the case where both variables exceed a certain value at the same time. For example, it might be interesting to know

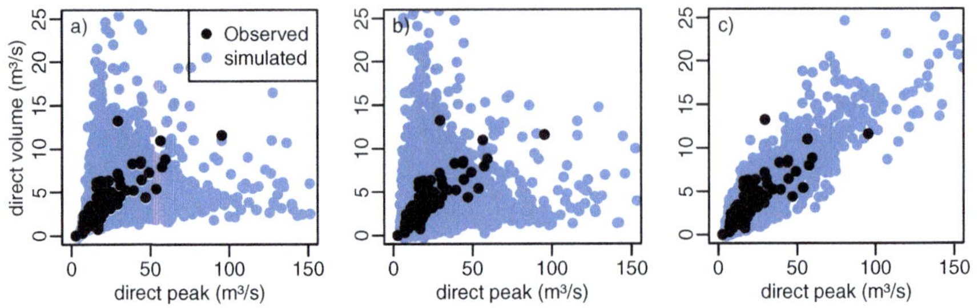

Fig. 3.1 Three copula models fitted to a sample of flood peaks and corresponding volumes (Kendall's $\tau = 0.671$) with GEV margins: **a** Clayton copula, **b** Frank copula and **c** Gumbel copula

when the flood peak exceeds a given threshold while the corresponding flood volume simultaneously exceeds a certain threshold for the volume. This is the so-called AND-case. It is based on the probability

$$\mathbb{P}(X > x \wedge Y > y) = 1 - F(x) - G(y) + C(F(x), G(y)). \tag{3.1}$$

On the other hand, for practical purposes it might also be interesting to know the probability that at least one of the two variables exceeds a given threshold, e.g., only the flood peak exceeds a critical threshold. This is the so-called OR-case. Then, the following probability is considered

$$\mathbb{P}(X > x \vee Y > y) = 1 - C(F(x), G(y)). \tag{3.2}$$

For both cases, the corresponding return periods can be estimated by

$$T_{AND} = \frac{\mu}{1-F(x)-G(y)+C(F(x),G(y))} \tag{3.3}$$

$$T_{OR} = \frac{\mu}{1+C(F(x),G(y))}, \tag{3.4}$$

where μ is the number of events per year, i.e., in the case of annual maxima $\mu = 1$.

Similarly, quantiles for given return periods for multivariate distributions do not consist of one pair of values only. Instead, there exist many combinations of the variables that occur with the same probability. For example, when considering flood peak and volume, there exist many combinations of those two characteristics that have the same probability. Therefore, quantiles for a given return period in the bivariate case always are described by isolines, which represent the many possible combinations of the two variables to result in the same probability. Again, it has to be differentiated between the AND and OR case (Fig. 3.2). It should be considered that the purely horizontal or vertical have less physical meaning when one moves away from the turning point of the isoline (see also Chap. 1). For example, it cannot be expected that a flood event with a very small volume will have an extremely large peak.

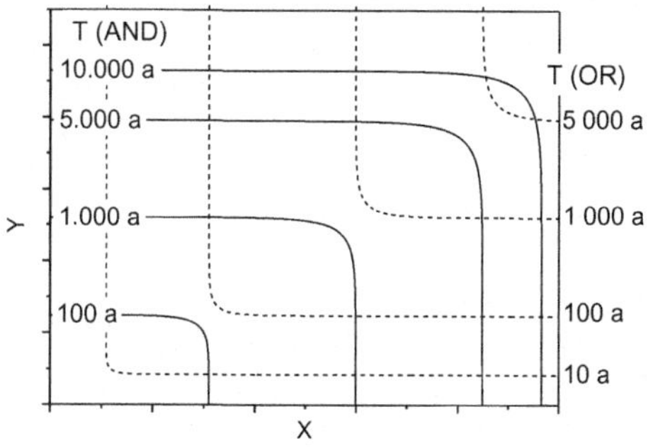

Fig. 3.2 Isolines for quantiles of a given return period of bivariate copulas for the AND and OR case

In cases where higher dimensional copulas are required, the above-mentioned concepts can be generalised to an arbitrary number of random variables. For example, if flood peak, volume and duration are considered jointly, a trivariate copula would be required to describe the joint distribution of all three characteristics:

$$H(x, y, z) = C(F_1(x), F_2(y), F_3(z)), \tag{3.5}$$

where F_1 is the marginal distribution of the flood peaks, F_2 is the marginal distribution of the flood volumes and F_3 is the marginal distribution of the flood durations. With increasing dimension, the complexity of the copula model increases, too. For example, a three-dimensional Gumbel copula with parameter θ has the form

$$C(u, v, w) = \exp\left(-\left((-\log u)^{\theta} + (-\log v)^{\theta} + (-\log w)^{\theta}\right)^{\frac{1}{\theta}}\right).$$

Therefore, the applicability of high-dimensional copulas always also depends on the underlying sample sizes.

3.3 Vine Copulas

In the previous section, bi- and multivariate copulas have been introduced. However, the larger the copula dimension, the more difficult becomes the handling of the copula. Vine copulas allow for multivariate dependence modelling without having to cope with multivariate copulas. Instead, they transform the multivariate dependence into a dendritic structure, where each knot can be modelled separately by a bivariate copula. This is advantageous, since there exist much more bivariate copula models than multivariate models and moreover the copula family of the applied bivariate model can be changed for each pair, allowing more flexibility (Daneskkhah et al. 2016). Vine copulas have been invented by Joe (1994). Theoretically, any multivariate distribution of n variables can be represented by a vine copula. They are applied in multiple fields like insurance or finance (Aas et al. 2009). In hydrology, they are used

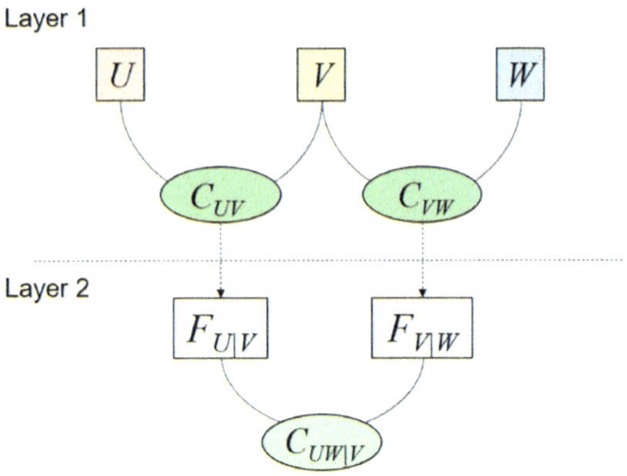

Fig. 3.3 Schematic overview of a 3D vine copula

for the multivariate modelling of, e.g., triples of peak, volume and duration of a flood, tuples of several peaks measured at neighbouring gauges or for drought frequency analysis (Graeler et al. 2013; Kao and Govindaraju 2008; Song and Singh 2010). In the context of this book, the application will be limited to 3D vines. This kind of vine consists of three initial time series which are combined via copulas in two layers. A schematic description is given in Fig. 3.3. In this case, the three variables u, v, w could be, e.g., the [0, 1]-transformed peak, duration and volume of a flood event. These are modelled pairwise by two copulas C_{uv} and C_{vw}, which do not necessarily have to be of the same family. To transit from the first layer to the second layer, the conditional distributions conditioned on the joint variable v have to be calculated. This can be done by partial derivation (Aas et al. 2009) such that

$$\begin{aligned} F_{U|V}(u|v) &= \frac{\partial C_{UV}(u,v)}{\partial v} \\ F_{W|V}(w|v) &= \frac{\partial C_{VW}(v,w)}{\partial v}. \end{aligned}$$

These conditional distribution functions are again uniformly distributed on [0, 1] such that they can be used to estimate the final copula $C_{UW|V}$. Combining these results, one obtains

$$c_{UVW}(u, v, w) = c_{UW|V}(F_{U|V}(u|v), F_{W|V}(w|v)) \cdot c_{UV}(u, v) \cdot c_{VW}(v, w),$$

the density of the 3D copula.

Based on this, different (conditional) probabilities can be calculated. For example,

$$\mathbb{P}(W \leq w|U = u, V = v) = \frac{\frac{\partial^2}{\partial u \partial v} c_{UVW}(u, v, w)}{\frac{\partial^2}{\partial u \partial v} c_{UV}(u, v)}. \tag{3.6}$$

As for the usual case of bivariate copulas, the bivariate copulas here can be fitted by different methods, e.g., Maximum Likelihood, which is chosen here (see Chap. 11). The best-fitting copula for each pair is determined by the AIC criterion.

Multivariate statistics allows the consideration of more than one random variable and hence more than one flood event characteristic in flood statistics. It is required when the focus is laid on the analyses of complete flood events instead of considering the flood peak only. Hence, it becomes more and more popular in hydrological sciences. Copulas are a simple multivariate model with a high flexibility regarding the underlying marginal distributions and the dependence structure. By using Vine copulas, a dendritic structure of the dependency of the variables can be applied to reduce complexity of high dimensional copula models.

References

Aas K, Czado C, Frigessi A, Bakken H (2009) Pair-copula constructions of multiple dependence. Insur: Math Econ 44(2):182–198

Daneshkhah A, Remesan R, Chatrabgoun O, Holman IP (2016) Probabilistic modeling of flood characterizations with parametric and minimum information pair-copula model. J Hydrol 540:469–487

Gräler B, van den Berg MJ, Vandenberghe S, Petroselli A, Grimaldi S, De Baets B, Verhoest NEC (2013) Multivariate return periods in hydrology: a critical and practical review focusing on synthetic design hydrograph estimation. Hydrol Earth Syst Sci 17(4):1281–1296

Joe H (1994) Multivariate extreme-value distributions with applications in environmental data. Can J Stat 22:47–64

Kao S, Govindaraju RS (2008) Trivariate statistical analysis of extreme rainfall events via the Plackett family of copulas. Water Resour Res 44(2)

Nelsen RB (2006) An introduction to copulas. Springer, Berlin

Salvadori G, De Michele C (2007) On the use of copulas in hydrology. Theory and practice. J Hydrol Eng 12(4):369–380

Salvadori G, De Michele C, Durante F (2011) Multivariate design via Copulas. Hydrol Earth Syst Sci Discuss 8(3):5523–5558

Sklar A (1959) Fonctions de repartition a n dimensions et leurs marges. Kybernetica 9:449–460

Song S, Singh VP (2010) Meta-elliptical copulas for drought frequency analysis of periodic hydrologic data. Stoch Env Res Risk Assess 24(3):425–444

Part III
Flood-Type-Specific Flood Statistics

Classical flood frequency analyses consider one flood characteristic, usually the flood peak, in one joint sample. However, these classical concepts lack the consideration of differences between certain meteorological drivers. Different flood-generating processes might lead to different flood types and a joint consideration hence to inhomogeneous samples. To overcome this issue, samples differentiated by their flood type should be considered. These flood type-specific samples then are analysed separately. However, the classical demands of flood frequency analyses, having one design flood for a given annuality, require the application of a mixture model to jointly consider the flood types for the specification of the contributions of several types to reach the same peak in one scenario. In this part, such a type-specific mixture model is introduced and compared to classical flood frequency approaches. The methodology introduced here forms the basis of type-based flood statistics and will be used in the following parts to define design hydrographs, to be extended to spatial consideration of floods and their generating processes and for regionalisation.

Chapter 4
Opportunities and Limitations of Specifying Flood Types

Abstract Flood types offer the opportunity to not only consider certain flood characteristics like the peak, shape or the volume jointly, but to incorporate information on the flood-generating processes into the analyses of flood events. Flood-generating processes encompass meteorological conditions as well as the catchment state. Both have significant impact on the resulting flood event. In hydrology, several flood type classifications exist, focusing on different aspects of flood-generating circumstances. In this chapter, the most common flood type classifications are introduced. When aiming to combine flood types and flood statistics, certain properties of flood classifications have to be fulfilled. These will be discussed at the end of this chapter.

4.1 What Is a Flood Type?

Apart from special natural events such as landslides, ice jams or the outburst of glacial lakes, and from anthropogenic causes such as dam breaches of reservoirs, floods are the result of meteorological phenomena associated with high precipitation amounts, rainfall intensities exceeding the infiltration capacity of soils, or an accumulation of snow and ice associated with their sudden thawing. The different meteorological causes are induced by different influences of the respective catchment characteristics. For example, snowmelt floods occur when an existing snow cover in a catchment melts due to increasing temperature. The larger the catchment and the larger the differences in elevation zones in a catchment, the longer is the process of snowmelt. It starts in the lower elevation parts since in these regions the snow cover usually is less thick, has a lower retention capacity for water and the temperature is higher than for higher elevations. As a consequence, floods caused by snowmelt are often characterised by a long duration. The water storage capacity of the snow cover as well as the soil conditions determine the release of water from snowmelt into the river network. When rain contributes to the increase of water content and to an additional energy input, the thawing is accelerated and the water release from the snowpack increases in short time. The hydrographs of such *rain-on-snow events* differ from other snowmelt events with less significant impacts of rain. Rain-on-snow events were responsible for many widespread floods in Central Europe (Krug et al. 2020).

S. Fischer and A. H. Schumann, *Type-Based Flood Statistics*, Water Science and Technology Library 124, https://doi.org/10.1007/978-3-031-32711-7_4

And even if a flood is caused only by rainfall, flood-inducing rainfall events can vary vastly in the spatial and temporal distribution of rain intensities. For example, rainfall events can have spatially limited extension but high intensities. Such heavy rainfall mainly occurs in summer in Central Europe and USA, e.g., accompanied by thunderstorms. In Germany, many of the recent extreme floods were caused by heavy rainfall. One example is the flood in July 2021 which hit large parts of western Germany, the Netherlands and Belgium causing extreme damage and high fatalities. For this flood, more than 150 mm of rainfall fell down in 15 h. Such storm events often dominate the flood regime in summer of small and medium-sized catchments in mountainous regions. Contrarily, due to the limited spatial extent of such events, in large catchments long-duration rainfall becomes more relevant. In combination with a high initial soil moisture, long rain events trigger extreme floods. Another rainfall-driven flood type can be observed when soil moisture has been increased by previous rainfall events. The catchment then responds to relatively small rainfall events and produces a series of flood peaks that often cannot be associated with individual flood waves because the discharges do not fall below certain thresholds used to estimate the end of individual flood events. Due to these differences between the rain types, the same flood peak can be observed for floods which differ significantly in their volumes and shapes of hydrographs.

It is therefore if high importance when aiming to understand floods to understand the atmospheric components and hence the meteorological causes in their interaction with the initial conditions (soil moisture and snow cover) of the catchment first. The temporally variable relationships between the wetness state of the soil, the capacities of infiltration and percolation in the deeper soil layers and the rainfall intensities determine the catchment response and the resulting flood hydrographs. If the rainfall intensity exceeds the infiltration intensity, surface runoff occurs. At the same time, if the water storage capacity of the upper zone is exhausted, i.e., the soil is saturated, rain cannot infiltrate any more and also becomes surface runoff. Inside the ground, differences between permeabilities of soil layers result in lateral flows of soil water and a speed-up of the drainage intensities. The hill slope of the catchment and the hydraulic conditions of the river network determine the flow velocities and the shape of the hydrographs. Both, meteorological and catchment processes that may cause a flood, are closely interlinked.

The complexity of the interplay between meteorological causes and catchment states can be easily illustrated by an example. Like many catchments of the Elbe and Danube river basins in Europe, eastern Germany was hit by two extreme floods in the years 2002 and 2013. Both flood events were characterised by unexpectedly high but almost identical flood peaks, leading to failures of many flood protection systems in the region. Though the impact of both flood events may have been similar, significant differences occurred in the generating mechanisms. The main difference lies in the temporal and spatial distribution of the rainfall events that caused these floods. This will be illustrated by the example of the Zwickau-Pölbitz gauge at the Zwickauer Mulde river in Saxony, eastern Germany, which has a catchment size of 1022 km^2. Both rainfall events almost resulted in the same amount of the average areal precipitation for the Mulde basin, 183 mm in 2002 and 161 mm in 2013. In 2002,

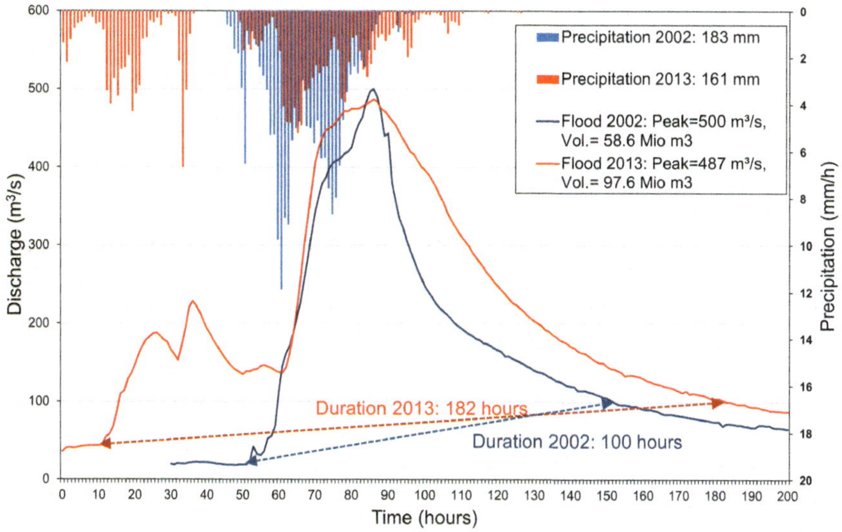

Fig. 4.1 2002 and 2013 flood events at the Zwickau/Zwickauer Mulde gauge in the Mulde river basin, eastern Germany, and the corresponding precipitation events

however, this amount of rainfall was unevenly distributed over three days: 16.8 mm at the first day, 141.8 mm at the second and 24.8 mm at the third day (Fig. 4.1a). This means that 77% of the rainfall causing the flood event fell down within one day, indicating high intensities of rainfall. In 2013, the rainfall that caused the flood events was distributed over eleven days. Within this time span, three partial rain events can be identified: the first three days with a total amount of 40.8 mm, the next five days with a total amount of 51.9 mm and the last four days with an amount of 97.0 mm (Fig. 4.1b). Compared to the year 2002, the 2013 flood is therefore characterised by rainfall with much longer duration and significantly smaller intensities as only 30% of the total amount of rainfall fell down within 24 h.

These different temporal distributions of rainfall are directly conveyed also to the hydrograph shapes of the floods. In 2002, the heavy rainfall with high intensity led to a fast and steep rising limb of the hydrograph. However, due to high flow velocities of surface runoff and the short duration, the recession of the hydrograph was also short with a total duration of only five days. In 2013, the different reactions of the catchment to each of the three partial rainfall events became visible. The first part of the rainfall only led to a small increase of the discharge. However, the soils got wetted. The second main interval of the rainfall event then led to a larger increase of discharge, but still a large amount of water was stored in the soils which became more and more saturated. The third part of the rainfall then led to a fast increase of discharge. This had two reasons: on the one hand, the recession of the hydrograph caused by the second partial rainfall was not finished when the third partial rainfall started and the newly generated runoff was overlaid. On the other hand, the soil was already saturated in large parts of the catchment by the first two partial rainfall events such that much less water could be stored. In total, not only the large flood peak proved to be disastrous for the Mulde basin but also the large volume of the 2013 flood event led to high damages and failing flood retention systems.

This short introduction offered only a glimpse on the complex topic of flood-generation. There exist much more detailed studies on how floods are caused and which components are the most relevant. However, this is beyond the scope of this book. Instead, this short introduction should demonstrate how complex flood processes can be. To understand the difference between normal floods and extreme floods and to better capture this in flood statistics, these flood-generating processes have to be considered. Due to the different flood-generating processes, flood events cannot be statistically treated alike. This is where the flood typology comes into play.

For many years now, in hydrological sciences, flood types have been analysed and different classification schemes have been introduced. A consistent classification of floods is not only relevant for flood statistics, where the use of inhomogeneous samples (due to different flood-generating processes) may cause uncertainty, but also to understand and identify changes in the flood regime of catchments which are often caused by changing meteorological or hydrological conditions. In most flood classifications, flood types are identified according to detailed catchment analyses and pre-assumed hydrological processes. Seldom, also data mining techniques are applied which require less knowledge on the hydrological processes in a catchment. Any attempt to consider deterministic processes to explain the differences between observed flood events by the interplay between meteorological drivers and catchment response is beneficial when aiming to not only obtain homogeneous samples but also link the statistical findings to hydrological processes.

4.2 How Detailed Can Floods Be Analysed?

Several flood typologies have been developed under the assumption of pre-existing hydrological processes. The main criteria for the classification can be grouped into four categories: Seasonal aspects (winter and summer flood events or hydro-meteorological phenomena such as, e.g., monsoon or snowmelt which happen more or less regularly in the course of the year) (Strupczewski et al. 2012), hydro-climatic classification (large-scale circulation patterns and atmospheric state at the time of the event) (Hirschboek 1988; Bardossy and Filiz 2005), causative classification of river flood events (catchment scale precipitation patterns and antecedent catchment state) (Tarasova et al. 2018; Stein et al. 2019), and hydrograph-based classification (indirectly considering generating mechanisms through their effects on hydrograph characteristics) (Fischer et al. 2016). The seasonal approach is widely applied as it does not require many additional data besides the day of occurrence of the flood. However, it has the limitation that different flood types within one time period cannot be differentiated. This is a drawback for most regions, where, e.g., different types of rainfall or a mixture of rainfall and snow(melt) occur during a season. Hydro-climatic classification focuses on large-scale climatological patterns and requires knowledge on the states of the atmosphere and the path of moisture transport. The catchment state, however, is usually ignored in this approach and the link to floods is only made by the day of occurrence. In the causative classification, instead, the precipitation event (usually limited to the catchment-scale) and the catchment state are considered. The most simple classification is based on a differentiation into rainfall-driven and snowmelt-induced floods only (Kampf and Lefsky 2016; Vormoor et al. 2016). This requires knowledge on the precipitation event as well as the snowmelt in the catchment. To extend the hydrological knowledge included in the classification, additional indices for, e.g., the soil moisture or the runoff time can be included. This knowledge can be increased further, if the spatial and temporal distribution of rainfall is linked with the catchment state to obtain pre-defined flood types (Tarasova et al. 2020). However, such an approach does not only require large amounts of additional data (soil moisture, high resolution rainfall data, temperature, soil and topography), it may also lead to many different flood types. For example, Tarasova et al. (2020) defined 16 different flood types. Though such a classification may include much hydrological information, it is a drawback for (statistical) analyses since the samples of each flood type are rather small. Another challenge for all classifications is the availability of data on the meteorological drivers and the states of watershed. Often, there is a lack of such data for long time series, which should be used to reduce the stochastic uncertainties of statistical analyses. Thus, instead of focusing on the many combinations of catchment states and drivers, the hydrograph-based classification focuses on the response of catchments, i.e., the flood event itself. Fischer et al. (2016) use the flood timescale to derive different flood types according to the hydrograph shape. Of course, also aspects of different classifications can be combined. Fischer et al. (2019), for example, combine the causative classification with a hydrograph-based classification. This approach has the advantage that on the one hand hydrological knowledge on the processes is included in the classification, but on the other hand no extensive additional data are required. This makes the approach applicable to long observation periods. For such observation periods, usually information on soil or atmosphere are not available. However, rainfall and temperature data have long records and can be included. The availability for long observation periods makes such hybrid classification attractive for flood statistics: Any classification of flood events with the aim to produce homogenous samples of flood events has to consider the need to subdivide the observed floods into groups which should be large enough for statistical analyses, the fitting of distribution functions and the establishing of a mixture model.

For more details on the categories of flood classification, the overview of Tarasova et al. (2019) is recommended.

4.3 Information Extension: Partial Duration Series

In the previous section, the need for extensive sampling for each flood type was pointed out if flood statistical approaches are to be applied. The methodology of the flood type classification can help to obtain meaningful and large subsets of flood types. But also the statistical approach of flood frequency analyses has an impact on the number of flood events available for estimating flood statistics. Obviously, the number of annual maxima is too small to be subdivided into several types, since only one event per year is considered. The partial duration series of floods (Sect. 2.2), instead, significantly increase the number of floods under consideration, as long as the threshold is defined appropriately. By considering several floods per year, the representativeness of the resulting samples and the reliable estimation of frequencies of event types within the statistical model are ensured. At the same time, the threshold which has to be applied to specify a flood event has to be selected in such a way that the number of events should not be too large to avoid that any increase of runoff is considered as a flood event. The definition of the threshold as well as the flood event separation play a crucial role here. In Chap. 5, this problem is discussed in greater detail and an opportunity to handle this problem is proposed.

Flood types allow for considering the flood-generating mechanisms, the atmospheric or the catchment conditions in the analyses of flood events. Yet, their applicability is limited by the data availability. Therefore, flood typologies can be of great use as long as they are applied in a meaningful way having in mind the purpose of the analyses. By combining flood types with an appropriate statistical model, the information content in flood frequency analyses can be increased significantly.

References

Bárdossy A, Filiz F (2005) Identification of flood producing atmospheric circulation patterns. J Hydrol 313(1–2):48–57

Brunner MI, Fischer S (2022) Snow-influenced floods are more strongly connected in space than purely rainfall-driven floods. Environ Res Lett 17(10)

Fischer S, Schumann A, Bühler P (2019) Timescale-based flood typing to estimate temporal changes in flood frequencies. Hydrol Sci J 64:457–470

Fischer S, Schumann A, Schulte M (2016) Characterisation of seasonal flood types according to timescales in mixed probability distributions. J Hydrol 539:38–56

Hirschboek KK (1988) Flood hydroclimatology. In: Baker VR, Kochel RC, Patton PC (eds.): Flood geomorphology: Wiley-Interscience, pp 27–49

Kampf SK, Lefsky MA (2016) Transition of dominant peak flow source from snowmelt to rainfall along the Colorado front range: historical patterns, trends, and lessons from the 2013 Colorado front range floods. Water Resour Res 52(1):407–422

Krug A, Primo C, Fischer S, Schumann A, Ahrens B (2020) On the temporal variability of widespread rain-on-snow floods. Meteorol Z 29(2):147–163
Stein L, Pianosi F, Woods R (2019) Event-based classification for global study of river flood generating processes. Hydrol Process 34:1514–1529
Strupczewski WG, Kochanek K, Bogdanowicz E, Markiewicz I (2012) On seasonal approach to flood frequency modelling. Part I: two-component distribution revisited. Hydrol Process 26:705–716
Tarasova L, Basso S, Zink M, Merz R (2018) Exploring controls on rainfall-runoff events: 1. Time-series-based event separation and temporal dynamics of event runoff response in Germany, Water Resour Res
Tarasova L, Basso S, Wendi D, Viglione A, Kumar R, Merz R (2020) A process?based framework tocharacterize and classify runoff events: The event typology of Germany. Water Resour Res 56:e2019WR02695
Tarasova L, Merz R, Kiss A et al (2019) Causative classification of river flood events. WIREs Water 6:e 1353
Vormoor K, Lawrence D, Schlichting L, Wilson D, Wong WK (2016) Evidence for changes in the magnitude and frequency of observed rainfall vs. snowmelt driven floods in Norway. J Hydrol 538:33–48

Chapter 5
Type-Based Mixture Model of Partial Duration Series (TMPS)

Abstract This chapter introduces the Type-based Mixture model of Partial duration Series (TMPS). The TMPS approach allows for directly incorporating different flood types in a statistical model and enables the combination of the different type-specific distribution functions in a multiplicative mixture model. The statistical formulas for deriving joint flood quantiles and probabilities on an annual basis are given, thus allowing to derive design flood estimates. Moreover, the flood-type specific distribution functions are defined by application of the classical Poisson-model for peak-over-threshold series (POT). The TMPS is compared to classical flood statistics based on annual maximum series or POT without differentiation between flood types. It is shown in which cases the application of the TMPS model is beneficial.

As introduced in Chap. 2, flood statistics build the foundation for deriving design floods. Usually, two common approaches are distinguished: the AMS and the POT approach. The POT approach offers the possibility to include more than one event per year and hence is preferable to the annual maximum series if different flood types occur in flood series, as all events can be considered in the model. It was demonstrated in the previous Chap. 4, how different flood event characteristics may result from different meteorological conditions. Floods caused by short but heavy rainfall events are significantly different in their peak-volume relationships and hydrograph shapes compared to floods caused by long-duration rainfall. Such differences indicate inhomogeneous samples, which violates the fundamental statistical assumption of identically distributed variables (Klemes 2000). Indeed, it is well-known practice in the meteorological sciences to distinguish between different processes and thus to account for inhomogeneity in the samples (Rossi et al. 1984; Gioia et al. 2008). In hydrology, several attempts to consider such differences failed and statistical analyses are still based on a common sample of all flood events. Hydrological statistical models that aim in a similar direction as the meteorologically differentiated models are seasonal statistics. When seasonal statistics are applied, the samples are divided according to different seasons. The classical differentiation is into winter and summer floods as proposed, e.g., by Waylen and Woo (1982), Strupczewski et al. (2012), but the exact time of seasonal splitting may vary over the different regions. There

S. Fischer and A. H. Schumann, *Type-Based Flood Statistics*, Water Science and Technology Library 124, https://doi.org/10.1007/978-3-031-32711-7_5

may also be reasons to define more seasons, e.g., midsummer with heavy rain events caused, e.g., by thunderstorms, or monsoon times, as long as there is sufficient data available. The idea behind this differentiation is always to obtain sub-samples that then can be assumed to be identically distributed such that requirements of the standard statistical theory are fulfilled. Additionally, also statistical approaches exist to define homogeneous sub-samples by statistical means in a sample. Often, the different sub-samples (periods) are then combined in an additive mixture model to obtain a joint distribution. The drawback of such a purely statistical approach is that the sub-samples are usually not hydrologically meaningful, i.e., there is no deterministic cause behind the division. The seasonal differentiation instead is a proxy for a hydrological understanding of the temporal variability of flood-generating processes. It is based on different time periods, where the majority of flood events originates from a specific flood-generation process, e.g., snowmelt. What is ignored is that even in one season different flood-generating processes may occur: the summer season does not only consist of heavy rainfall events, but includes also synoptic rainfall events. Especially the spring season in Central Europe and Northern America may be quite diverse regarding the flood-generating processes, including snowmelt as well as rainfall. Even in winter time, not all floods are affected by snowmelt in Central Europe. The relevance and frequency of the different flood types may vastly depend on the catchment characteristics. Hence, a seasonal differentiation may not be meaningful since inhomogeneity of the samples still persists in many cases.

Instead, a flood typology should be combined with a statistical mixture model to allow for a consideration of different flood-generating processes in flood statistics. Such an approach is introduced in this chapter.

5.1 Statistical Mixture Models in Hydrology

When considering different samples, either due to seasonal, statistical or process-based aspects, the different distributions of these samples have to be combined statistically to obtain joint probabilities, which are the basis for design floods. This can be achieved by the application of mixture distributions.

In statistics, a mixture distribution is defined as the probability distribution (pdf) of a random variable that has been drawn from more than one probability distribution, the so-called component distributions. The classical mixture distribution function H (in the continuous case) consists of a convex distribution of other pdfs assigned with weights:

$$H(x) = \sum_{i=1}^{k} \omega_i F_i(x), \tag{5.1}$$

where k is the number of component pdfs, $F_i(x)$ is the pdf of component distribution i and $\omega_i \geq 0$ is the weight of component distribution i with $\sum_{i=1}^{k} \omega_i = 1$. Analogously, the mixture probability density is defined as

$$h(x) = \sum_{i=1}^{k} \omega_i f_i(x), \tag{5.2}$$

where f_i is the probability density corresponding to F_i. Theoretically, k can converge to ∞, though this is of less practical use in hydrological statistics.

In flood statistics, mixture distributions have been applied to model different components in a sample of flood events. For AMS in the USA, multiple approaches with combinations of two or three distributions of the same distribution family such as Gumbel respectively GEV, Log-PearsonIII or Normal distributions exist, e.g., in Singh (1968), Singh and Sinclair (1972). Similarly, Szulczweski (2018) modelled the AMS by a mixture model consisting of GEV and Gamma distributions under the assumption that floods and maximum flows (which are defined as high runoffs but not in the range of floods) are both included in the AMS. These approaches, however, do not use a deterministic distinction into different samples but optimise the mixture parameters by criteria such as least squares. In contrast, often the sample of floods is split first into sub-samples, e.g., according to season or other deterministic causes, and the component distributions are fitted to these sub-samples before combining them in a mixture model. Waylen and Woo (1982), Woo and Waylen (1984) differed between rainfall and snowmelt floods in Canada according to meteorological data and fitted a Gumbel distribution to each sub-sample. Again, it is assumed that at least within one season floods are identically distributed. Under the same assumption, (Todorovic and Rousselle 1971) apply a two-component mixture distribution consisting of Gumbel distributions, commonly denoted as the two-component extreme value distribution (TCEV). Other applications of the TCEV can be found in Cunderlik et al. (2004), Allamano et al. (2011). Rossi et al. (1984) use the same model but instead of a seasonal distinction they differ between basic and outlying floods in the AMS in Italy.

What is missing so far in statistical hydrology, is the incorporation of deterministically derived flood types in a statistical mixture model. Such an approach is described in the following section.

5.2 The TMPS Model

To overcome the inhomogeneity of the flood event sample and at the same time obtain hydrologically meaningful sub-samples, the flood events can be differentiated according to their flood type (Chap. 4). Each type-specific sample then has its own distribution. Statistically speaking, suppose that there exist n_t flood types resulting in samples of peaks $X_{1;j}, \ldots, X_{N_j;j}$, $j = 1, \ldots, n_t$, where N_j is the number of events of type j. To omit a too high uncertainty, a minimum sample size for each flood type should be available. Fischer and Schumann (2023) propose $N_j > 10$. Flood types which do not fulfill this requirement can be merged with the most similar flood type, e.g., R3 floods with R2 floods, to not loose too much information. Similar to the POT approach, not all events should be included in the statistical analyses but only the large ones referring to floods. Thus, for each flood type j, a threshold u_j is defined, for which each exceedance indicates a flood event. That is, if $X_{i;j} > u_j$, then $X_{i;j}$ is a flood peak. The type-specific threshold is required as the absolute size of flood peaks may vary according to the baseflow regime. In some cases, alpine rivers which have higher baseflow in summer due to the impact of glaciers, have relative small flood peaks in other seasons. A distribution for these type-specific exceedances arises naturally from statistical theory: the GPD distribution (Sect. 2.2). The distribution of all exceedances of type j, $G_j(x;\theta_j,u_j)$, is modelled by the GPD distribution for each flood type separately with parameter set $\theta_j = (\kappa_j, \beta_j)$. Analogous to the POT-approach, the distribution of each flood type can be generalised to an annual distribution $\tilde{G}_j$ by application of the total probability theorem to obtain

$$\begin{aligned}\tilde{G}_j(x) = &\sum_{k=0}^{\infty} \mathbb{P}_j(l=k)(G_j(x,\theta_j,u_j))^k \\ = &\sum_{k=0}^{\infty} \mathbb{P}_j(l=k)\left(1-\left(1+\kappa_j\left(\frac{x-u_j}{\beta_j}\right)\right)^{\frac{-1}{\kappa_j}}\right)^k,\end{aligned} \tag{5.3}$$

where $\mathbb{P}_j(l=k)$ is the probability that the annual number l of flood peaks of type j above the threshold u_j is equal to k and can be described by the Poisson distribution with parameter λ_j

$$\mathbb{P}_j(l=k) = \frac{\lambda_j^k}{k!}e^{-\lambda_j}$$

as shown in Sect. 2.2. The resulting distribution $\tilde{G}_j$ is once again a GEV distribution with parameters $\xi_j' = \kappa_j, \sigma_j' = \beta_j\lambda_j^{\kappa_j}$ and $\mu_j' = u_j - \beta_j(1-\lambda_j{}^{\kappa_j})/\kappa_j$. Although having a separate distribution for each flood type may reveal interesting information on the flood regime, it does not serve the aim of flood frequency analyses: the assessment of a design flood for a given return period. Practitioners require such design floods for the planning and operation of flood protection systems. An example of such an application would be: if an inundation problem arises by a flood peak of x m^3/s, the probabilities of such a peak can be estimated for each flood type separately. But, for design purposes, the joint probability related to the combination of all flood types is required. Hence, the different flood types once again have to be combined. However, this time not the flood events are considered jointly in one sample, but the different distributions are considered jointly. This can be reached statistically by application of a mixture model.

Such an approach that explicitly considers flood types in statistics was first described by Fischer (2018), Fischer et al. (2019). In their model, which is based on the approach of Zhou et al. (2016), the type-specific distribution of the exceedances,

G_j, is combined with the type-specific probability of exceedance of the threshold, $1 - F_j(u_j)$. F_j in this case is the distribution of all flood peaks of type j below and above the threshold u_j. Fischer et al. (2019) propose to use a GEV distribution for F_j. Due to this model concept, the probability of all large floods above the threshold u_j is weighted with the probability of exceedance of the threshold.

The pdf of the type-based mixture model of partial duration series (TMPS) is thus defined as

$$
\begin{aligned}
H(x) &= \prod_{j=1}^{n_t} \left(G_j(x;\theta_j,u_j)(1 - F_j(u_j)) + F_j(u_j)\right) \qquad (5.4)\\
&= \prod_{j=1}^{n_t} \left(\left(1 - \left(1 + \kappa_j \left(\frac{x - u_j}{\beta_j}\right)\right)^{\frac{-1}{\kappa_j}}\right) \left(1 - \exp\left(-\left(1 + \xi_j \frac{u_j - \mu_j}{\sigma_j}\right)^{-\frac{1}{\xi_j}}\right)\right)\right.\\
&\quad \left. + \exp\left(-\left(1 + \xi_j \frac{u_j - \mu_j}{\sigma_j}\right)^{-\frac{1}{\xi_j}}\right)\right).
\end{aligned}
$$

It has to be noted that in the TMPS model no convex combination of the mixture components is used but instead the product is considered. This multiplicative approach is reasoned by the fact that the maximum flood event per year is of interest. Since the distribution function of the maximum of n independent random variables is simply the product of the n distribution functions, the product is applied here to the mixture components as well. If the MQ varies distinctly between the seasons, this is automatically mirrored in the type-specific defined parameters mu_j and in σ_j, which also result in seasonally varying parameters.

Due to the complex equation to derive the distribution function of the TMPS model, no analytical function can be derived to estimate the quantiles. To nevertheless obtain flood quantiles for given return periods, which is essential for flood statistics, the quantiles have to be derived by numerical optimisation. For this purpose, classical optimisation algorithms like the Broyden-Fletcher-Goldfarb-Shanno algorithm (BFGS, Fletcher 1987) or variants of it can be used to minimise the squared difference between the given return period and the return period derived for any flood discharge in a pre-defined range (typically this range is between 0 and 100000, but it can be reduced for computational efficiency according to the catchment size). The derived flood quantile then is the discharge that results in a return period with minimal squared distance to the given return period. Alternatively, one could estimate the return periods for given flood discharges and fit a spline or polynomial function to these supporting points. For this purpose, the number and values of the supporting points have to be selected carefully to represent the full range of flood quantiles. Moreover, the fitted polynomial should neither have a too high nor a too low degree to optimally fit to the supporting points. As a rule of thumb, no less than ten supporting points for a fitting between return periods of 2 and 200 years and polynomials of degree two up to four should be considered.

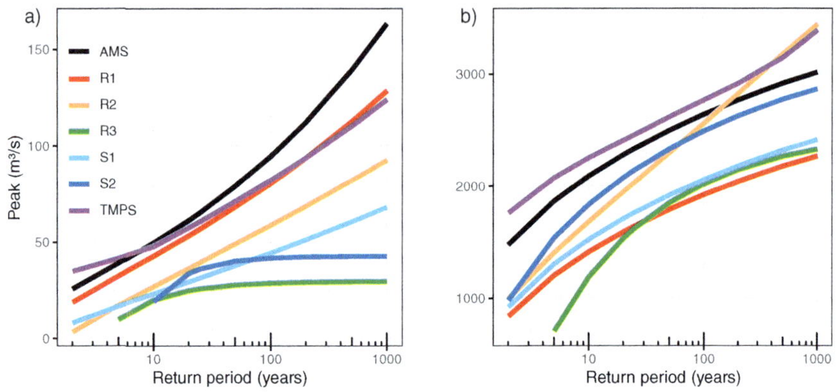

Fig. 5.1 Type-specific distributions and TMPS model for Achsheim/Schmutter and Schwabelweis/Danube gauges in Bavaria, Germany

As an example for the application of type-based statistics, the distribution of each flood type as well as the TMPS model are given for the two gauges Achsheim at Schmutter river and Schwabelweis at the Danube river (Fig. 5.1). Both gauges are located in Bavaria, Germany, the former belonging to a rather flat catchment of size 359.1 km^2 and the latter belonging to the large catchment of the Danube with 35476 km^2. As a comparison, also the distribution function of the AMS (with L-Moments and GEV) is given. The typology used here will be described in more detail in Chap. 6. The results demonstrate that for the small catchment, the highest quantile estimates for large annualities ($T > 20$ years) are derived from the rainfall-induced flood types R1 and R2. Especially the flood type R1 dominates the right tail, leading to a heavy tail behaviour and a convex shape of the TMPS distribution, which follows the dominant flood type, i.e., the one resulting in the largest quantiles. The snow-impacted flood types S1 and S2 and the long-duration rain-floods R3 instead only contribute little to the right tail. The TMPS distribution of the large Danube catchment instead is affected by more flood types. Especially flood types R2 (medium rainfall-floods) and S2 (snowmelt floods) contribute to the right tail. However, the right tail is much less pronounced than for the smaller catchment, with a tendency to a concave shape of the distribution function. Heavy-rainfall does not play a crucial role for the generation of large floods in this catchment. A more detailed comparison of the distribution of the flood types and the TMPS model for different gauges in Germany is given in Part IV. In the next Section, the TMPS model is compared in greater detail to the classical FFA approaches, AMS and POT. Already Fig. 5.1 revealed that the TMPS may deviate from the AMS distribution. This depends mainly on the flood types and the shape of their distribution. It will be investigated in how far the flood types have an impact on the resulting TMPS model.

5.3 Comparison of TMPS to Common Approaches Based on Monte-Carlo Simulations

When applying complex models such as the TMPS with distinctly more parameters than classical approaches such as AMS and POT, the question of over-parametrisation and increased uncertainty arises. The higher the number of parameters, the higher is the uncertainty in estimation under the assumption of the limited same sample size, as pointed out already in the previous section. The TMPS model with five flood types has 30 parameters, compared to two or three parameters for the classical AMS approach. At the same time, a larger data basis is considered and more information is incorporated in the model. This makes detailed investigations of the suitability of the proposed model and the uncertainty introduced by its application necessary. In statistics, such comparisons are usually made by extensive simulation studies. In these so-called Monte Carlo simulation, repeatedly random samples are drawn from a known distribution. A large number of repetitions ensures the asymptotic validity of the result. Typically, such simulations are performed for different combinations of parameters and sample sizes, to understand the impact of the sample moments as well as the sampling effect. For example, Madsen et al. (1997) compare the POT and the AMS model for different shape parameters of the GEV model, different sample sizes as well as different return periods. This has the advantage that, in contrast to the real-life data comparison in the previous section, the true underlying distribution is known and can be used as comparison value. For the TMPS model, the simulation approach is built on the benchmark methodology by Madsen et al. (1997) and the procedure is extended to make it suitable for a comparison of AMS, POT and the TMPS model as follows.

The aim of the analyses here is to compare AMS, POT and TMPS models concerning bias and Root Mean Squared Error (RMSE), typical measures for systematic over-/underestimation. Different return periods and sample sizes are compared that lead to recommendations on when to use which model. However, it is not straightforward to design such Monte Carlo simulations. Already the comparison of POT and AMS model led to several problems that were addressed by Madsen et al. (1997): the simulation of a POT sample from a given distribution such as GPD does not provide sufficient information about the AMS and vice versa. When simulating a POT sample, the probability that one event of the AMS is below the POT-Threshold is $\exp(-\lambda)$ when λ is the mean annual number of events above the threshold. Similarly, the simulation of an AMS sample will exclude several POT events, especially if $\lambda > 1$. To overcome this problem, they made use of the threshold-stability of the GPD, i.e., the above-excess distributions are invariant to the choice of u.

Here, a similar technique is used to design the Monte Carlo simulation. However, for the TMPS model an additional component has to be considered compared to the POT model: it makes use of the non-exceedance probability of the threshold u_j. This probability cannot be derived from a POT-sample and hence it is impossible to know the true value when performing a Monte Carlo simulation. Therefore, this probability is fixed for each flood type. Of course, this reduces the complexity of the TMPS model and hence a comparison with the AMS model might not be totally fair, but it is the only way to design such an experiment. Moreover, it only has small impact on the TMPS results, which are mainly influenced by the GPD distribution (Fischer 2018). This non-exceedance probability of the threshold is denoted with $p_{0;j}$ for each flood type $j = 1, \ldots, n_t$. Though p_0 is somehow related to the threshold u, it cannot be estimated from u, since p_0 requires more information, also from the values below the threshold.

Additionally, the average number of events per flood type and year has to be defined. This number is denoted by λ_j. It is not required for the TMPS model but to derive the AMS model from the sample, as shown later on.

With these assumptions, first a sample of size n years is drawn from the GPD distribution for each flood type. For this purpose, $n \cdot \lambda_j$ values $X_{1;j}, \ldots, X_{n\lambda_j;j}$, the exceedances of the threshold, are generated from the GPD using the parameter set $\theta_j = (\kappa_j, \beta_j)$ and the threshold u_j.

The assumed distributions are used to generate annual samples

$$X_{1;1}, \ldots, X_{n\lambda_1;1}, \ldots, X_{1;n_t}, \ldots, X_{n\lambda_{n_t};n_t}$$

which are distributed according to the TMPS model with parameters $\kappa_1, \ldots, \kappa_{n_t}$, $\beta_1, \ldots, \beta_{n_t}, u_1, \ldots, u_{n_t}$ and $F_j(u_j) = p_{0;j}$.

Following (Madsen et al. 1997), the corresponding AMS then can be derived by taking the maximum of each year and adding the threshold parameter u. However, since n_t different flood types are considered, this has to be extended. In this case, the AMS of year i, Y_i, is defined as the maximum of the maximum value of each flood type per year plus the type-specific threshold u_j:

$$Y_i = \max_{1 \le j \le n_t} \left(\max_{(i-1)\lambda_j + 1 \le k \le i\lambda_j} X_{k;j} + u_j \right), i = 1, \ldots, n.$$

In a next step, the three models, TMPS, POT and AMS, are fitted to the generated samples respectively. For the TMPS, the GPD distribution is fitted for each type separately using the L-moments. For the AMS model, the GEV is fitted to the AMS sample using the L-moments. Both scenarios are also compared to the POT model as described in (Madsen et al. 1997), where a GPD distribution is fitted to the POT sample. Then, the respective quantiles for a given return period T are estimated. This procedure is repeated N times and normalised Bias and RMSE are derived:

$$Bias(T) \quad = sum_{i=1}^{N} \frac{q_T - \hat{q}_T}{q_T} \tag{5.5}$$

$$RMSE_T = \sqrt{\frac{1}{N} \sum_{i=1}^{N} (\frac{q_T - \hat{q}_T}{q_T})^2}, \tag{5.6}$$

where q_T is the true quantile and $\hat{q}_T$ is the estimated quantile either from the TMPS, POT or the AMS model for return period T. The whole simulation procedure is summarised in Fig. 5.2.

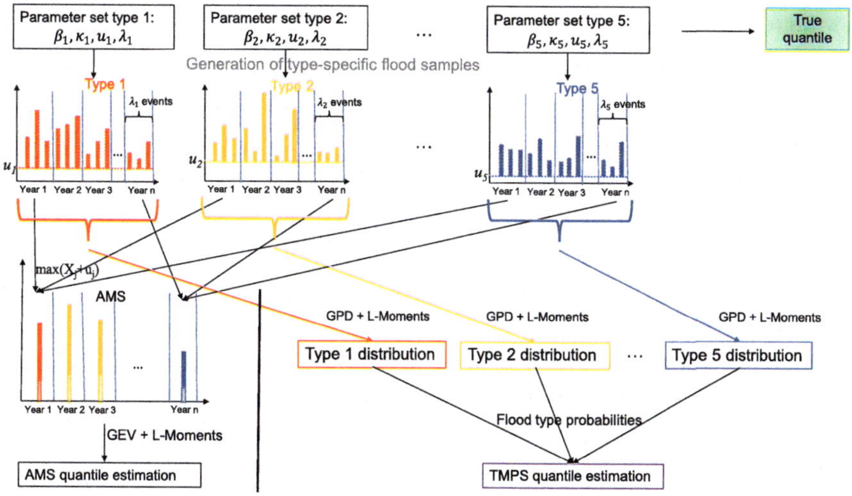

Fig. 5.2 Simulation procedure to obtain the quantiles of AMS and TMPS estimation (POT is analogous)

To limit computational efforts, only *L*-Moments are considered for parameter estimation. For right-skewed samples, referring to a shape parameter larger than 0.2, which is often the case for samples of flood peaks, Madsen et al. (1997) show that for the AMS as well as the POT series the *L*-moment estimates are preferred regarding RMSE for small and moderate sample sizes as well as return periods. Similar results have been reported by Hosking (1986), Hosking et al. (1985) for POT and AMS series. The MoM estimates, instead, often led to a large negative bias for a shape parameter larger than 0.3, while the ML-estimator does not converge for small sample sizes and small shape parameters (Madsen et al. 1997). Thus, reduction of the setting to *L*-Moments only, reduces the complexity of the simulation study and, since small and large sample sizes and return periods are considered, reliable estimates for all these ranges are guaranteed.

For the simulation study, different settings of the assumed TMPS model are considered. For each setting, 1000 runs are performed. A higher number of runs would be computationally too expansive due to the optimisation procedure used to derive the TMPS quantiles. Return periods T are varied between $T = 2, 5, 10, 20, 25, 50, 100, 200, 500$ years to consider the full spectrum of small and extreme quantiles and therefore possible heavy tails in the data. Sample sizes n are varied between $n = 30, 50, 100, 200, 500$ years and the case where asymptotic behaviour should be valid, $n = 10000$. A number of five flood types is considered to have a reasonable variation between the flood events. The probability of exceedance of the threshold is chosen as $p_{0;1} = 0.24$, $p_{0;2} = 0.19$, $p_{0;3} = 0.17$, $p_{0;4} = 0.26$ and $p_{0;5} = 0.14$. These values have been obtained by averaging all probabilities of exceedances of the respective flood types for a large sample of gauges in Bavaria, Germany. The variability in the probability is considered in the analysis of the results.

All remaining parameters are varied in different scenarios.

1. Scenario: Variation of shape, one event per year
   The number of events per year is chosen as $\lambda_j = 1$ for all $j = 1, \ldots, 5$, the scale parameter is chosen as $\beta_j = 5$ for all $j = 1, \ldots, 5$ and the threshold parameter is set to $u_j = 10$, $j = 1, \ldots, 5$. The shape parameter is first set to $\kappa_j = 0.2$, but the number of flood types with $\kappa_j = 0.6$, corresponding to a heavy tail, is increased successively from $n_E = 0$ to $n_E = 5$ in different simulation runs.
2. Scenario: Variation of shape, two events per year
   Same as Scenario 1, but here the number of events per year is chosen as $\lambda_j = 2$ for all $j = 1, \ldots, 5$.
3. Scenario: Variation of scale
   Same as Scenario 2 but with a constant shape parameter $\kappa_j = 0.2$ and a constant threshold parameter $u_j = 10$ for all $j = 1, \ldots, 5$. Here, the scale parameter is first set to $\beta_j = 5$, $j = 1, \ldots, 5$, but the number of flood types with $\beta_j = 20$, corresponding to a high variation, is increased successively from $n_E = 0$ to $n_E = 5$ in different simulation runs.
4. Scenario: Variation of location
   Same as Scenario 2 but with constant shape parameter $\kappa_j = 0.2$ and scale parameter $\beta_j = 5$ for all $j = 1, \ldots, 5$. Here, the threshold parameter is first set to $u_j = 10$, $j = 1, \ldots, 5$ and the number of flood types with a threshold $u_j = 50$, corresponding to a high mean value, is increased successively from $n_E = 0$ to $n_E = 5$ in different simulation runs.
5. Scenario: Variation of shape and scale
   Same as Scenario 2 but both the shape and the scale parameters are varied for an increasing number of flood types. The shape parameter is first set to $\kappa_j = 0.2$, but the number of flood types with $\kappa_j = 0.6$, corresponding to a heavy tail, is increased from $n_E = 0$ to $n_E = 5$. The scale parameter is first set to $\beta_j = 5$, but the number of flood types with $\beta_j = 20$, corresponding to a high variation, is increased successively from $n_E = 0$ to $n_E = 5$ in different simulation runs.

The scenarios are summarised in Table 5.1.

Results in terms of normalised bias are given for different return periods and sample sizes in Figs. 5.3, 5.4, 5.5, 5.6 and 5.7 for all scenarios. Mean normalised bias as well as RMSE are given for selected Scenarios and conditions in Table 5.2 to obtain information on the variability of the estimation for 1000 repetitions. Due to the low performance results of the POT model in Scenario 1 (Fig. 5.3), this model is no longer considered for the remaining scenarios.

For the first scenario, where the shape parameter is increased from $\kappa_j = 0.2$ to $\kappa_j = 0.6$ for a varying number of flood types j and one flood event of each flood type is considered per year, the TMPS model results in a normalised bias around zero (Fig. 5.3). Only for small sample sizes of $n = 30$ up to $n = 100$, return periods greater than 50 years and no extreme shape parameter for any of the flood types, the bias increases to a maximum of 0.4. Therefore, the TMPS is well applicable for the estimation of the mixture distribution, even for small sample sizes and large

Table 5.1 Overview of the five simulation scenarios

Scenarios	Number of events per year λ_j	Shape parameter κ_j	Scale parameter β_j	Threshold parameter u_j
1	1	0.2, but gradually replaced by 0.6 for a number of 1–5 flood types	5	10
2	2	0.2, but gradually replaced by 0.6 for a number of 1–5 flood types	5	10
3	2	0.2	5, but gradually replaced by 20 for a number of 1–5 flood types	10
4	2	0.2	5	10, but gradually replaced by 50 for a number of 1–5 flood types
5	2	0.2, but gradually replaced by 0.6 for a number of 1–5 flood types	5, but gradually replaced by 20 for a number of 1–5 flood types	10

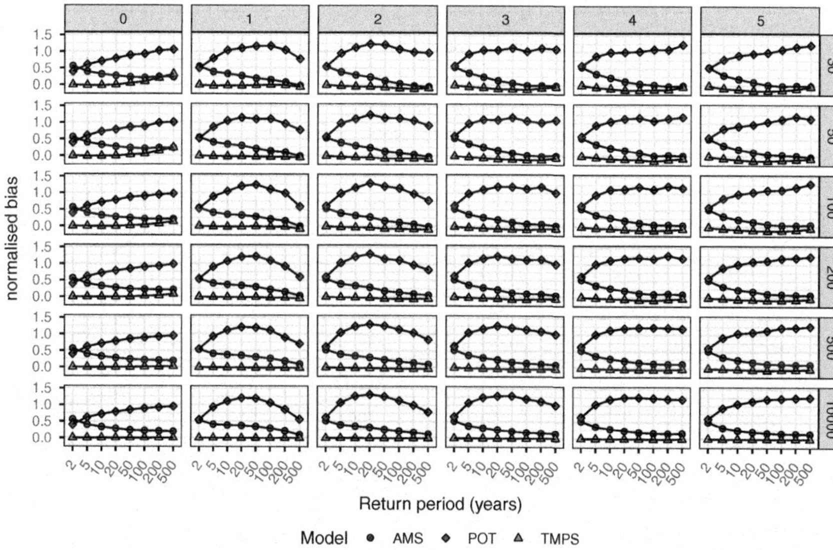

Fig. 5.3 Mean normalised bias for different return periods stratified by sample size n (rows) and number of flood types with extreme parameters n_E (columns) for AMS, POT and TMPS models in Scenario 1

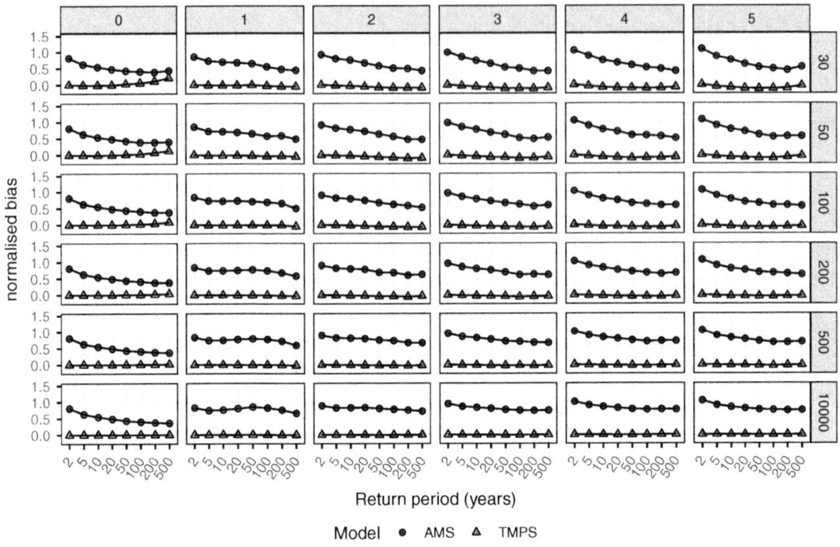

Fig. 5.4 Mean normalised bias for different return periods stratified by sample size n (rows) and number of flood types with extreme parameters n_E (columns) for AMS and TMPS models in Scenario 2

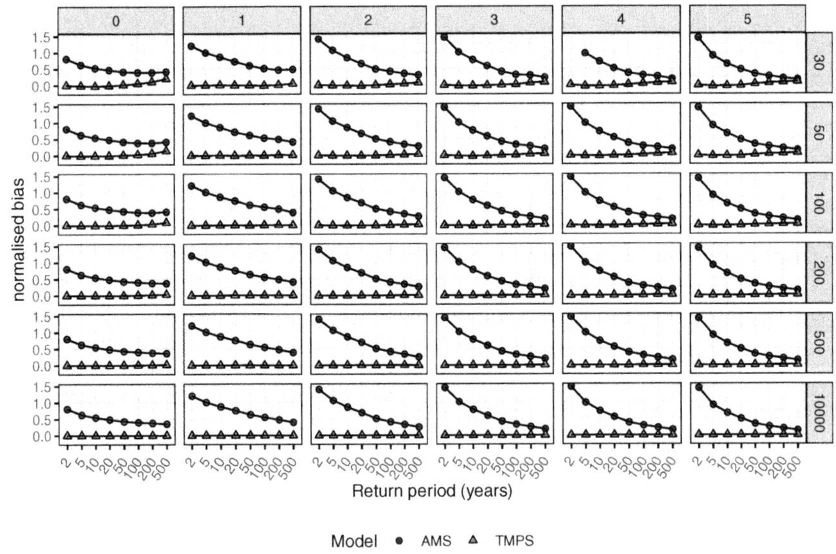

Fig. 5.5 Mean normalised bias for different return periods stratified by sample size n (rows) and number of flood types with extreme parameters n_E (columns) for AMS and TMPS models in Scenario 3

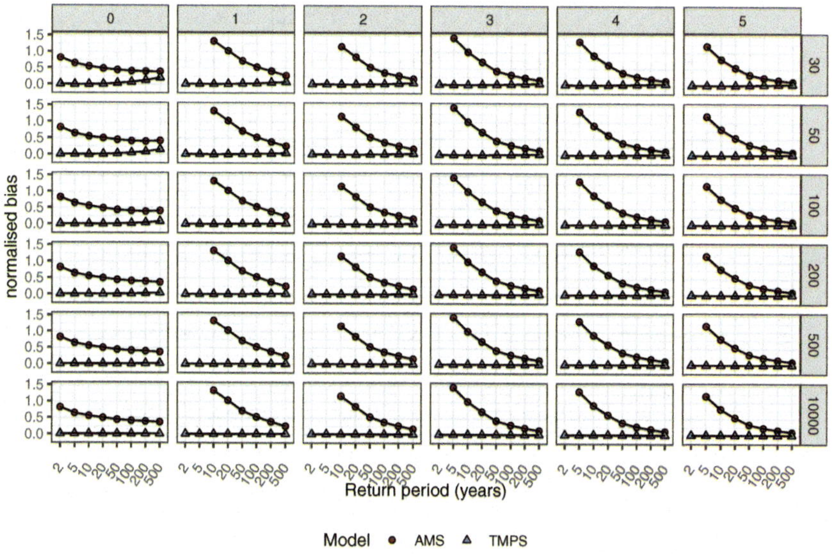

Fig. 5.6 Mean normalised bias for different return periods stratified by sample size n (rows) and number of flood types with extreme parameters n_E (columns) for AMS and TMPS models in Scenario 4

Table 5.2 Normalised bias and RMSE for selected scenarios and return period $T = 100$ with the AMS or TMPS model

	Bias				RMSE			
Length of time series	n = 30 years		n = 100 years		n = 30 years		n = 100 years	
Model	AMS	TMPS	AMS	TMPS	AMS	TMPS	AMS	TMPS
Scenario 1: Variation of shape, one event per year (varying κ and $\lambda_j = 1$)								
$n_E = 0$	0.210	0.099	0.211	0.036	0.293	0.250	0.182	0.137
$n_E = 2$	0.043	−0.096	0.118	−0.074	0.592	0.434	0.436	0.285
Scenario 2: Variation of shape, two events per year (varying κ and $\lambda_j = 2$)								
$n_E = 0$	0.403	0.055	0.408	0.021	0.386	0.186	0.217	0.098
$n_E = 2$	0.501	−0.095	0.617	−0.061	0.861	0.323	0.598	0.219
Scenario 3: Variation of scale (varying β and $\lambda_j = 2$)								
$n_E = 2$	0.395	0.024	0.393	0.006	0.491	0.252	0.292	0.130
Scenario 4: Variation of threshold (varying u and $\lambda_j = 2$)								
$n_E = 2$	0.645	0.017	0.664	0.009	0.206	0.118	0.132	0.068
Scenario 5: Variation of shape and scale (varying κ, varying β and $\lambda_j = 2$)								
$n_E = 2$	0.519	−0.110	0.612	−0.079	0.893	0.411	0.630	0.241

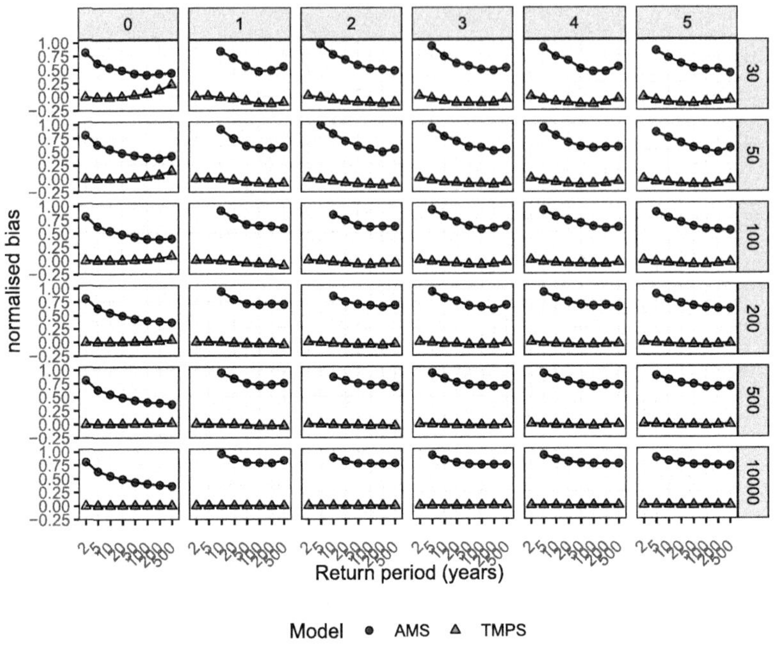

Fig. 5.7 Mean normalised bias for different return periods stratified by sample size n (rows) and number of flood types with extreme parameters n_E (columns) for AMS and TMPS models in Scenario 5

return periods, especially if one or more flood types have a high shape parameter. The AMS model instead results in constantly high bias of about 0.5 for small return periods of 2–5 years and all sample sizes. However, with increasing return periods, the bias decreases and for return periods of about 500 years the bias of the AMS model is comparable to that of the TMPS model. Similarly to the TMPS model, the bias for large return periods is smaller if one or more flood types have a large shape parameter. The POT is only comparable to the AMS model for very small return periods up to 5 years. For larger return periods and despite of the sample size, it leads to much higher bias. The POT model performance suffers most from the increasing number of flood types with large shape parameters, where especially the bias of the large return periods greater than 50 years increases. The very poor performance of the POT model in this case makes it unsuitable for the consideration of flood types. Therefore, it is no longer considered in the following scenarios.

For the second scenario, the number of flood events per year for each flood type is increased to $\lambda_j = 2$, while the remaining setting stays the same as in Scenario 1. In this case, the bias of the TMPS model still stays low around zero and is comparable to the first scenario (Fig. 5.4). The bias of the AMS model, however, increases significantly. Only for the case where none of the flood types has a large shape parameter, the bias of the AMS and the TMPS model are comparable for large return periods and small sample sizes. The more flood types are assigned to large shape parameters and the larger the sample size, the higher becomes the bias of the AMS model. For a number of five flood types with large shape parameter and a sample size of $n = 100$, for example, the bias is never smaller than 0.5.

A distinct bias of the TMPS model occurred in the third scenario, where again the number of flood events per flood type is set to $\lambda_j = 2$ per year, but instead of the shape parameter the scale parameter is altered from $\beta_j = 5$ to $\beta_j = 20$ (Fig. 5.5). For this scenario, the bias of the TMPS model increases for small sample sizes up to $n = 100$ years, large return periods and for those scenarios, where (almost) all flood types have equal scale parameter, independent of whether it is high or low. For all other cases, the bias of the TMPS model is still around zero. The case where none of the flood types has increased parameter values is of course identical to the second scenario, when also none of the parameters was altered. The AMS model in general results again in higher biases than the TMPS model, especially for small return periods. This bias reduces with increasing return periods, independent of the sample size. The largest bias occurs for those cases, where only one up to three flood types have large scale parameters.

Similar results can be observed for the fourth scenario, where instead of the scale parameter the threshold parameter is increased from $u_j = 10$ to $u_j = 50$ (Fig. 5.6). While the TMPS model has in general lower bias in this scenario, the AMS model results in even higher bias compared to Scenario 3.

Based on these results, for the final scenario, the two parameters with largest impact on the results are varied at the same time: the shape and scale parameter. On the one hand, the shape parameter is increased from $\kappa_j = 0.2$ to $\kappa_j = 0.6$ for an increasing number of flood types and similarly the scale parameter is varied between $\beta_j = 5$ and $\beta_j = 20$ with again a number of two flood events per flood type and year. This way, high bias can be expected for both AMS and TMPS models since the TMPS model is most affected by the shape parameter and the AMS model by the scale parameter. Except for small sample sizes of $n = 30$ and $n = 50$ years and high return period of more than 100 years together with small values for both parameters and all flood types, the TMPS model is only slightly affected by the changes in parameters (Fig. 5.7). The bias is constantly low and around zero, even for small sample sizes. The AMS model instead is highly affected by this scenario constellation and the bias is mostly above 0.5, no matter which return period or sample size is considered.

The results so far gave the bias of the estimation with either the AMS or the TMPS model. However, it is also of interest how high the variability of the estimation is to quantify the uncertainty. In Table 5.2, this variability is given in terms of the RMSE. The RMSE for the TMPS model and small sample sizes is comparable to that of the AMS for almost all scenarios. Interestingly, despite of the higher number of parameters of the model, the RMSE of the TMPS model decreases much faster for increasing sample sizes compared to the AMS model. Yet, for both models the variability is high for the given special cases.

The results reveal that there are indeed scenarios, where the AMS delivers comparable results in terms of bias to the TMPS model. This is especially the case if only one flood event per year and flood type occurs and for large return periods. For these cases, the value of the distribution parameters of the flood types is irrelevant for the estimation error. The AMS model is able to capture the distribution from the distribution of the dominant flood type if only one or two flood types have large shape parameters, since in this case the annual maximum flood will probably be of this flood type. The large shape parameter leads to extreme floods of this type which then will be also the annual maxima. However, if all flood types have similar parameters, there is no difference between the samples and it does not make a difference which flood type is represented in the AMS. Due to the simplicity of the AMS model compared to the TMPS model, the application of the AMS model is recommended in this case, especially for the estimation of design floods of large return periods. The scenario of only one flood event per year and type may occur in arid regions and therefore the AMS model may be sufficient there. However, the results revealed also that an increase of the number of flood events per year leads to large differences between the AMS and the TMPS model and that the TMPS model results in much smaller bias close to zero. These differences increase for increasing shape parameters of the distribution of exceedances of the flood types. This behaviour of the AMS model implies that it cannot incorporate the increased information that is obtained when having more than one flood event per year and flood type. Since the AMS approach only considers one event per year, it loses information: the larger the shape parameter and the more flood types are assigned with high shape parameters, the more large and extreme floods are generated. However, the annual maximum series can only consider one event per year and therefore several large events are ignored. This leads to higher bias of the AMS model. This performance of the AMS model is worsened when not the shape but the scale and location parameters are increased. These two parameters increase the variability and the general level of the flood peaks and the information loss by only considering one event per year is highest. The bias of the TMPS model, instead, increases most when having increasing shape parameters. The uncertainty introduced by the increased shape parameters multiplies due to the many parameters that have to be estimated for this model. Therefore, the TMPS model is very sensitive to high shapes. Nevertheless, the bias is still below the bias of the AMS model. The more parameters are increased and therefore the higher the large floods and especially the variability are, the worse is the performance of the AMS model. On the other hand, the results also reveal that the AMS model has problems in modelling highly different flood types. The less equal the distributions of the flood types are, the higher is the bias of the AMS model. Again, the selection of only one event per year and therefore only one flood type does not mirror the complex distribution of many flood types correctly. Therefore, for regions with highly unequal distributions of flood types and where comparably large parameters for one or two flood types can be expected, it cannot be recommended to use the AMS model since the bias is high. This leads to a large underestimation (as the bias is positive), especially of extreme flood peaks, and may have serious impact on the estimation of design floods and flood protection. This is the case for many regions around the globe, e.g., in Central Europe where floods caused by heavy rainfall seem to become more extreme than other flood types. Changing climate can increase this problem.

Often, when applying more complex distributions such as the TMPS model, it is argued that the smaller bias is only due to over-parametrisation and comes at the cost of high variability of the estimation. However, the results demonstrate that the performance of both models, AMS and TMPS, is almost equal for most scenarios. On the contrary, for large sample sizes, the TMPS model has smaller RMSE. The AMS model has constantly high uncertainty since it only considers the samples partially and the consideration of floods depends much on their respective occurrence in single years.

It has to be noticed that the proposed simulation scenarios only give a limited view of the variability for varying parameters due to the limited number of variations of parameters. However, because of the complex TMPS model with its many parameters, a more detailed analysis is more complicated. The proposed scenarios mirror the most important variations and therefore provide information on almost all important cases. Still, the simulation is limited and, e.g., the assumption of fixed probabilities of non-exceedance of the threshold in the TMPS model may reduce the RMSE and the bias of the TMPS model. A variation of these parameters, too, is not possible. The large differences between the bias of AMS and TMPS cannot be explained by this limitation and therefore the results remain significant.

The TMPS model is available in the R-package FloodR, downloadable from GitHub. At the end of this chapter, a short example is presented on how to apply the TMPS model in R (R Core Team 2022) and to obtain the distribution plot with the quantiles for the TMPS model, each flood type as well as the AMS model.

Program Code

```
devtools::install_github(repo = "PhilippBuehler/FloodR")

library("FloodR")

# Use examplatory data for daily mean discharges
data("Discharge")
head(Discharge)

# Load a table with flood events and their respective flood
# types (see next part of the book on how derive these)
data("Floods_typed")

# Select date of flood peak, flood peak discharge and flood
  type
Floods_typed <- Floods_typed[, c("Peak_date","HQ", "Type")]

# Estimate flood quantiles for return periods of 2,5,... years
TMPS_quantiles <- qTMPS(p = c(2, 5, 10, 20, 25, 50, 100,
    200, 500, 1000),
Flood_events = Floods_typed, Daily_discharge = Discharge,
return_TMPS = c("TMPS", "R1", "R2", "R3", "S1", "S2"))
```

```
TMPS_quantiles <- as.data.frame(round(TMPS_quantiles,1))
TMPS_quantiles[TMPS_quantiles < 0] <- NA
TMPS_quantiles

##          2     5    10     20     25     50    100
## TMPS 549.5 708.3 897.9 1259.6 1397.7 1911.5 2586.8
## R1    68.0 236.4 403.4  623.7  709.3 1033.1 1468.9
## R2   162.7 402.2 616.3  876.4  972.3 1316.2 1743.8
## R3      NA    NA    NA     NA  165.2  457.7  571.7
## S1   382.8 472.9 518.2  553.2  562.9  588.9  610.0
## S2      NA    NA 286.4  525.6  571.6  661.7  705.6

##         200    500   1000
## TMPS 3479.1 5122.4 6853.7
## R1   2056.7 3160.7 4341.5
## R2   2277.1 3190.5 4083.8
## R3    616.5  637.2  642.4
## S1    627.2  645.1  655.8
## S2    727.2  740.0  744.2

# For comparison, the quantiles based on AMS are estimated.
# This is not essential for the TMPS model.

# Load the annual maximum series
data("AMS")

# Estimation of AMS parameters with L-Moments
AMS_params <- gevFit(AMS, type="pwm")@fit$par.ests
AMS <- qgev(1-1/c(2,5,10,20,25,50,100,200,500,1000),
mu=AMS_params[2], xi=AMS_params[1],
beta=AMS_params[3])

# Now, the results are condensed in a plot.

# Merge the results in one data.frame
Results <- rbind(TMPS_quantiles, AMS=AMS)
Results <- cbind(Results, "Method" = rownames(Results))
Results_melt <- melt(Results, id.vars = "Method",
    variable.name = "ReturnPeriod")
Results_melt$ReturnPeriod <- as.numeric(as.character
  (Results_melt$ReturnPeriod))

# Definition of colors for plot
cols <- c("AMS"="black",c(R1 = "#D7191C", R2 = "#FDAE61",
R3 = "#1A9641", S1 = "#6BAED6", S2 = "#2171B5"),
"TMPS"="darkorchid1")

# Plot
ggplot(Results_melt)+
theme_bw()+
geom_line(aes(x=ReturnPeriod, y=value, colour = Method ),
     size=1.5)+
scale_color_manual(values=cols)+
```

```
scale_x_log10()+
labs(x="Return period (years)",y="Peak discharge (mÂ³/s)")+
annotation_logticks(sides = "b")

# Obtain probabilities of the TMPS for given peak discharges
# Here: use AMS as peak discharges
# In practice, any discharges can be used
TMPS_probs <- pTMPS(q = round(AMS),
Flood_events = Floods_typed, Daily_discharge = Discharge,
return_TMPS = c("TMPS", "R1", "R2", "R3", "S1", "S2"))
```

The TMPS model offers the opportunity to consider different flood types directly in flood statistics. It makes use of type-specific partial duration series which are combined in a multiplicative mixture model, to the cost of a larger number of parameters compared to classical FFA models. The dominating flood type defines the right tail of the TMPS model. There exist spatial differences between the dominating flood types and they vary with the magnitude of the flood peak. The more different the distributions of the flood types, the better the performance of the TMPS model compared to AMS. It is recommended to use the AMS model only for regions where the pdfs of different flood types are similar in their shape and scale parameters and where approximately only few events per year occur. A large number of flood events per year leads to a bias of the AMS since only one event per year can be considered in this model. Moreover, if one or two flood types differ significantly from the remaining ones in terms of their distribution parameters, the AMS cannot be recommended, too. In this case, the AMS cannot incorporate the inhomogeneity of the samples and the statistical assumptions of the model are violated. This leads to significant under-estimation of design floods. Especially when having in mind recent large floods caused by heavy rainfall events in Central Europe, which clearly show that the distribution of this flood type differs significantly from, e.g., snowmelt floods, the wide-spread use of the AMS for the estimation of design floods should be re-considered.

References

Allamano P, Laio F, Claps P (2011) Effects of disregarding seasonality on the distribution of hydrological extremes. Hydrol Earth Syst Sci 15(10):3207–3215

Cunderlik JM, Ouarda TBMJ, Bobee B (2004) Determination of Flood seasonality from hydrological records. Hydrol Sci J 49(3):511–526

Fischer S (2018) A seasonal mixed-POT model to estimate high flood quantiles from different event types and seasons. J Appl Stat 45:2831–2847
Fischer S, Schumann A, Bühler P (2019) Timescale-based flood typing to estimate temporal changes in flood frequencies. Hydrol Sci J 64:457–470
Fischer S, Schumann AH (2023) Generation of type-specific synthetic design flood hydrographs. Hydrol Sci J
Fletcher R (1987) Practical Methods of Optimization, 2nd edn. Wiley, New York
Gioia A, Iacobellis V, Manfredo S, Fiorentino M (2008) Runoff thresholds in derived flood frequency distributions. Hydrol Earth Syst Sci 12:1295–1307
Hosking JRM (1986) The theory of probability weighted moments. IBM Research Report, IBM, p 1986
Hosking JRM, Wallis JR, Wood EF (1985) Estimation of the Generalized Extreme-Value distribution by the method of probability-weighted moments. Technometrics 27(3):251–261
Klemes V (2000) Tall tales about tails of hydrological distributions. J Hydrol Eng 5:227–231
Madsen H, Rasmussen PF, Rosbjerg D (1997) Comparison of annual maximum series and partial duration series methods for modeling extreme hydrologic events: 1 At-site modeling. Water Resour Res 33(4):747–757
R Core Team (2022) R: A language and environment for statistical computing. R Foundation for Statistical Computing, Vienna, Austria. https://www.R-project.org/
Rossi F, Fiorentino M, Versace P (1984) Two-component extreme value distribution for flood frequency analysis. Water Resour Res 20:847–856
Singh KP (1968) Hydrologic distributions resulting from mixed populations and their computer simulation. Publication No. 81, International Association of Scientific Hydrology, pp 671–191
Singh KP, Sinclair RA (1972) Two-distribution method for flood frequency analysis. J Hydraul Div 98(1):29–44
Strupczewski WG, Kochanek K, Bogdanowicz E, Markiewicz I (2012) On seasonal approach to flood frequency modelling. Part I: two-component distribution revisited. Hydrol Process 26:705–716
Szulczewski W, Jakubowski W (2018) The application of mixture distribution for the estimation of extreme floods in controlled catchment basins. Water Resour Manag 32(10):3519–3534
Todorovic P, Rousselle J (1971) Some problems of flood analysis. Water Resour Res 7(5):1144–1150
Waylen P, Woo MK (1982) Prediction of annual floods generated by mixed processes. Water Resour Res 18:1283–1286
Woo MK, Waylen PR (1984) Areal prediction of annual floods generated by two distinct processes. Hydrol Sci J 29(1):75–88
Zhou XY, Schmidt F, Toutlemonde F, Jacob B (2016) A mixture peaks over threshold approach for predicting extreme bridge traffic load effects. Prob Eng Mech 43:121–131

Part IV
Application of Type-Based Flood Statistics

Part IV of this book demonstrates the variety of possible applications of the type-based statistical model TMPS. First, the required data processing is described: separation of flood events and typing of floods. An automated statistical separation procedure is introduced together with a hybrid causative-hydrograph based flood typology that is suitable to be used in the context of flood statistics. More precisely, three different rainfall-driven flood types are defined that differ in their spatial and temporal characteristics of rainfall. Moreover, two snow-impacted flood types are defined, namely rain-on-snow floods and snowmelt-induced floods. Based on these data preparations, the application of the TMPS model for estimating design floods is demonstrated and the benefits of incorporating flood types in flood statistics are shown for practical examples. Moreover, changes of flood types are analysed and their impact on the flood statistics is demonstrated, once again making use of the TMPS model. Finally, synthetic type-specific flood scenarios are obtained, i.e., flood hydrographs for a given return period with assigned occurrence probability depending on the flood type. Such scenarios are most relevant for practical purposes as they might help to better consider the fact, that extreme floods with a similar peak could differ in their remaining flood wave characteristics. It can be applied to differentiate between more or less probable flood scenarios. For example, flood retention systems could be operated seasonally differentiated, depending on the occurrence probability of the flood type that is most relevant to cause large floods with high volumes.

Chapter 6
Data Processing

Abstract The application of type-based statistics requires an appropriate data basis. In contrast to classical flood frequency analyses, it is no longer sufficient to consider only one event per year, let alone only the flood peak. In this chapter, methods to separate flood events and the corresponding precipitation events are introduced. The methods are based on statistical threshold criteria and define flood events from daily discharges. In a second step, a hybrid causative-hydrograph-based flood typology is proposed that is explicitly designed for the use in the context of flood statistics and thus requires only few input data available for a long time period. As a result, samples of separated flood events that are classified into five flood types—three rainfall-driven flood types, two snowmelt-induced flood types—are derived that then can be used for application in type-based flood statistics.

6.1 Flood Event Separation

For the application of classical univariate flood frequency analyses, knowledge on the flood peak alone is sufficient, i.e., on the flood peak magnitude and, if applicable, on the date of the peak occurrence. When aiming at considering more flood characteristics than the peak, additional information is required. If, for example, flood peak and volume are considered jointly, usually the corresponding flood event to the given flood peak has to be defined by its beginning and end (Tang and Carey 2017; Bezak et al. 2015). The definition of the beginning of a flood event and its end is not straightforward and often includes subjectivity. The task becomes even more difficult when all flood events in observed discharge series should be included in the analyses. In this case, the whole time series of discharge has to be scanned and flood events must be defined according to a certain convention. One possibility for such a convention is to change the perspective to the analysis of precipitation series to define precipitation events first. This can be done, e.g., by defining minimum dry spells (Dunkerley 2008) or thresholds for rainfall intensity (Loukas and Quick 1996). Despite the subjective choice of thresholds, which may depend vastly on the region of application, these approaches do not put the focus on flood events

S. Fischer and A. H. Schumann, *Type-Based Flood Statistics*, Water Science and Technology Library 124, https://doi.org/10.1007/978-3-031-32711-7_6

but on rainfall events and analyse the related increase of discharge. Flood events that are caused by snowmelt or have a large peak because of a high initial soil moisture with an only small amount of accompanying rainfall might be overlooked. Another challenge is how to deal with baseflow, i.e., the runoff that cannot be associated with a hydro-meteorological cause. It depends on the initial wetness conditions of the catchment and can enhance or weaken its response to precipitation. Hence, when aiming to analyse flood events, the focus should be laid on runoff that can be associated with a hydro-meteorological cause. Most approaches aim to separate baseflow first. The flood event is then defined as the runoff that exceeds the baseflow, the so-called direct runoff component or quick flow. There exist various methods for baseflow separation, ranging from digital filters over recession methods (where the quick flow is assumed to have a higher declining rate than the baseflow) to minimum location and tracer methods (Chapman and Maxwell 1996; Eckhardt 2005; Piggott et al. 2005; Blume et al. 2007; Aksoy et al 2009; Mei and Anagnostou 2015), leading to quite subjective results. The combination of a search for peak discharges exceeding a threshold and the subsequent separation of the direct discharge is problematic in that it requires an iterative approach. The direct runoff waves that have been determined again need to be tested for their characteristic as flood events. Otherwise, the separation of baseflow does not guarantee that indeed flood events are considered only. Instead, when considering any deviation from baseflow, even small increases of runoff might be separated. This is contra-productively when aiming to analyse flood events. More recent studies hence often avoid the separation of baseflow but instead develop methods to directly identify flood events, e.g., by the speed of increase of runoff or by machine-learning approaches (Oppel and Mewes 2020; Towler and McCreight 2021). However, such approaches require test data sets to train the algorithms or to define timing errors. Another constraint on separation algorithms when aiming to use the results for statistical analyses is the data resolution. Hourly discharge data, which are required for some approaches (Mei and Anagnostou 2015), are often not available for long observation periods. Instead, mostly only daily mean discharges are available.

One approach for an automated separation of flood events that does neither require baseflow separation nor training sets but considers a variance-based threshold is proposed by Fischer et al. (2021a). This approach is introduced in more detail in the following section, as it is especially applicable in the context of flood types, where separated flood events are required. However, no automated approach can truly identify all the complex flood events that may occur due to an overlap of several flood-generating processes. Especially those events that are characterised by multiple flood peaks, caused by several successive rainfall events, cannot be easily identified, as will be shown. Hence, a manual inspection of the results is always recommended. The approaches introduced in this chapter have been developed for daily data, i.e., daily mean discharges and daily areal precipitation. In theory, they can also be applied to hourly data, though in this case the parameters should be adjusted.

6.1.1 A Variance-Based Separation of Flood Events

In this section, the separation algorithm by Fischer et al. (2021a) is introduced. Due to its variance-based flood separation that only requires few variables that are available for long time periods, it is particularly suitable for flood statistics. Fischer et al. (2021a) state that their separation algorithms is based on two simple assumptions

1. A flood event is a temporally limited exceedance of normal discharges (which have to be defined based on averages of discharges). For each flood event, beginning and end can be defined.
2. A flood event is characterized by a significantly increased dynamic of discharges. Especially around the peak, the variance of discharges is high for a limited period of time.

Especially the second assumption implies a possibility to define flood events: by the variance of discharges.

Considering the variance of daily discharges Q_i over a given time period allows for an estimation of the discharge dynamics. For this purpose, consider

$$\mathrm{Var}_{d_{\mathrm{Var}}}(i) = \mathrm{Var}(Q_{i-(d_{\mathrm{Var}}-1)}, \ldots, Q_i),$$

for $i \geq d_{\mathrm{Var}}$. The parameter d_{Var} defines the time window length that is considered for the variance estimation. It controls, how long respectively how strong a certain variation has to occur in the discharges to be considered as the beginning of a flood event, and is catchment specific. It depends on the catchment size but mainly on the general runoff dynamics in a catchment. If the dynamics are high, d_{Var} is small. For all catchments in Central Europe considered in several studies Fischer and Schumann (2021a), Fischer and Schumann (2021b), Brunner and Fischer (2022) as well as in the sensitivity analysis by Fischer et al. (2021a), $d_{\mathrm{Var}} = 3$ proved to be sufficient. The variance alone cannot indicate unusual variability of discharge and hence the beginning of a flood event. Instead, the variance has to be in a range that can be seen as extraordinary. In statistics, such ranges are used quite often to define outliers or ranges of high quantiles, e.g., in a boxplot, where the whiskers are defined by the Interquartile Range (IQR). A similar threshold is the σ-range. In statistics, values are often treated as outliers when they are $\theta > 0$ times σ away from the mean, where σ is the standard deviation. A classical example is the 2σ-rule in case of the normal distribution. Transferring this concept to the variance of discharges, Fischer et al. (2021a) developed a threshold

$$TH_{\mathrm{Var}} = \overline{\mathrm{Var}_{d_{\mathrm{Var}}}} + \theta \cdot \sqrt{\mathrm{Var}(\mathrm{Var}_{d_{\mathrm{Var}}})}.$$

Roughly speaking, a significant increase of discharge and hence the beginning of a flood event occurs when the 3-day variation of discharge is larger than the long-term mean of the 3-day variance plus θ times the standard deviation of the 3-day variance (Fig. 6.1).

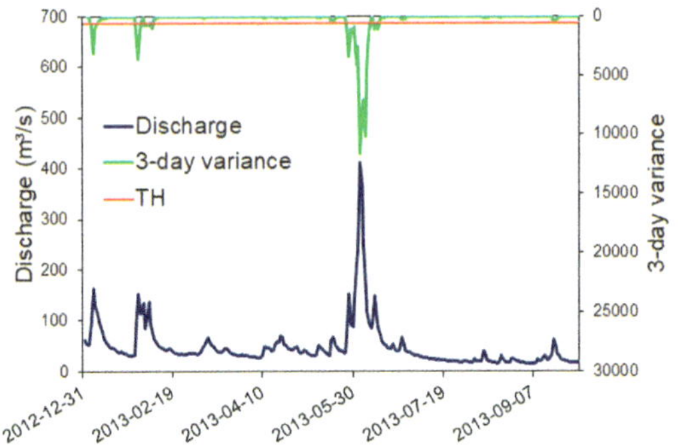

Fig. 6.1 Daily discharges with moving variance of a three-day window Var_3 and variance threshold TH_{Var} of the Marienthal/Regen gauge

With this threshold TH_{Var}, periods of significantly increased variance are detected. Each period is a potential flood event with the flood peak defined as the maximum discharge at time t_p during the period of variance above the threshold. The starting point t_s of the flood event is defined as the time step before occurrence of the flood peak for which the discharge does not increase for the last time:

$$t_s = \max\{2 \leq i < t_p | (Q_i - Q_{i-1}) < 0\}.$$

This definition ensures that from the start of the event on the discharges increases strictly monotonously until the peak is reached. However, Fischer et al. (2021a) point out that

> not every increase of runoff automatically defines the beginning of a flood event.

Small increases of discharge might be part of the naturally variable discharge with its daily up-and-down-behaviour. Of course, such discharges should not be included in the flood event separation. Contrarily, due to breaks within the rainfall event, the increase of the discharge might not be strictly monotone between beginning and peak but with short recession phases in between. Yet, even short initial increases of discharge might contribute to the flood event and thus should be included in the event separation (Fig. 6.2). To take both cases into account, Fischer et al. (2021a) propose the following extension of the definition of the flood beginning.

1. As long as $|(Q_{t_s} - Q_{t_s+1})/Q_{t_s+1}| < \eta$, $\eta \in (0, 1)$, set

$$t_s = t_s + 1.$$

2. If

$$\max_{1 \leq i \leq \gamma, 2 \leq j \leq \gamma+1, i<j} (Q_{t_s-i} - Q_{t_s-j}) > \kappa \cdot (Q_{t_p} - Q_{t_s}), \tag{6.1}$$

set

$$t_s = t_s - m$$

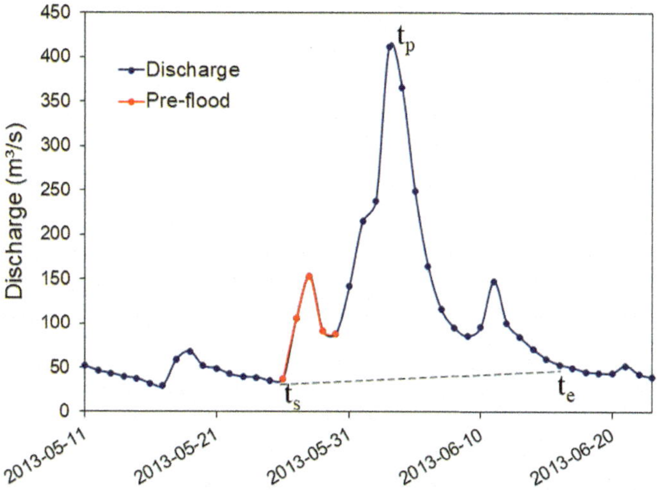

Fig. 6.2 Example of an adjustment of the event begin t_s due to a pre-flood for the 2013 flood at Marienthal/Regen gauge

where

$$m = \arg\max_{2 \le j \le \gamma+1} \left\{ (Q_{t_s-i} - Q_{t_s-j}) > \kappa \cdot (Q_{t_p} - Q_{t_s}), 1 \le i \le \gamma, i < j \right\}$$

and $\gamma \in \mathbb{N}^{>0}$.

The first step ensures that the increase of the discharge at the beginning is at least of a certain given magnitude. The second step searches for a local maximum of peak magnitude of at least $\kappa \in (0, 1)$ times the flood peak within $\gamma + 1$ days before the beginning of the flood event. Fischer et al. (2021a) denote such a local maximum as pre-flood.

Compared to the definition of the beginning of the flood event, the definition of the end is much more complicated. While the beginning is characterised by a sudden and steep increase of discharge, the end of a flood event is usually a long recession with decreasing gradient. One possibility would be to define the end as the point where the gradient is zero, but this might lead to very long flood events due to the variability of runoff. Instead, another idea is to define the end of a flood as the point for which the sum of the 1-day differences of discharge is zero. Mathematically speaking, the end of the flood event would then be defined as the time step $t \ge t_p$ for which

$$\sum_{i=t_s+1}^{t} (Q_i - Q_{i-1}) \approx 0.$$

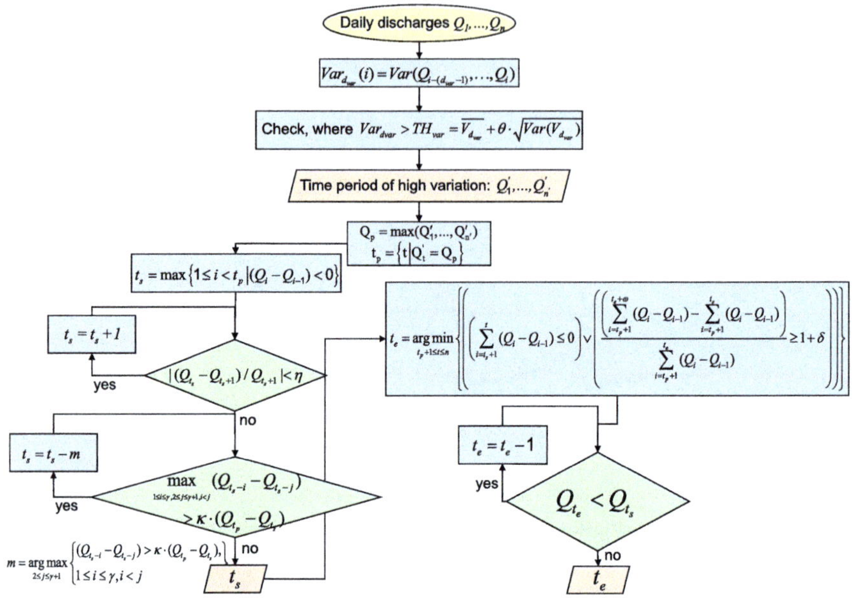

Fig. 6.3 Schematic overview of the flood event separation (modified from: Fischer et al. 2021a)

This is a hydrologically meaningful definition, though it does not fully consider the nature of floods. After a flood event, often the baseflow increases due to the increasing runoff from groundwater. A strict cut of the flood event only when the baseflow from the beginning of the flood event is reached might therefore again lead to much too long flood events. Fischer et al. (2021a) thus propose a criterion to consider increasing baseflow in the definition of the flood end:

$$t_e = \arg \min_{t_p+1 \le t \le n} \left\{ \left(\left(\sum_{i=t_s+1}^{t} (Q_i - Q_{i-1}) \le 0 \right) \vee \left(\frac{\sum_{i=t_e+1}^{\omega} (Q_i - Q_{i-1})}{\sum_{i=t_p+1}^{t_e} (Q_i - Q_{i-1})} \ge 1 + \delta \right) \right) \right\}. \tag{6.2}$$

The first criterion checks for the baseflow to be at the same level as at the beginning of the flood event. The second criterion ensures that within the next ω days after the defined end of the flood event less than $\delta \in (0, 1)$ of the volume of the falling limb is manifested as discharge. Since one of the criteria has to be met, either the assumption of equal baseflow or an increased baseflow is fulfilled.

A full schematic overview of the algorithm is given in Fig. 6.3. A sensitivity analysis of the parameters used in the separation can be found in Fischer et al. (2021a).

With these definitions for beginning and end, the flood event is fully characterised and the corresponding flood volume and duration can be calculated. To estimate the direct runoff, the baseflow has to be separated. However, since beginning and end of the flood event are already known at this point, the simplest way to obtain the baseflow is the straight-line method which is applied here. For this approach, the baseflow is modelled by a straight line between beginning and end of the flood event.

The procedure for flood event separation described so far implicitly assumes that flood events have a more or less regular shape with a rising and falling limb. However, there exist flood events which cannot be classified like this: the multiple-peaked floods. Such events usually are caused by intermittent rainfall events with breaks between them. In such a case, the recession of the previous flood event is not back at the baseflow level when the next rainfall event begins, the direct runoff components overlay and the discharge rises again. Yet, here again different types of floods may occur. Fischer et al. (2021a) identify three critical cases of multiple-peaked flood events:

1. single floods with unsteady rising and/or falling limbs,
2. the superposition of two flood events which can be separated and
3. floods which are superimposed on high and long lasting baseflow, following each other in a short time span and cannot be separated. Here, the rainfall event consists of several partial events that fall on a wet catchment area, each causing a further increase in direct runoff.

The first two cases differ significantly from the third, as they are of much shorter duration and can be clearly linked to rainfall events. The third type of floods however is different. Due to the increased baseflow, even rainfall that normally might not result in a flood event now causes significant increases of discharge due to the wet catchment state. Such multiple events cannot be separated into single floods, as they are clearly interconnected (Fig. 6.4b). Hence, they have to be treated differently than the first two types of multiple-peaked floods. Fischer et al. (2021a) propose to distinguish between the first two and the third type by the flood duration d, with the latter type expected to have a longer duration. The parameter d is catchment specific, depending on the catchment size and climatic conditions, e.g., the duration of rainfall events. For Central Europe, a duration of 30 days proved to be applicable for most catchments.

For the first category of multiple-peaked floods with comparably short duration, a separation is possible, as usually the rainfall events, and to a large degree also their contribution to each partial flood, can be identified. Fischer et al. (2021a), for example, propose to apply an independence criterion to separate both flood events at the minimum discharge between both peaks and to reconstruct the recession of the first event (Fig. 6.4a). This way, each flood event can be considered separately together with the corresponding rainfall. Superimposed flood events, however, cannot be separated. They should be treated as a separate category. In any case of multiple-peaked flood events, a manual inspection of the separation events is recommended.

The methodology by Fischer et al. (2021a) introduced in this chapter was explicitly developed to be applied to daily discharges. However, by adjusting the parameters, it also applicable to data with higher temporal resolution. For example, it has been successfully applied to mean hourly discharges, as the basic assumptions on floods remain valid. When using mean daily discharges, one aspect has to be mentioned:

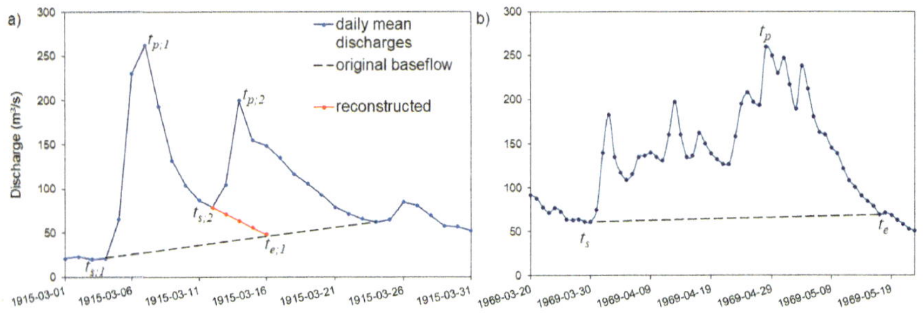

Fig. 6.4 **a** Superpositioned floods, which are differentiated into two separated events; **b** superimposed floods leading to a single event with multiple peaks which is characterized by a long-lasting high baseflow. (modified from: Fischer et al. 2021a)

the flood event separation delivers a flood event with a flood peak derived from daily mean discharges. Especially for small catchments, daily mean discharges underestimate the flood peak significantly, as they consider the mean discharge of one day, while the flood peak might occur for only few hours and hence is smoothed in the mean. An example is given in Fig. 6.5, where the daily mean discharges, the hourly mean discharges and 15-min discharges for the 2013 flood in Goeritzhain/Chemnitz in Saxony, Germany are given. The catchment of Goeritzhain with an area of 532 km^2 is strongly influenced by the urban runoff from a city (Chemnitz) leading to a fast catchment reaction. Therefore, the resolution of the discharge data plays a considerable role when analysing the flood event. Due to the fast reaction and hence fast rise and recession, the daily mean discharges smooth the 15-minute discharges. This leads to an underestimation of the flood peak of about 20% when considering daily mean discharges. Small events like the first pre-wave even totally disappear in the daily mean discharges. For the hourly discharges, no such differences occur (<1%) and hence these are suitable to estimate the flood peak. For the flood volume, the resolution of the discharge data is not relevant, as it is captured by both daily and hourly mean discharges correctly.

This demonstrates that for catchments of the mesoscale, additionally to the separated flood events, at least hourly values of the flood peak should be used. The availability of monthly maxima limits the number of flood events to one per month, but having in mind that only flood events are separated, this is a sufficient number. For larger catchments of the macroscale, such a consideration of the hourly flood peaks is not frequently required, since the flood wave is much more smoothed.

An application to more than 160 catchments in Bavaria, Germany (see Appendix A.1) demonstrates that this separation procedure is meaningful (Fig. 6.6). Parameters for this separation have been chosen as $\theta = 0.5$, $\kappa = 0.8$ and $\eta = 0.05$. In mean, two up to four events per year were separated, a reasonable number when considering flood events. No spatial differences can be observed, indicating that the separation performs equally well for all catchment sizes and topographies. Moreover, this means that the separation is consistent within a river and thus for consecutive gauges. Considering the critical multiple-peaked events, i.e., superpositioned and

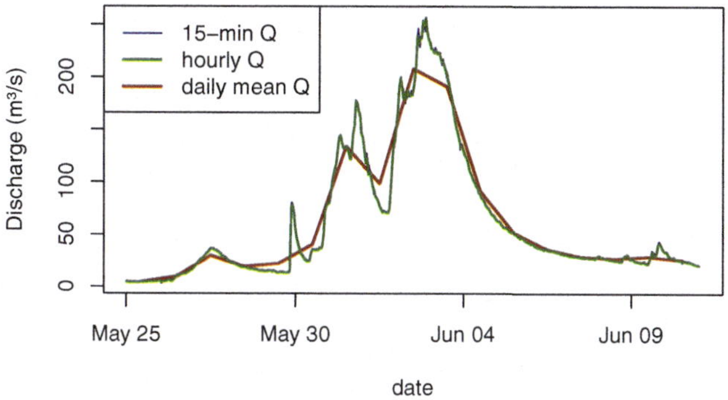

Fig. 6.5 Daily mean discharges, hourly mean discharges and 15-minute discharges of the 2013 flood event at Goeritzhain/Chemnitz gauge in the Mulde river basin, Germany

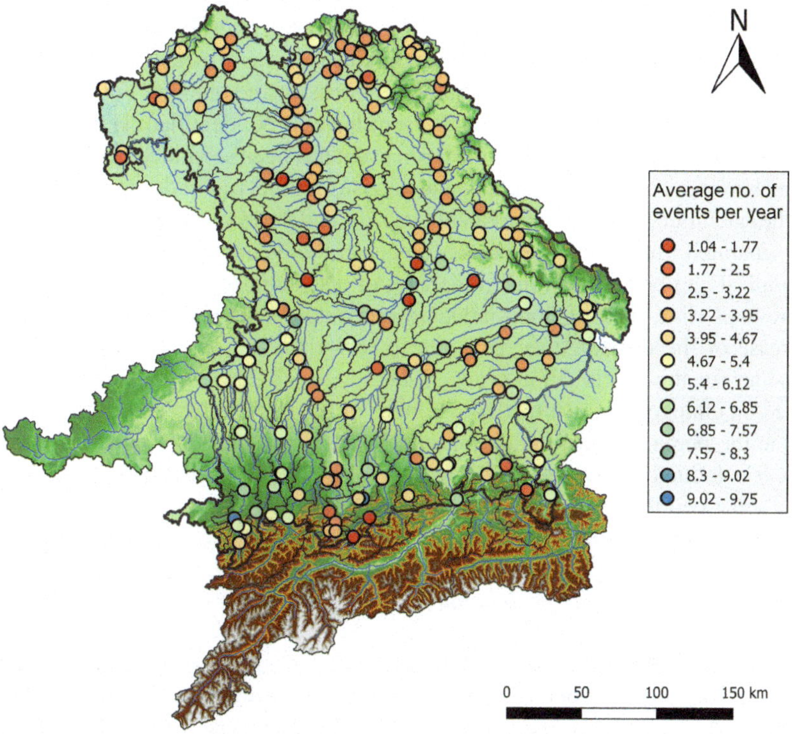

Fig. 6.6 Mean number of separated flood events per year for catchments in Bavaria, Germany

superimposed floods, there appears a clear spatial pattern regarding their frequency. In the low and flat parts of the region, the proportion of such events on all floods is below 1%. Often, none such events occur at all. However, in the more mountainous regions, in the uplands and the Alps, the proportion increases significantly up to more than 10%. For these catchments, a manual inspection of the flood event separation is highly recommended. A reason for this high proportion of multiple-peaked flood events might be the snowmelt that is much higher in mountainous regions than for the flatlands. Snowmelt in spring might increase baseflow and hence superimposed floods might occur more frequently. This is also illustrated later in this chapter.

6.1.2 Separation of the Flood-Generating Precipitation

In the introduction of this chapter, it was pointed out that when aiming to analyse flood events, first these flood events have to be separated and afterwards the corresponding rainfall event should be identified. Since rainfall is the main meteorological driver for floods, it is important to know its duration, intensity and amount. For this purpose, the beginning and end of the rainfall event have to be identified. However, this is not straight-forward, even when the flood event has been separated before. The time delay between beginning of rainfall and the rise of direct runoff does not only depend on the catchment size but on the catchment state and the rainfall event itself. However, the separated flood event can be used to detect the beginning of the flood-inducing rainfall: it can be assumed that a increase of discharge is most probably related to a previous time-span of flood-inducing precipitation. Hence, the starting time of increase of the discharge serves as an indicator for the required time window to consider for the precipitation event. Fischer et al. (2021a) propose a methodology based on change-point estimation of slopes of the curve of accumulated rainfall to define the beginning of rainfall under these assumptions. More precisely, a buffer time b is defined to consider the delay between precipitation and flood begin. Hence, for a separated flood event starting at time t_s and ending at time t_e, the precipitation is considered in the time window $t_s - b, \ldots, t_e$, with

$$b = \max(\xi, t_p - t_s + 1).$$

The parameter ξ depends on the catchment under consideration and represents a minimum buffer time. For mesoscale catchments in Europe, it was set to $\xi = 3$ (days). Then, the time series of accumulated precipitation of this time window, $P_{t_s - b}, \ldots, P_{t_e}$, is split into two sub-series at a point

$$\tau \in \{t_s - b + 2, ..., t_e - 3\}.$$

The idea is to compare the first sub-series and the second sub-series regarding their increase to find the beginning of the precipitation as the point where the difference between both increases is largest (Fig. 6.7). For this purpose, a linear regression is fit to both sub-series

$$\sum_{i=t_s-b}^{j} P_i = \alpha + \beta_{t_s-b:\tau} t_j, \quad \text{for } j = t_s - b, \ldots, \tau$$

$$\sum_{i=\tau+1}^{j} P_i = \alpha' + \beta_{\tau+1:t_e} t_j, \quad \text{for } j = \tau + 1, \ldots, t_e.$$

Each sub-series has to consist of a minimum number of values, otherwise the linear regression is not meaningful. For example, Fischer et al. (2021a) recommend a minimum number of three.

With these assumptions, the beginning of the flood-inducing precipitation $\hat{\tau}$ can be derived by taking the time step τ, for which the difference between both regression slopes is largest:

$$\hat{\tau} = \arg\max_{t_s-b+2 \leq \tau < t_e-2} \{\beta_{\tau+1:t_e} - \beta_{t_s-b:\tau}\}.$$

The change-point methodology is based on the assumption that only one precipitation event with distinct breaks before and after the event occurs. However, precipitation events are not always that simple. Often, pre-event precipitation occurs that would lead to a falsified precipitation beginning when using the above-mentioned methodology. To overcome this problem, Fischer et al. (2021a) propose a second approach that should be applied jointly with the first: Instead of considering one change-point τ, the series of precipitation is split into three sub-series according to two change-points τ_1 and τ_2. With the three-fold split, the beginning can be defined as the part where the slope is steepest while considering small increases before and after. Hence, the beginning of the precipitation event is estimated by

$$\hat{\tau}_1 = \arg \max_{t_s-b \leq \tau_1 < \tau_2 | t_s-b+3 < \tau_2 < t_e-2} \left| (\beta_{t_s-b:\tau_1} + \beta_{\tau_2+1:t_e}) / (2\beta_{\tau_1+1:\tau_2}) \right|.$$

When applying method 1 and method 2, the beginning of the precipitation is chosen that is closer to the beginning of the flood event. If there is no significant precipitation before the flood event, the flood probably is snowmelt-induced. For flood events impacted by snowmelt, instead of precipitation the sum of precipitation and snowmelt, determined from a simple snowmelt model as the day-degree-method Howard (1996), can be considered and the methodology for detection of the beginning of the flood-inducing process can be applied again.

The methodology is illustrated by an example taken from Fischer et al. (2021a). Figure 6.7 shows the estimation of precipitation beginning corresponding to a flood event. In this example, both approaches result in different estimates for precipitation beginning: if only one change-point is considered, the beginning of the precipitation would be dated to February 24th. If two change-points are assumed, the beginning would be dated to February 27th. In this case, the later date is chosen as this is closer to the beginning of the flood event which was estimated before.

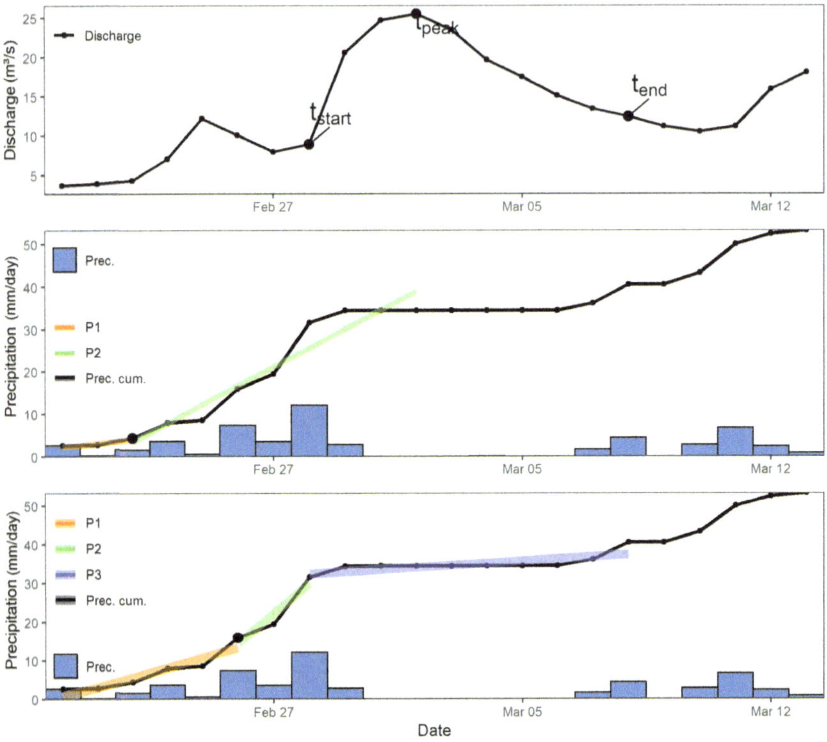

Fig. 6.7 One- (centre) and two-change-point method (bottom) to define event precipitation for a flood event (top) with $\xi = 7$. The black lines in the two bottom figures represent accumulated precipitation. (from: Fischer et al. 2021a)

The end of the flood-inducing precipitation is much easier to define as the beginning, as it can be set simply to the end of the flood event. At a later point in time, the precipitation cannot contribute to the flood event. Of course, it might occur that the precipitation ends before the flood event ends due a longer recession of the flood. However, in this case the amount and intensity of the precipitation event is not altered when defining the precipitation end as the end of the flood event, as no significant amount of additional precipitation falls down in this period.

6.2 A Hybrid Causative-Hydrograph-Based Flood Typology

The previous section allows to separate flood events and the corresponding meteorological drivers. These separated events then form the basis for the application of a flood typology. The flood typology proposed in this section has been developed

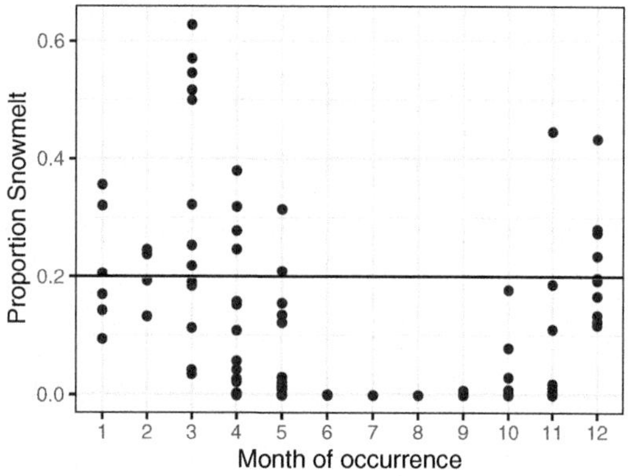

Fig. 6.8 Proportion of snowmelt on the runoff-generating water differentiated by month for flood events at the Munich/Isar gauge in Bavaria, Germany. The horizontal line shows the threshold for distinguishing between rainfall-driven and snowmelt-induced floods

by Fischer et al. (2019). It consists of two aspects: the causative classification by taking into account the amount of precipitation and snowmelt of each flood event and the hydrograph-based classification by defining flood types according to the flood timescale. It requires separated flood events with flood peak and volume and the corresponding areal precipitation and snowmelt. In a first step, flood events are distinguished according to their meteorological causes by differentiating between pure rainfall-driven and snowmelt-impacted flood events. The impact of snow often cannot be estimated from observed snow data as such data do not exist in many cases. To fill this gap, Fischer et al. (2019) propose to apply a degree-day model to assess the accumulation and ablation of the snow cover which only requires precipitation and temperature data. Rainfall-driven floods are then defined as those flood events for which the amount of snowmelt on the total volume of runoff-generating water stays below a threshold. Consequently, snowmelt-induced floods have a proportion of snowmelt above this threshold. The use of a threshold to distinguish between rainfall-driven and snowmelt-induced floods is a basic approach of causative flood classification. For example, Kampf and Lefsky (2016) use a threshold of 0.2, while Vormoor et al. (2016) define the threshold as 0.33. The choice of the threshold greatly depends on the region where the classification is applied. For Europe and USA, a threshold of 0.2 proved to be a valid choice. In the summer months, no snowmelt occurs, while the highest proportion of snowmelt on the flood-generating water can be observed in spring (Fig. 6.8). This first step of the classification provides a rough distinction according to the main meteorological drivers of the flood event. However, what is not taken into account are the specific characteristics of the event. To include further deterministic knowledge and to deepen the understanding of the flood-generating processes, Fischer et al. (2019) propose a further distinction of the flood events. This distinction is introduced in the following two subsections.

6.2.1 *Flood Types of Rainfall-Driven Floods*

In the case of rainfall-driven floods, the temporal distribution of rainfall is directly linked to the shape of the hydrograph. While short, heavy rainfall will lead to floods with steep hydrographs and short duration, synoptic rainfall will result in flood waves with large peak-volume relationships. Long intermitted rainfall consisting of several successive rainfall events will lead to multiple-peaked flood events with long duration. According to these differences, instead of analysing the rainfall patterns which is difficult due to the often low time resolution, the hydrographs are analysed. A measure describing the shape of the hydrograph is the flood timescale. This characteristic was originally proposed by Gaal et al. (2012) and was found to differ significantly between the flood types and to be closely linked to meteorological drivers and the catchment state. The flood timescale of a flood event, which is usually given in hours, is defined as

$$TQ = \frac{Vol}{Q_p}, \tag{6.3}$$

where Vol is the flood volume and Q_p is the flood peak. TQ has the unit of time. It is recommended to not use the total flood volume and the flood peak but instead the direct volume and direct peak only, i.e., the flood volume and peak minus the respective baseflow components. By subtracting the baseflow, only the direct runoff is considered and the peak and volumes of the separated event are directly related to the flood-inducing precipitation. Baseflow can have a large impact on both the total flood peak and volume, especially in catchments where relatively high seasonal baseflow can be expected. This will be discussed in more detail in Sect. 6.2.3.

By using the flood timescale, the rainfall-driven floods can be subdivided into flood types according to thresholds of the TQ-values. However, it is not straight-forward to define such thresholds, as the flood timescale depends much on the catchment size and drainage characteristics. Fischer et al. (2019) propose an empirical definition of thresholds according to linear regression. For this approach, the peak and volume of the rainfall-driven floods are modelled by a simple linear regression through origin

$$Vol = \beta Q_p.$$

The flood events are ordered according to their flood timescale and divided into a pre-defined number of sub-samples. For each of the sub-samples, the linear regression is performed and the coefficient of determination is estimated. This is repeated for each possible distinction of the timescale-ordered flood events. Finally, the partition is chosen which results in the highest sum of coefficient of determination. More precisely, for a sample $(x_1, y_1), \ldots, (x_n, y_n)$ of pairs of peak x_i and volume y_i and a number of three rainfall-driven flood types, the timescale thresholds between the types are estimated by maximising the overall sum of coefficients of determination r^2:

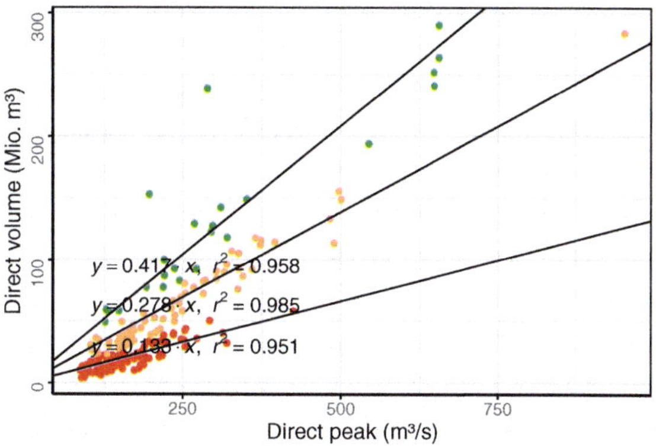

Fig. 6.9 Example of the classification of rainfall-driven floods according to the flood timescale for the Munich/Isar gauge in Bavaria, Germany

$$TQ_0 = \arg\max_{h_1,h_2} \{r^2(1,h_1) + r^2(h_1+1,h_2) + r^2(h_2+1,n) \mid$$
$$[n\varepsilon] \le h_1 \le n - 2[n\varepsilon] + 1, 2[n\varepsilon] \le h_2 \le n - [n\varepsilon] + 1, h_1 < h_2\},$$

where $r^2(i, j)$ is the coefficient of determination of the linear regression of the sample $(x_{|TQ_{(i)}}, y_{|TQ_{(i)}}), \ldots, (x_{|TQ_{(j)}}, y_{|TQ_{(j)}})$ and ϵ determines a minimum sample size of each subsample. Fischer et al. (2019) recommend to choose ϵ according to the available number of flood events and suggest an $\epsilon = 0.20$ for a number of two up to four events per year. As peak-volume pairs which are located very distant from the mean of the sample may affect a regression strongly, the linear regression should be estimated without extreme events first and then the extremes should be classified according to their flood timescales into the most appropriate group.

As illustrative example, in Fig. 6.9 the rainfall-driven flood events are given for a catchment in Bavaria, Germany (2838.4 km^2). A number of three flood types has been pre-defined for this catchment. The peak-volume pairs ordered by their flood timescale are divided into three subsamples. For example, for flood type R1 (red dots), the regression is estimated as

$$Vol = 0.133 Q_p,$$

where the volume is given in million m^3 and the flood peak is given in m^3/s, with a coefficient of determination $r^2 = 0.951$. The sum of all coefficients of determination is 2.894, which is the maximum possible summed coefficient of determination that could be reached in this example. Hence, the given distinction of rainfall-driven floods is the optimal distinction for this catchment. Fischer et al. (2019) and Fischer and Schumann (2021a) found a number of three rainfall-driven flood types sufficient for catchments in Europe, as this number allows for a comprehensive understanding of the flood-generating processes and results in samples of sufficient size to be analysed statistically. The three flood types are labelled with R1, R1 and R3 from small to large flood timescales.

Hence, flood type R1 is associated with small flood timescales and thus steep hydrographs with large peaks but small volumes and short duration. Flood type R2 is associated with hydrographs of medium duration with a balanced relation between flood peak and volume. Flood type R3 is characterised by large flood timescales and thus by large volumes and long duration. However, the maximum peaks of R3-events are usually smaller than those of the two other rain-induced flood types, though the events often have multiple peaks. For information on how the defined flood types are related to their meteorological drivers is given in Sect. 6.2.4.

6.2.2 Flood Types of Snowmelt-Floods

Similar to the rainfall-driven flood events, also the snowmelt-induced flood events are differentiated further. Snowmelt-induced floods are usually considered as either being mainly driven by snowmelt, i.e., large amount of the snowpack in a catchment melt and result in high discharges, or as rain-on-snow floods (Nied et al. 2014; Kampf and Lefsky 2016; Sikorska et al. 2016). This differentiation is difficult especially in mountainous study regions, where snowmelt periods are often accompanied with warm fronts, which are in many cases connected with rain. The resulting floods depend strongly on the water storage capacity of the snow cover at the beginning of the rain event. If the snow height is low and the snow density is high, rain induces a fast increase of runoff which emerges in a relatively short time as the falling rain provides additional energy for snowmelt and the water storage capacity of the snow is exceeded fast. In difference to such rain-on-snow floods, another type of snowmelt-induced floods is dominated by runoff that is retarded by a higher snow cover where the snow storage is more relevant for the duration of the flood event.

Under this assumption, Fischer et al. (2019) propose a distinction based on a kMeans clustering approach. As input data, the sum of precipitation, the sum of snowmelt and the runoff coefficient are considered. These data are normalised to the interval [0,1] by a simple min-max-scaling approach. The kMeans clustering algorithm Hartigan and Wong (1979) then defines a pre-defined number of clusters such that the sum of squares of the deviation of the data points to the cluster centroid is minimised (Fig. 6.10). Theoretically, besides the two above-mentioned snowmelt-induced flood types, additional types such as glacial floods could be considered, if the data basis is appropriate. However, the more types are considered, the smaller is the sample size for each flood type and hence the larger the uncertainty of statistical analyses.

The whole flood typology is summarised in Fig. 6.11. It has been applied in several studies for Europe (Fischer et al. 2019; Fischer and Schumann 2021a; Fischer et al. 2021b; Fischer and Schumann 2021b; Brunner and Fischer 2022). After the application of the flood typology, it is recommended to check the flood types of the largest floods in a catchment. For example, Fischer and Schumann (2018) propose a classification for extreme events which could be the basis for such a check.

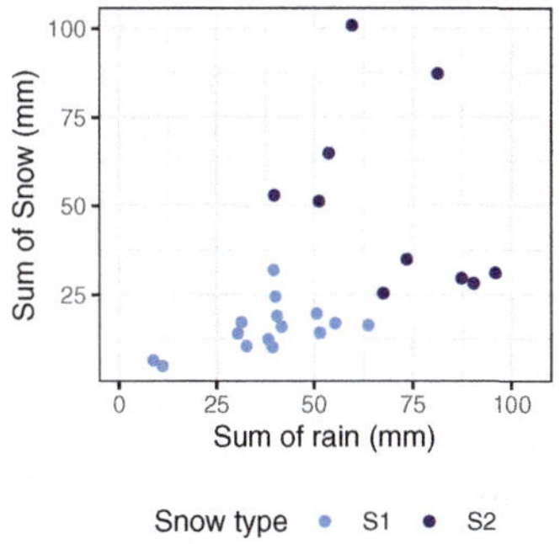

Fig. 6.10 kMeans clustering of snowmelt, rainfall and runoff coefficient (on average 0.2129 for S1 and 0.1779 for S2) for distinction of snowmelt-induced flood events for the Munich/Isar gauge

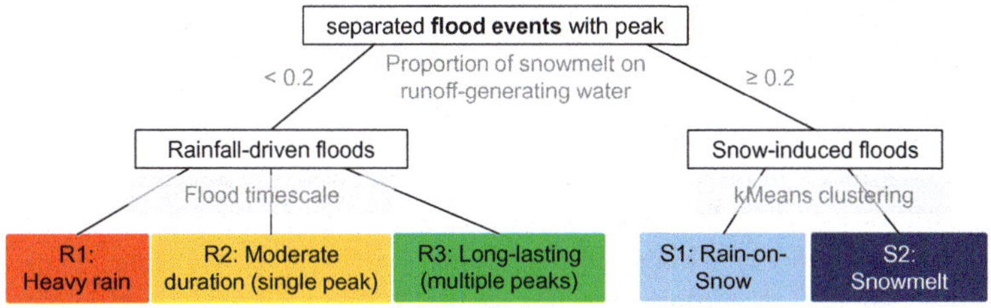

Fig. 6.11 Hybrid flood typology proposed by Fischer et al. (2019). (Modified from Fischer and Schumann 2021a)

6.2.3 The Role of Baseflow

Runoff can be separated into two main components: direct runoff and baseflow. While direct runoff is directly associated with a hydrometeorological event and reflects the immediate response of the catchment, the majority of baseflow cannot be associated with a hydrometeorological cause. Baseflow, thus, is usually seen as an indicator of the wetness state of the catchment or disregarded in flood analyses. Often, floods are only identified as such because direct runoff overlays with high baseflow. This is especially the case for approaches where floods are specified according to discharge-level or discharge-quantile thresholds. For example, for partial duration series, usually thresholds of discharge are applied to identify those flood peaks that are considered in the partial duration series. Often, these thresholds are based on the minimal observed annual discharge maximum within the AMS or by certain quantiles of flood peaks. Due to this definition, flood peaks that result from only small amounts of direct runoff but occur during a period of high baseflow are considered as flood peaks while flood peaks resulting from large amounts of direct runoff but only low baseflow are ignored. Knowledge on the baseflow is therefore required when analysing flood events. Moreover, baseflow might provide valuable information on the flood-generating mechanisms: high baseflow results mostly from wet

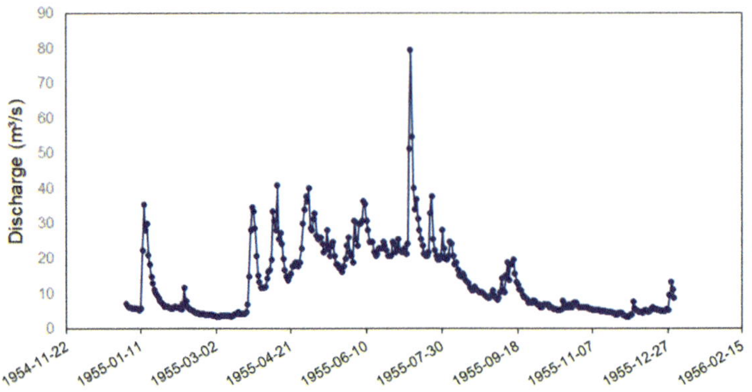

Fig. 6.12 Daily discharges at the Farchant/Loisach gauge (catchment size 424.3 km^2) with a distinct period of high baseflow in spring

initial conditions and hence saturated soils in a catchment. For example, in spring the snowpack melts in different elevation zones successively and feeds runoff to the river over a long period. In such situations, the catchment quickly generates runoff from rainfall (Fig. 6.12). This results in flashy hydrographs with large peaks, even though the amount of rainfall may not be high. Often, several flood events occur successively due to several short rainfall events. The flood timescale of such single events is similar to those of heavy rainfall events (R1). Therefore, the flood timescale alone is not sufficient to distinguish between both flood types. Instead, Fischer and Schumann (2021b) propose to additionally consider the baseflow of the event as an indicator for the flood type and to define a baseflow threshold below which the floods can be associated with heavy rainfall (type R1) and above which a new flood type is defined that is associated with high baseflow but comparably small amounts of rainfall (type R0).

Fischer and Schumann (2021b) show by an example of the Rhine river (see Sect. A.3 for more details), how such a baseflow threshold can be derived: When considering the histogram of baseflow of all flood events with a small flood timescale, a bimodal distribution becomes visible (Fig. 6.13a). Defining the baseflow threshold as the discharge that separates both modes, the two flood types R0 and R1 become easily differentiable.

Such an additional flood type considering the baseflow for small flood timescales is recommended for river basins with distinct periods of high baseflow. Usually, such periods are related to snowmelt in the high-elevation parts of the catchment. Hence, river basins with mountainous parts or thawing glaciers are probably affected by this phenomenon.

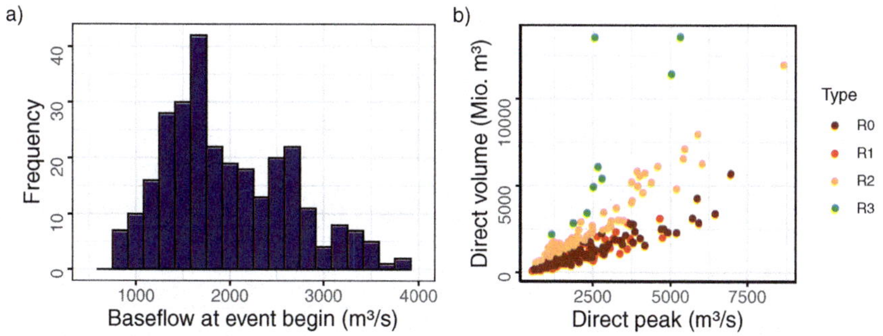

Fig. 6.13 **a** Histogram of the baseflow at the beginning of all events with small flood timescale and **b** peak-volume relationship of the rainfall-driven flood types R0, R1, R2, and R3 at Andernach/Rhine gauge. (modified from: Fischer and Schumann 2021b)

6.2.4 Coherence Between Flood Types and Flood-Generating Processes

The typology introduced in this chapter is based on causative as well as hydrograph-based considerations of the assumed flood-generating processes. For example, it is assumed that a certain flood timescale is linked to a certain type of rainfall event. By analysing the event-specific characteristics, links between the flood-generating processes and the flood type can be verified and thus the information contained in the flood types can be demonstrated. For a case study, once again the large data basis of catchments in Bavaria, Germany, is used (Appendix A.1). The more than 160 catchments offer the opportunity to consider diverse topographical and hydrological attributes ranging from alpine catchments to flatland catchments of various size.

The analyses of the flood types demonstrate that these are linked clearly to different flood-generating processes (Fig. 6.14). For example, flood type R3 is characterised by the longest duration and largest volume caused by the highest amount of precipitation. However, the rainfall intensity (defined as the proportion of the maximum one-day-rainfall sum on the total precipitation event) is lowest for all rainfall-driven floods. In contrast, for flood type R1, a rainfall that mostly falls down within one to three days is responsible for a fast catchment reaction where the resulting flood volume is small in relationship to the peak. Often, the flood duration is about five days. For the snowmelt-induced floods, it can be seen that the rainfall intensity for rain-on-snow floods is much higher than for snowmelt floods, though the sum of rainfall usually is not. Hence, the energy that starts the melting of the snowpack originates from rather heavy rainfall but not from a large amount of it. For the flood peak, no significant differences are observed, meaning that in theory each flood type might lead to large flood peaks. However, in the boxplots these large events have been removed for better readability. If the most extreme events would be included, the range of the R1 and R2 floods would clearly exceed those of the remaining flood types.

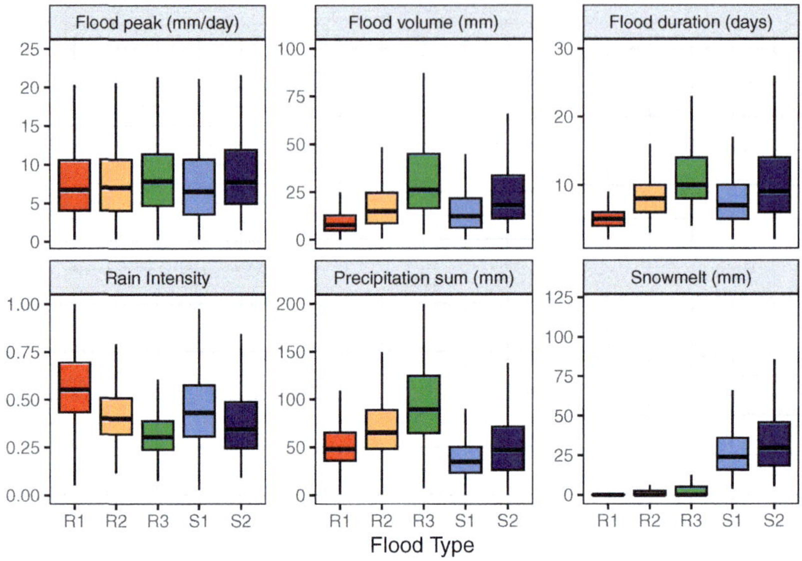

Fig. 6.14 Boxplots of several flood event characteristics differentiated by flood types for catchments in Bavaria, Germany

Similar results on the coherence of the flood types with flood-generating processes were observed by Fischer et al. (2019), Fischer and Schumann (2021b), Brunner and Fischer (2022).

The flood event separation as well as the flood typology proposed in this chapter are available in the R-package FloodR, downloadable from GitHub. At the end of this chapter, a short example is presented on how to apply these methods in R (2022).

Program Code

```
devtools::install_github(repo = "PhilippBuehler/FloodR")

library("FloodR")

# Load daily mean discharge data
# here: examplatory generated data set
dailyMQ <- data.frame(Date=seq(from=as.Date("01.01.2000",
 format="%d.%m.%Y"),
to=as.Date("01.01.2004", format="%d.%m.%Y"), by="days"),
discharge=rbeta(1462,2,20)*100)

#Run the separation
Flood_events <- eventsep(dailyMQ)
```

```
# The Separation might still contain overlaid flood events
# which need to be corrected
head(Flood_events)

#Run the Web separation to manually check the results

Run_WebFlood()

#### Separation of precipitation ####

# After the separation of the flood events,
# the corresponding precipitation events have
# to be estimated. For both functions, a daily
# precipitation timeseries is required,
# as well as the parameter indT, which consists
# of the position indices of begin, end and peak
# of the flood event.

# A sample dataset for testing:
dailyprec <- data.frame(Date=seq(from=as.Date("01.01.2000",
 format="%d.%m.%Y"),
to=as.Date("30.04.2000", format="%d.%m.%Y"), by="days"),
precipitation=rbeta(121,2,20)*100)

# Create indices of beginning, end and peak of
# the flood event, e.g. for first flood event
indT <- which(dailyprec$Date %in% Flood_events[1,c(1:3)])

# Run the separation for both methods
Date1 <- PreconeCP(dailyprec, indT = indT)
Date2 <- PrectwoCP(dailyprec, indT = indT)
print(c("Method1" = Date1, "Method2" = Date2))

##       Method1      Method2
## "2000-01-13" "2000-01-04"

# the later date is chosen

##### Flood Typology ####
 # For the typing of the flood event, multiple
 # characteristics for each flood event have
 # to be calculated beforehand (e.g. by separation):

 # Sum_SM: Sum of snowmelt during the floodevent in mm
 # Sum_N: Sum of precipitation during the floodevent in mm
 # dir_Volume: Direct volume of the flood event in Mio. m³/s
 # HQ_dir: Direct peak in m³/s
 # PSI_SM: Runoff coefficient WITH snowmelt + precipitation


# Open the sample flood event data
data("Sample_Flood_events")
head(Sample_Flood_events)
```

```
# Run the event typing
Floods_typed <- Flood_typology(Floods = Sample_Flood_events,
    n_G = 3, Type_3_min_samplesize = 10)

table(Floods_typed$Type)
# Check if S1 and S2 are assigned correctly by clustering
# approach, i.e., S2 should have higher mean snowmelt

# Plot the event typing of rainfall-driven floods
Floods_Rain <- Floods_typed[Floods_typed$Type
%in% c("R1", "R2", "R3"),]

library(ggplot2)

ggplot(Floods_Rain)+
geom_point(aes(x=dir_Volume, y=HQ_dir, fill=Type),
colour="black", shape=21, size=2)+
scale_fill_manual(values = c("R1"="#D7191C",
"R2"="#FDAE61", "R3"= "#1A9641"))
```

A thorough preparation of the data basis is the key to type-based flood statistics. For this purpose, flood events have to be separated and classified according to their flood-generating processes. The proposed separation based on a variance-threshold allows for considering flood events with high dynamics as well as the corresponding rainfall event. Combined with a hybrid causative-hydrograph-based typology using the flood timescale and clustering, a large number of catchments can be analysed with comparably small effort. The hydrograph-based perspective has the advantage that the catchment reaction instead of the meteorological causes is used for the flood type definition.

References

Aksoy H, Kurt I, Eris E (2009) Filtered smoothed minima baseflow separation method. J Hydrol 372:94–101

Bezak N, Horvat A, Sraj M (2015) Analysis of flood events in Slovenian streams. J Hydrol Hydromech 63(2):134–144

Blume T, Zehe E, Bronstert A (2007) Rainfall-runoff response, event-based runoff coefficients and hydrograph separation. Hydrol Sci J 52(5):843–862

Brunner MI, Fischer S (2022) Environ Res Lett 17(10):104038

Chapman TG, Maxwell AI (1996) Base flow separation-Comparison of numerical methods with tracer experiments. In Paper presented at hydrological and water resources symposium. Institution of Engineers, Hobart, Australia

Dunkerley D (2008) Identifying individual rain events from pluviograph records: a review with analysis of data from an Australian dryland site. Hydrol Process 22:5024–5036

Eckhardt K (2005) How to construct recursive digital filters for baseflow separation. Hydrol Process 19(2):507–515

Fischer S, Schumann A (2018) A distribution-free ordinal classification of floods based on moments. Hydrol Sci J 63(11):1605–1618

Fischer S, Schumann AH (2021) Regionalisation of flood frequencies based on flood type-specific mixture distributions. J Hydrol X 13:100107

Fischer S, Schumann AH (2021) Multivariate flood frequency analysis in large river basins considering tributary impacts and flood types. Water Resour Res 57:e2020WR029029

Fischer S, Schumann A, Bühler P (2019) Timescale-based flood typing to estimate temporal changes in flood frequencies. Hydrol Sci J 64:457–470

Fischer S, Schumann A, Bühler P (2021) A statistics-based automated flood event separation. J Hydrol X 10:100070

Fischer S, Bühler P, Schumann A (2021) Impact of flood types on superposition of flood waves and flood statistics downstream. J Hydrol Eng 26(7)

Gaál L, Szolgay J, Kohnová S, Parajka J, Merz R, Viglione A, Blöschl G (2012) Flood timescales: understanding the interplay of climate and catchment processes through comparative hydrology. Water Resour Res 48:W04511

Hartigan JA, Wong MA (1979) Algorithm AS 136: A K-means clustering algorithm. Appl Stat 28:100–108

Howard C (1996) Revisiting the degree-day method for snowmelt computations - Discussion. Water Resour Bull 32:411–413

Kampf SK, Lefsky MA (2016) Transition of dominant peak flow source from snowmelt to rainfall along the Colorado front range: historical patterns, trends, and lessons from the 2013 Colorado front range floods. Water Resour Res 52(1):407–422

Loukas A, Quick MC (1996) Spatial and temporal distribution of storm precipitation in southwestern British Columbia. J Hydrol 174:37–56

Mei Y, Anagnostou EN (2015) A hydrograph separation method based on information from rainfall and runoff records. J Hydrol 523:636–649

Nied M, Pardowitz T, Nissen K, Ulbrich U, Hundecha Y, Merz B (2014) On the relationship between hydro-meteorological patterns and flood types. J Hydrol 519:3249–3262

Oppel H, Mewes B (2020) On the automation of flood event separation from continues time series. Front Water 2:18

Piggott AR, Moin S, Southam C (2005) A revised approach to the UKIH method for calculation of baseflow. Hydrol Sci J 50:911–920

R Core Team (2022) R: A language and environment for statistical computing. R Foundation for Statistical Computing, Vienna, Austria. https://www.R-project.org/

Sikorska AE, Viviroli D, Seibert J (2015) Flood-type classification in mountainous catchments using crisp and fuzzy decision trees. Water Resour Res 51(10):7959–7976

Tang W, Carey SK (2017) HydRun: A MATLAB toolbox for rainfall-runoff analysis. Hydrol Process 31(15):2670–2682

Towler E, McCreight JL (2021) A wavelet-based approach to streamflow event identification and modeled timing error evaluation. Hydrol Earth Syst Sci 25:2599–2615

Vormoor K, Lawrence D, Schlichting L, Wilson D, Wong WK (2016) Evidence for changes in the magnitude and frequency of observed rainfall vs. snowmelt driven floods in Norway. J Hydrol 538:33–48

Chapter 7
Estimation of Flood Probabilities and Further Applications of Type-Based Statistics

Abstract The estimation of design flood peaks, i.e. of flood peaks for given probabilities respectively return periods, is the main practical application of flood frequency analyses. In this chapter, the TMPS model is applied to estimate design flood peaks for the joint mixture distribution as well as for each flood type separately. It is shown how the flood-type-specific model can be used to estimate the best- and worst-case scenarios under consideration of the flood types and by assigning a type-specific probability to each of the scenarios. Moreover, the impact of the different flood types on the mixed distribution, especially the right tail, is shown. The practical implications of such a type-specific consideration are explained and recommendations for hydrological practice are given.

The TMPS model extends classical flood frequency analyses by considering the flood-generating processes in flood statistics. It can therefore be classified as a multivariate approach. In contrast to other multivariate approaches, the TMPS model still allows to estimate one design flood peak for a given return period by combining the different flood types in a mixture model. However, the TMPS model provides much more information on the design flood peak than classical approaches such as AMS and POT. It takes into account that a given flood peak can result from different meteorological causes and allows the estimation of flood volumes and hydrographs for these different event types. In this chapter, the various possibilities to use the information content of the TMPS model are introduced. By the consideration of catchments in Bavaria, Germany, practical applications are given.

7.1 Estimation of Flood Probabilities

As a first example, the Marienthal catchment is considered. This catchment is part of the Regen basin, a tributary of the Danube river. It is located in Bavaria, Germany, and has a catchment size of 2613 km^2. With a mean elevation of 597 m a.s.l. it belongs to the eastern uplands. Currently, design floods are estimated by their peaks

S. Fischer and A. H. Schumann, *Type-Based Flood Statistics*, Water Science and Technology Library 124, https://doi.org/10.1007/978-3-031-32711-7_7

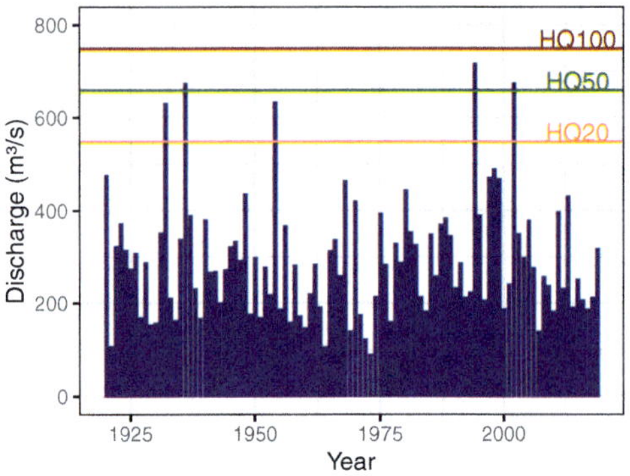

Fig. 7.1 AMS at Marienthal/Regen gauge with design flood peak values for different return periods T denoted with HQT

with the AMS approach. For example, for a return period of 100 years, this results in a design flood peak of HQ100 = 750 m^3/s (Fig. 7.1).

When estimating such design floods, the flood-generating process and hence the flood type is usually not taken into account. However, by application of type-based statistics, it can be considered retrospectively. For example, for a given design flood peak of return period 100 years, one can estimate the probability that such a flood peak originates from a given flood type. This way, it can be assessed how probable the occurrence of such a peak is during the period of snowmelt or in summer, when heavy rainfall is expected. Depending on the flood type, the exceedance of a given flood peak might be more probable for some flood types while others most likely will never reach or exceed it. Since hydrographs differ among flood types by their volumes and shapes, this allows the diversity of these characteristics to be estimated for a given flood peak as that peak may occur due to different flood types.

Coming back to the example of the Marienthal gauge, where a flood peak of 750 m^3/s might be used for the estimation of, e.g., flood inundated areas. This flood peak has quite different exceedance probabilities for the different flood types. Considering the flood typology proposed in Chap. 6, flood events are identified and classified. A total number of 297 flood events has been separated for the period 1930–2018 (the period for which precipitation data were available) based on daily discharges and the events were assigned with an instantaneous flood peak. Of these events, 116 are of type R1, 36 are of type R2, 33 are of type R3, 92 are of type S1 and 19 are of type S2. Considering these flood types and their respective distributions (see Chap. 5), the flood peak of 750 m^3/s has a return period of 180 years, if it caused by a heavy rainfall event, but of 8100 years, if it would result from a rain-on-snow flood event. Hence, it is highly improbable for the peak to be exceeded under snow conditions. In summer, however, when heavy rainfall events occur most frequently, the design flood peak is more likely. For the Marienthal gauge, the return periods of the flood types are higher than the joint return period derived from the AMS model (100 years) in this quantile range, but the total return period derived from the TMPS model is smaller (62 years). This is because the distributions of the rainfall-driven floods do not differ much for this range and hence several flood types contribute similarly to the joint distribution. A few examples on how the flood types differ in their return periods for given flood peaks are given in Table 7.1. There, probabilities of exceedance of selected flood peaks are given. For the single flood type distributions, always the annual distribution is considered, i.e., the one where the distribution function of the flood peaks is combined with the Poisson model to consider the number of floods per year of the respective flood type (Eq. 2.8). For example, take a flood peak of 500 m^3/s. With the classical AMS approach, this flood peak would have a return period of 14 years. For each of the flood types, the return period would be much higher, ranging from 44 years (R1) to 80 years (R3 and S1). How can this be explained? The annual distributions of the flood types only consider one specific flood type and do not consider other flood events. Hence, a return period of 44 years for heavy-rainfall only means that a heavy-rainfall flood of this peak is exceeded every 44 years in statistical mean. Still, any other type of flood can reach or exceed the peak of 500 m^3/s much more often. For example, having a return period of 44 years for heavy-rainfall floods and 72 years for R2 floods, this means when considering both flood types jointly, the flood peak of 500 m^3/s would be exceeded roughly every 30 years. If then all flood types are considered jointly, again the return period is reduced further. This is what is captured in the TMPS model, which results in a return period of 11 years for this flood peak. The TMPS model includes the probability of each flood type to exceed the flood peak under consideration of the probability of the occurrence of such a flood type. The more distinct the distributions of the flood types, the more the TMPS model depends on one single flood type (the one with smallest return period) when estimating the joint return period.

This can be explained by a short calculus example. Consider the case of a flood peak of 1000 m^3/s at Marienthal gauge (Table 7.1). Using the AMS model, this peak would have a return period of about 1000 years. Using the TMPS model, the same flood peak only has a return period of 210 years. This seems to be much smaller and one could question the applicability of the TMPS model. However, when having a look at the flood types, the return period of the TMPS model can easily be explained. First, there are the heavy-rain floods (R1). For these kind of floods the flood peak of 1000 m^3/s has a return period of about 500 years. This means, in statistical mean there

Table 7.1 Type-specific return periods of selected flood peaks for the Marienthal/Regen gauge

	Return period (years)						
Flood peak (m^3/s)	AMS	TMPS	R1	R2	R3	S1	S2
400	5.7	2.9	22	37	35	21	25
500	14	11	44	72	80	81	∞
600	35	24	82	130	186	397	∞
700	86	47	142	221	426	2663	∞
750	100	62	183	283	642	8086	∞
1000	1084	216	544	830	4851	$63 \cdot 10^6$	∞

will be two flood events with such a flood peak (or higher) in 1000 years. Similarly, for the long-duration-rain floods (R2) about one event exceeding this flood peak per 1000 years might occur. For the remaining flood types, the occurrence of such flood peaks is more improbable, only R3 flood peaks might occur once in 5000 years with such a magnitude. In summary, this means that we have three flood events with a flood peak as high or even higher than 1000 m^3/s in 1000 years. This would result in a return period of 330 years, when considering all events jointly. This is already quite similar to the result of the TMPS model with 210 years. The difference can be explained by the time period that was considered. Having only 1000 years, the impact of the flood types R3, S1 and S2 is neglected. If the time period in this toy example is increased, the empirical return period derived from simply adding up the expected number of events per flood type becomes more and more similar to the one estimated by the TMPS model, as the expected number of flood events per flood type is estimated more precisely. Asymptotically, the return period of the TMPS model will be obtained. This toy example emphasises the large difference between the AMS and the TMPS models: obviously, the two options (R1 and R2) reduce the exceedance probability of such a flood peak. The AMS model does not make such a distinction. Instead, the R1 and R2 events were considered as members of one sample. This leads to an underestimation of the exceedance probabilities here.

From the example of the Marienthal gauge, first conclusions on when large differences between the AMS and the TMPS model occur can be drawn. The difference mainly depends on three factors: the skewness of the distribution of each flood type and hence the heaviness of tail, the frequency of each flood type and the difference between the flood types. The difference of the TMPS model to the AMS model is larger, the larger one or two flood types deviate from the remaining flood types in their right tail. The heavier the tail, the larger the difference. However, such extreme deviation of flood types is not always the case. In the following, two additional examples are presented where different cases of the interplay of the flood types are illustrated.

As a second example, the Fürstenfeldbruck gauge at the river Amper is considered. Its catchment has a size of 1230 km^2 and belongs to the alpine foreland in Bavaria, Germany. With a mean catchment elevation of 744 m a.s.l. it has a higher elevation than Marienthal. For Fürstenfeldbruck, the design flood peak for a return period of 100 years was estimated as 134 m^3/s using the AMS approach (GEV, L-Moments) (Fig. 7.2).

For the same time period as for the Marienthal gauge, 308 flood events have been separated, 212 thereof of flood type R1, 44 of flood type R2, 5 of flood type R3, 38 of flood type S1 and 9 of flood type S2. Hence, flood types R3 and S2 are very rare and can not be considered statistically. As a rule of thumb, which was validated for more than 1000 catchments in Europe, flood types that represent less than 2% of all flood events in a catchment should be excluded from the statistical analyses, as the statistical modelling of these flood types might result in high uncertainty. Of course, these flood types should not contribute to the large flood events, i.e., have no flood peaks among the 25% largest flood peaks in the catchment. If these two criteria are fulfilled, the contribution of the respective flood type on the TMPS model

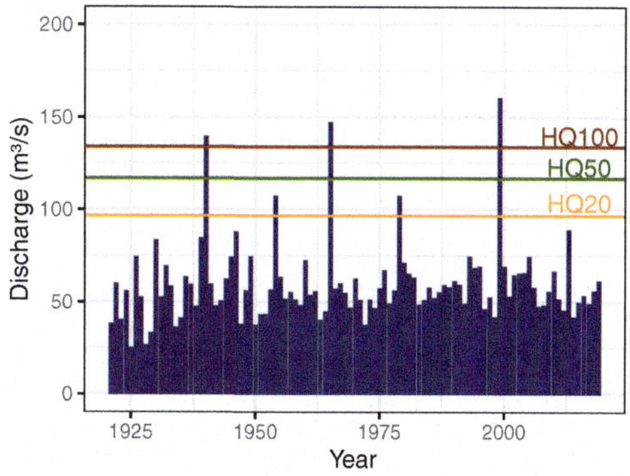

Fig. 7.2 AMS at Fürstenfeldbruck/Amper gauge with design flood peak values for different return periods T denoted with HQT

is negligible and the flood type should be excluded from the TMPS model and the respective flood events should be merged with the most similar flood type, e.g. flood type R3 with R2 and S2 with S1.

For the Fürstenfeldbruck catchment, the probability distributions of the flood types differ much (see also next section). Here, the R1 floods are mainly responsible for large flood peaks. This becomes obvious when considering once again the type-specific probability of the estimated flood peak with return period 100 years based on the AMS approach. If such a flood peak occurs due to a heavy rainfall event and hence is of type R1, its return period is only 69 years. Hence, it occurs much more frequently than expected. For the remaining flood types, the return periods are once again higher than 100 years. For the TMPS model, the return period of the 100-year flood of the AMS model is 96, so basically the same as for the AMS approach. The consideration of flood types for this catchment reveals that the design flood peaks which are used may be too small and that the probability of a failure of a flood protection system would be higher than expected by using the AMS-based design flood. Such large floods are caused by heavy rainfall, but when mixed up with other, smaller floods in a sample, as it is done in the AMS approach, they are underestimated. Type-based statistics can reveal such issues and help to prepare large floods, as the flood-generating processes are considered.

The last example is that of the Kautendorf gauge at the river Southern Regnitz in Bavaria, Germany. This catchment has a size of 92.4 km^2 and is located in the eastern uplands. With a mean catchment elevation of 586 m a.s.l. it is comparable to the Marienthal catchment in location and elevation, though much smaller. An extraordinary large flood occurred in 2013 in this catchment, which was almost twice as large as the largest flood peak in the records so far. For the period of 1958–2018, 166 flood events have been separated, 45 thereof of flood type R1, 34 of type R2, 7 of type R3, 55 of type S1 and 25 of type S2. For Kautendorf, the design flood peak for a return period of 100 years was estimated as 34 m^3/s using the AMS approach (GEV, L-Moments) (Fig. 7.3). When comparing the return periods derived with AMS and TMPS models for selected flood peaks (Table 7.2) it becomes obvious that for this catchment AMS and TMPS models do not differ much. This can be explained by

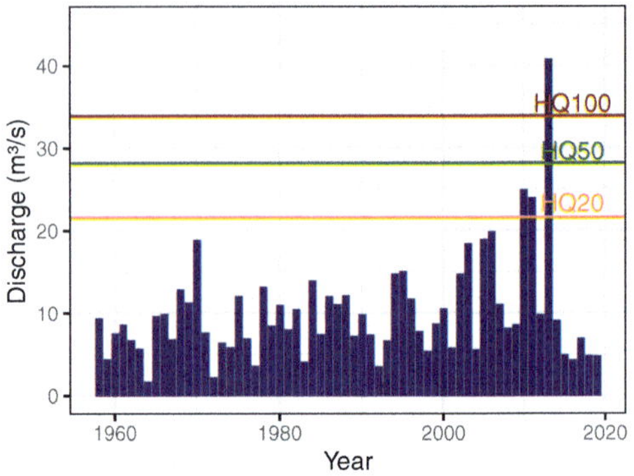

Fig. 7.3 AMS at Kautendorf/Southern Regnitz gauge with design flood peak values for different return periods T denoted with HQT

Table 7.2 Type-specific return periods of selected flood peaks at the Kautendorf/Southern Regnitz gauge

	Return period (years)						
Flood peak (m^3/s)	AMS	TMPS	R1	R2	R3	S1	S2
10	2.6	1.3	4.4	5.9	72	6.2	4.0
15	6.5	2.8	15	13	∞	30	9.3
20	15	11	131	25	∞	207	62
25	32	36	137892	48	∞	2140	∞
30	63	67	∞	88	∞	40950	∞
50	484	516	∞	665	∞	∞	∞

the fact that for Kautendorf mainly one flood type is responsible for the right tail, the R2 floods. On the first sight, it might be a similar case as for Fürstenfeldbruck and one could wonder why the deviation of AMS and TMPS models is different for the two gauges. However, this becomes clearer when having a closer look at both distributions. For the Fürstenfeldbruck gauge, the AMS and TMPS probabilities have also been similar for peaks of up to 150 m^3/s. Only then the return periods started to deviate. For Kautendorf gauge, the return periods are almost equal for both AMS and TMPS model for basically all peak levels except the small peaks, where the TMPS model results in smaller return periods (see also next section). This difference in the right tail can be explained by one factor: the frequency of the dominant flood type in the AMS sample. When one flood type clearly dominates certain parts of the TMPS distribution, i.e., it results in much smaller return periods for a given flood peak than any other flood type, the TMPS model follows the distribution of this flood type and hence results in similar return periods as this flood type. At the same time, also the AMS model results in similar return periods, as these floods will likely be included in the AMS due to their size. This applies mainly to the medium to large flood peaks. The return periods of small and very large flood peaks, however, are impacted by another factor: the frequency of the flood types in the AMS sample. For the Fürstenfeldbruck gauge, 42 of the 88 annual maxima belong to the dominating flood type R1. However, usually the R1 floods consist of many floods with a small flood peak and only few with a large flood peak. In the AMS sample, all these small R1 floods are replaced by larger flood peaks of other flood types in those years, especially R2, such that the overall skewness of the distribution is reduced compared to the skewness of the R1 floods, which results in higher return periods for large flood peaks compared to the TMPS model. For the Kautendorf gauge, only 10 of the 60 annual maxima are of the dominating flood type R2. These are all large flood peaks. The several smaller flood peaks of R2 are not considered in the AMS sample as other flood types (mainly S1 and S2) often have somewhat larger peaks. Since the larger R2-floods are retained in the AMS, the total skewness does not change significantly compared to the sample that includes only the R2 floods. This sample then results in a similar distribution of AMS and TMPS model, which is dominated by the R2 events, except for the small quantiles, where the TMPS model considers the large frequency of floods with comparable high peak of flood types R2, S1 and S2.

This comparison shows how the TMPS model can change the perception of flood risk. If a combination of flood causes is considered, the exceedance probabilities determined by the TMPS model may be significantly higher, but if one flood type clearly dominates the high quantile range, the TMPS and AMS return periods (or quantiles of equal probability) will differ less from each other.

7.2 Estimation of Design Flood Peaks

In the previous section, it was demonstrated how the flood probabilities for a given flood peak might differ between flood types and AMS (or POT) model and TMPS model. Of course, the type-based statistics can also be used vice-versa: flood peaks can be estimated for given return periods. This is most relevant for the purpose of assessing design flood peaks for flood risk assessment and management, e.g., by dimensioning of flood protection systems.

Considering once again the example of the Marienthal gauge from the previous section, flood peaks can be derived for given return periods by considering quantile functions (Fig. 7.4). It can be seen that the TMPS model does not differ much from the AMS model for return periods up to 20 years. For example, the HQ20 (the design flood peak for a return period of 20 years) is 550 m^3/s when applying the AMS and 577 m^3/s when applying the TMPS model. However, for increasing return periods, both models differ significantly. The distribution function of the TMPS has a steeper course and thus resulting in higher flood quantiles for large return periods than the AMS. Hence, the right tail of the TMPS is heavier than that of the AMS model. The HQ100 for both models already differs by 88 m^3/s (being 838 m^3/s for the TMPS model). This difference can be explained when considering the distributions of the flood types. Flood types R1 and R2, associated with heavy and long rains, clearly have a similarly heavy tail which makes them different compared to the remaining flood types. This implies that these two flood types more probably will result in extreme

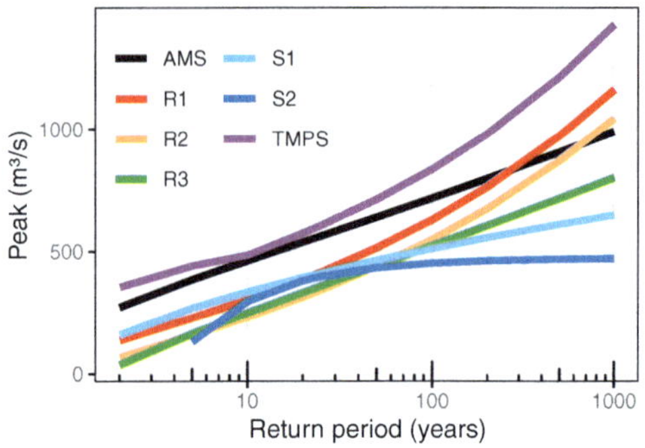

Fig. 7.4 Quantiles of the TMPS model and the different annual flood-type distributions at Marienthal/Regen gauge. For comparison, also the AMS quantiles are shown

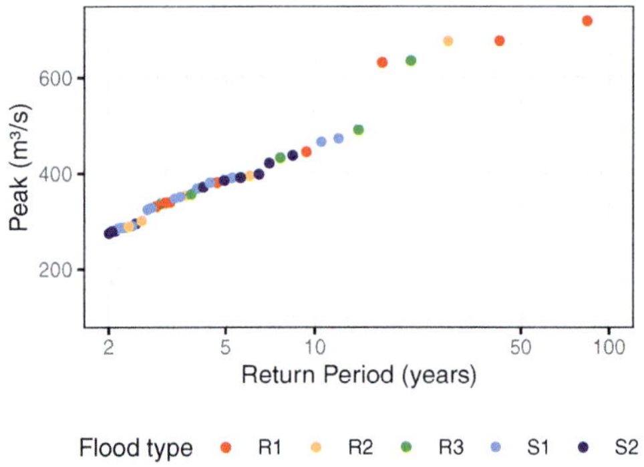

Fig. 7.5 Annual maximum discharges and corresponding empirical return periods (plotting positions) with respective flood type of the Marienthal/Regen gauge

flood events with high peaks. The heavy tail of the flood types is then inherited by the mixture model. The remaining flood types, especially the snowmelt-induced types, only play a minor role for the right tail of the TMPS model. Instead, these define the smaller quantiles. The AMS model mixes these flood types up, which leads to a reduction of the highest quantiles which are levelled out by those flood types resulting only in small flood peaks (Fig. 7.5).

Another interesting aspect of the flood type distributions becomes obvious for this gauge: the distribution of the snowmelt floods (S2) has a concave shape in Fig. 7.4. This implies that there exists an upper limit for the flood peak of this flood type. In this case, a flood peak of 500 m^3/s cannot be exceeded by snowmelt-floods, which results in return periods equal to ∞. Such an upper limit is related to the physical conditions that come along with a snowmelt event. More precisely, the snow cover that originally leads to the snowmelt also limits the amount of free water that generates the steepness of the hydrograph and the peak by its retention capacity. The water storage capacity of a thick snow cover itself balances the release of water. Basically, this means that an infinite amount of snow does not lead to an infinite high flood peak as it would dampen the reaction of the catchment during snowmelt.

The role of the flood types for the TMPS model cannot only be quantified visually. Instead, the type-specific distributions can be used to derive the probability that the design flood peak occurs for the respective flood type. Take as an example the

Table 7.3 Type-specific return periods of design flood peaks at the Marienthal/Regen gauge

		Return period (years)					
Return period TMPS (years)	Quantiles TMPS (m^3/s)	AMS	R1	R2	R3	S1	S2
5	443	8.4	30	50	50	37	63
10	486	12	40	66	71	66	∞
20	577	28	71	114	153	269	∞
50	714	97	153	238	478	3590	∞
100	838	283	276	425	1317	82600	∞
200	983	945	509	777	4236	∞	∞

HQ100. This was estimated as 838 m^3/s with the TMPS model. The same flood peak has a return period of 276 years when considering heavy-rainfall floods (R1) only (Table 7.3). For snowmelt-induced floods (S2), such a flood peak cannot be expected at all. The Marienthal catchment is located in the uplands in Bavaria. In this region, snowmelt-floods cannot be expected to cause large flood events. However, this flood type is frequent when considering small floods with return periods up to ten years. This implies that flood protection systems for extreme floods should focus on heavy-rain events. But the flood typology provides even more information. Since a hydrograph-based approach was applied, it can also be concluded that these heavy-rain events come along with extreme flood peaks but comparably small volume, because they have small flood timescales. Such information can be used to assess the performance of flood retention systems where the storage capacity is limited by local conditions.

As was pointed out before, for the example of the Fürstenfeldbruck gauge, the difference between the type-specific distributions is much larger (Fig. 7.6). Here, the heavy-rain floods clearly dominate the right tail of the distribution and hence also the shape of the distribution of the TMPS model. The remaining flood types are mostly irrelevant for large floods (Table 7.4). Hence, when estimating design floods, one should focus on heavy rain floods instead of considering all floods jointly and thus underestimating the large quantiles, as does the AMS model.

The case of quantiles of one flood type being larger than those of the TMPS model is generally rare. It can be observed for about 20% of all catchments in Bavaria (164 analysed catchments) and occurs almost only for the flood type R1, the heavy-rain floods. This phenomenon can be read in advance from the data: if the large floods at a gauging station are dominated by the R1 floods and at the same time there are only few small R1 floods in the sample, the higher quantiles of the distribution of the R1 floods will exceed those of the TMPS model. This occurs mostly from a return period of about 100 years on. Below this return period, the TMPS model provides the largest flood quantiles. For large return periods, the distribution of the R1 floods delivers the highest flood quantiles. However, it has to be considered that only the

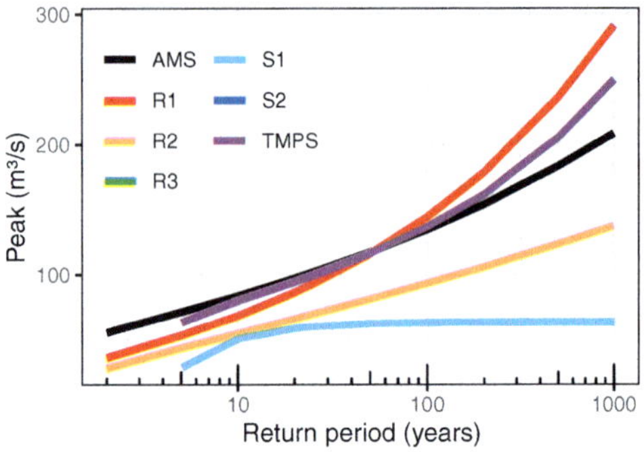

Fig. 7.6 Quantiles of the TMPS model and the different annual flood-type distributions at Fürstenfeldbruck/Amper gauge. For comparison, also the AMS quantiles are shown

Table 7.4 Type-specific return periods of design flood peaks for the Fürstenfeldbruck/Amper gauge

		Return period (years)					
Return period TMPS (years)	Quantiles TMPS (m^3/s)	AMS	R1	R2	R3	S1	S2
5	64	3.3	7	15	–	∞	–
10	80	8.1	14	40	–	∞	–
20	95	18	23	94	–	∞	–
50	117	49	45	307	–	∞	–
100	137	109	74	846	–	∞	–
200	162	260	127	2753	–	∞	–

distribution of this single flood type and the number of events per year of this type is considered in the Poisson model (Eq. 2.8). The relative frequency compared to other flood types is ignored. Therefore, one cannot draw any direct conclusions about the impact on the TMPS model. The TMPS model, instead, considers the relative frequency of each flood type and hence may down-weight single distributions such as the one of R1 floods. This is not a drawback of the model but a realistic assessment of the relevancy of each flood type. It has to be considered, though, that for the TMPS model constant frequencies of flood types are assumed over the full range of observed peaks. However, this is not necessarily the case in reality. In fact, depending on the catchment area, it can often be observed that extreme floods occur almost exclusively due to a specific flood type (Fig. 7.7). To consider this aspect, it would be necessary to estimate the frequency of the flood types quantile-dependently and to build a multi-stage model for different quantile ranges. However, the observation series are usually too short and contain only few extreme floods, such that a reliable estimate of the frequency of the flood types for large quantiles is not possible. For this reason, this is not done here.

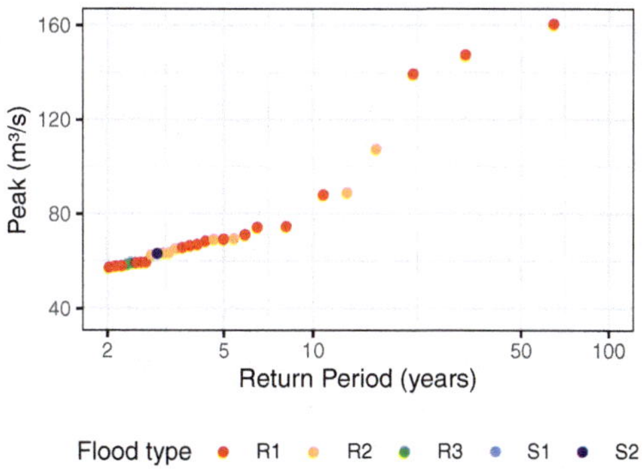

Fig. 7.7 Annual maximum discharges and corresponding empirical return periods (plotting positions) with respective flood type of the Fürstenfeldbruck/Amper gauge

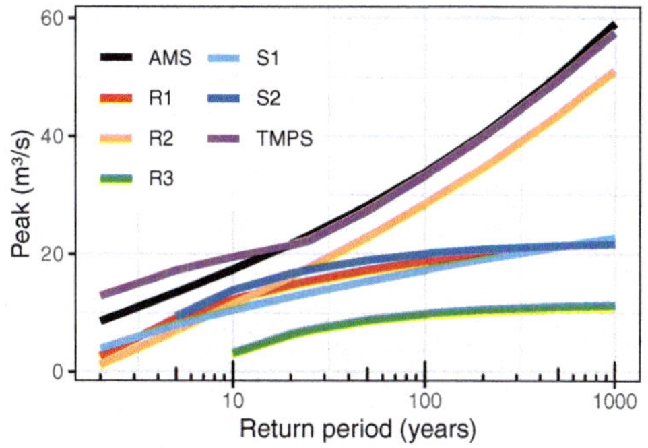

Fig. 7.8 Quantiles of the TMPS model and the different annual flood-type distributions at Kautendorf/Southern Regnitz gauge. For comparison, also the AMS quantiles are shown

Finally, the case is illustrated where the AMS and the TMPS models result in quite similar flood quantiles, at least for larger return periods, in this case for the Kautendorf gauge from return periods of 10 years on (Fig. 7.8). This was already explained in the previous section by the even distribution of flood types in the AMS sample, where no flood type clearly dominated and all were represented well. Only for small return periods, where many flood types result in similar flood peaks and due to the limitation of one event per year in the AMS their frequency is underestimated, the TMPS model results in higher quantiles.

The examples given here demonstrate that the TMPS model is able to capture the different flood type distributions. It is impacted by the flood types that result in the largest flood quantiles, especially in the right tail. Therefore, differences to classical FFA models like the AMS might occur. The TMPS model results in similar quantiles as the AMS model in the case where the frequency of the flood types is captured well by the AMS sample. This requires not only a long record of floods but also a balanced occurrence of the flood types as annual maximum floods. If one flood type dominates one part of the AMS and the frequency of this flood type is not captured in the AMS sample, as it was the case in Fürstenfeldbruck (Fig. 7.7) where the R1 flood events dominated the right tail, this part of the TMPS model will deviate from the AMS. Similarly, as was shown for the Marienthal gauge, some flood types might become irrelevant for the right tail. In this case, again the frequency of these flood types will not be captured correctly in the AMS model and the TMPS model will result in larger quantiles than the AMS model.

7.3 Analyses of Evidence of Flood Types

The analyses of flood types and their impact on flood statistics may reveal important local and spatial patterns and can provide information on the relevance of the different flood types for the generation of extreme events. Flood types occur with different frequency for different regions. Moreover, they are not uniformly distributed over all peak magnitudes, as shown in the previous section. As was demonstrated in Fig. 7.4, different flood types may be associated with different peak magnitudes. Large floods may be caused by different flood types than small floods. This also has an impact on the flood samples. When considering the POT approach, different flood types may be included than in the AMS. For example, in Bavaria in Germany, the largest floods in the southern catchments close to the Alps are most often caused by heavy rainfall (R1 floods), while in the north snowmelt often leads to large floods (type S2) (Fig. 7.9). Indeed, the southern part of this basin is known to be the area in Germany with the most frequent heavy-rainfall events (Kaiser et al. 2020). Such differences are important to know for the operation of flood protection systems, since for snowmelt floods the reservoir in winter should be empty while for heavy-rainfall floods the reservoir more probably in summer should be empty. Such differences are hidden for the POT and AMS samples, where all flood types are mixed. Similar differences have been observed for the US as well as all over the globe (Stein et al. 2019).

Since especially the largest flood events are relevant for flood protection measures, it is of interest to have a closer look at the right tail of the TMPS model. The tail behaviour of the TMPS model is an especially interesting indicator for the tail heaviness of the flood types and moreover helps to identify those flood types that result in the most extreme flood peaks. For example Merz et al. (2022) point out the role of mixing distributions in the context of heavy tails. They state that the most dominant component tail is inherited by the tail of the mixture distribution. Thus, it is of interest which flood type defines the tail of the joint distribution. For all catchments of the exemplary data set of Bavaria, Germany (see Sect. A.1), the dominating flood type for the quantiles with return period between 2 and 1000 years, i.e., the quantile of the flood type distribution with smallest deviation to the quantile of the TMPS model, is given in Fig. 7.10.

The observed difference between northern and southern catchments remains valid for the medium and large floods with return periods between 10 and 100 years. The northern catchments are mostly dominated by snowmelt-floods (S2) while the southern catchments are dominated mostly by heavy-rainfall-floods (R1) in this quantile

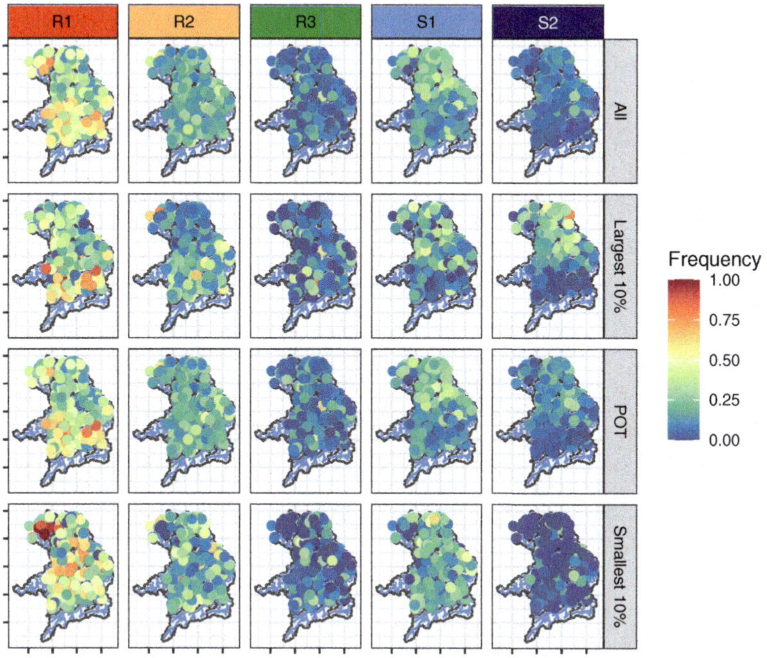

Fig. 7.9 Frequency of each flood type for different flood samples in Bavaria, Germany (from: (Fischer and Schumann, 2021))

range. For small floods with return periods between 2 and 5 years, in the north rain-on-snow floods (S1) play a major role for the flood-generation, but for floods with return periods from 20 to 50 years, the snowmelt floods (S2) become more relevant here. For extreme floods with return periods of 200 years and more, however, the difference between north and south is no longer visible. Instead, also the north of the basin is dominated quite often by heavy-rain-floods (R1) in this quantile range, while catchments close to the river Danube are most affected by long-duration rainfall (R3). The latter may be related to the catchment size: the gauges located directly at the river Danube belong to catchments of several thousand km^2. In such large catchments, the spatially limited heavy-rainfall-floods are not capable of generating extreme floods since they will only hit a small part of the catchment. Long-duration rainfall or several successive rainfall events, though, might hit the whole large catchment, leading to flood generation at several locations in the catchments and thus an extreme flood event downstream. Such regional characteristics encourage the use of the TMPS model for regionalisations (Part VI), taking into account differences in flood frequencies and distribution functions.

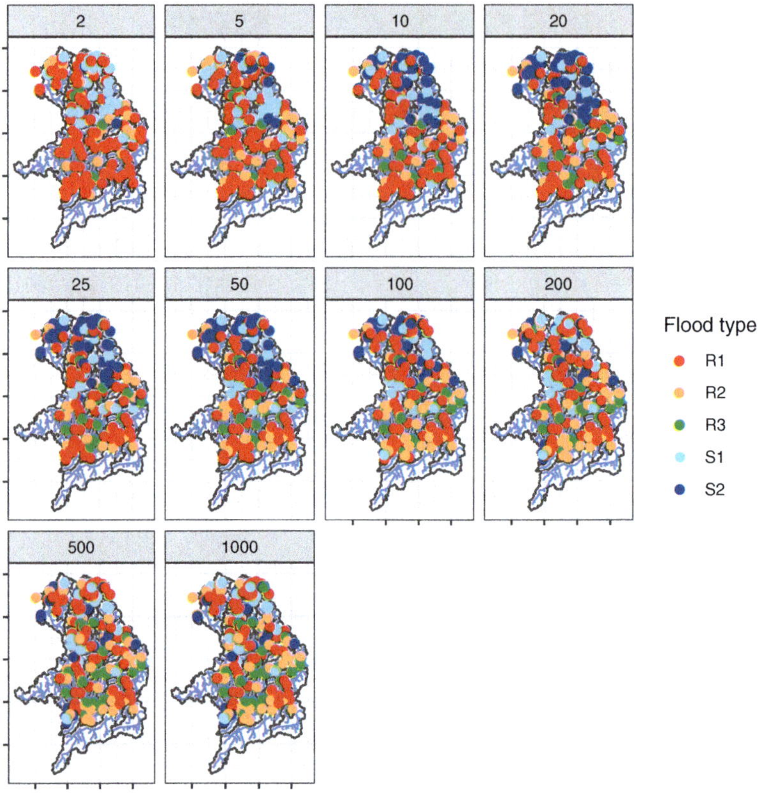

Fig. 7.10 Dominating flood type for the quantiles of different return periods $T = 2, 5, \ldots, 1000$ years in Bavaria, Germany

The TMPS model allows for the consideration of flood types in flood statistics. The resulting design flood peak might differ from those obtained by traditional FFA using the AMS approach. AMS and TMPS models result in similar flood quantiles for the same return period if the sample size of the AMS is sufficient and all flood types are represented uniformly in it. AMS and TMPS models deviate if the frequency of each flood type is not captured correctly in the AMS model and distinct flood types result frequently in very large floods. In such a case, the AMS model under-represents this flood type and will usually result in smaller design flood peaks. Flood types should not be included in the statistics, if their sample size is smaller than 2% of the whole sample of all flood events and they do not include floods with a peak among the largest 25% of all flood peaks. If there are only very few but large flood events of one flood type, these might be included in the sample of a similar flood type, e.g., floods of type R3 in the sample of floods of type R2.

References

Kaiser M, Borga M, Disse M (2020) Occurrence and characteristics of flash floods in Bavaria (Germany). In: Leal Filho W, Nagy G, Borga M, Chávez Muñoz P, Magnuszewski A, (eds) Climate change, hazards and adaptation options. Climate Change Management. Springer, Cham

Merz B, Basso S, Fischer S, Lun D, Blöschl G, Merz R, Guse B, Viglione A, Vorogushyn S, Macdonald E, Wietzke L, Schumann A (2022) Understanding heavy tails of flood peak distributions. Water Resourc Res 58:e2021WR030506

Stein L, Pianosi F, Woods R (2019) Event-based classification for global study of river flood generating processes. Hydrol Process 34:1514–1529

Chapter 8
Temporal Changes of Flood Types and Their Impact on Flood Statistics

Abstract Standard flood frequency analysis is based on the assumption of stationarity of the sample of floods, i.e., no change of the distribution over time. However, long-term variability in climate and anthropogenic impacts call this assumption into question. As a consequence, more and more non-stationary models are applied in flood frequency analyses. Yet, most of them only consider a change or trend in the magnitude of flood peaks while ignoring changes in the underlying flood geneses. In this chapter, flood types are applied to detect such changes in the flood geneses. It is investigated if and how certain flood types change in frequency. Moreover, the effects of such changes on the flood frequency analyses are demonstrated. Here, a special focus is laid on the benefits of the TMPS model, which is able to consider changes of each flood type separately.

The climate of our world is subject to change. These changes can have natural causes, such as previous ice ages, or anthropogenic causes, as is currently suspected. Current climate research (Kundzewicz et al. 2006; Burt et al. 2016) as well as the IPCC report (IPCC 2021) agree that significant changes in the climate are to be expected in the coming decades, regardless of the actual global warming. In part, the changes can already be observed now and are reflected in the observations of temperature and precipitation data of the last decades. One assumed change is the increase of convective rainfall events, especially heavy rainfall, for Central Europe and the USA (Butler 2020). It is also assumed that large-scale snow covers that remain for several days will become significantly rarer. These changes in the meteorological boundary conditions also have a direct influence on the generation of floods. Due to an increase in heavy rainfall events, it is likely that flash floods and floods of comparatively short duration with very high peaks will also increase in frequency. This can already be observed in part in Central Europe (Blöschl et al. 2019). A decrease in large-scale snow cover, on the other hand, is more likely to lead to the absence of long-lasting floods in spring, which used to cause extensive damage, due

S. Fischer and A. H. Schumann, *Type-Based Flood Statistics*, Water Science and Technology Library 124, https://doi.org/10.1007/978-3-031-32711-7_8

to a lack of snowmelt. These changes have important consequences for hydrology and water management. The seasonality of the flood regime is shifting (Blöschl et al. 2017) and management plans for dams and flood protection concepts must be adapted. However, it is difficult to specify these changes, especially with regard to statistically sound results (significance). In the scientific community, for example, it is highly controversial whether a long-term trend can be assumed that could be extrapolated years into the future, or whether cyclical fluctuations (long-term persistence), which are only present in parts of the observation series, are wrongly regarded as such monotonous changes. Numerous statistical methods exist to investigate such changes: Change-point and trend tests (Madsen et al. 2014; Kimuli et al. 2021), studies on long-term dependence (Szolgayova et al. 2014), or tests for cyclical behaviour (flood-rich and flood-poor periods, Lun et al. 2020). A major disadvantage of most studies of changes in flood time series is that they do not take into account the different flood geneses. Mostly, investigations are carried out on the basis of annual maximum flows or POT series. However, if opposite changes occur, such as the above-described decrease in snowmelt-related floods with a simultaneous increase in heavy rainfall floods, these changes can be disguised by mutual compensation in series mixed by geneses.

Considering flood types and studying them separately offers significant advantages in such cases. In this way, it can be determined which geneses have changed and to what extent. The results can then be concretely compared with corresponding climate studies and thus be given a deterministic justification. In this chapter, a procedure is presented on how significant changes in the frequency of certain flood types can be statistically detected. It is based on the variance of the temporal intervals between flood events. In a second step, the consequences of these changes for flood statistics are demonstrated. Here, too, it is of great importance to consider the flood types separately in the statistics, as otherwise effects can be balanced or weighted disproportionately.

8.1 Changes in the Frequency of Flood Types Over Time

To detect changes in the frequency of occurrence of certain flood types, again the flood typology described in Sect. 6.2 is applied. This typology captures the relevant meteorological drivers of floods for which a change of the climatic conditions is assumed, such as heavy rainfall or snowmelt. Moreover, the long records of flood events allow for a meaningful investigation of changes. The aim is to detect changes in the frequency of occurrence. For this purpose, a change-point test is applied. Alternative investigations for changes, e.g., by using a trend test, mainly focus on the magnitude of the flood events and often have low explanatory power due to comparably short record lengths and unfavourable trend-to-noise relationships. Using a change-point test and only the time of occurrence, instead, allows for an investigation independent of the flood (peak) magnitude. How the results are linked to current results on flood-rich and flood-poor periods, which also focus on the frequency of flood events, will be explained later on. One major difference is that the proposed change-point test does not rely on a definition of flood events except for the separation algorithm.

8.1.1 A Statistical Test to Detect Changes in Frequency of Events

Change-point test are designed to detect a certain point in time where some statistical characteristic of the distribution of the sample changes abruptly. In such a case, the characteristic is on a different level before and after the assumed change-point. This is contrary, e.g., to trend tests, where a continuous monotone trend is investigated. To detect changes in the frequency of occurrence of events, a change-point test for the variance is applied to the time series of temporal distances between two consecutive events of the same flood type. More precisely, for the sample of flood events of a given flood type, the time difference (in days) between two consecutive flood events is derived, resulting in a new sample that consists of these time spans. If the change-point test in variance is applied to this sample, it can be detected if there is a significant change in the variability of time spans. For example, if an increase of frequency of a certain flood type is assumed, this would lead to smaller time spans between the flood events and hence a smaller variability. If such a change-point is detected, the relative frequency of events before and after the change-point can be used to evaluate whether there is an increase or decrease in frequency. A change-point test in mean, instead, would focus on the observed time spans and hence being highly impacted by single time spans.

For the given task, the robust and non-parametric variance test based on Gini's Mean difference is applied (Fischer et al. 2019). The use of a robust test reduces the impact of single extraordinary large time gaps between two events, which could be caused by anthropogenic impacts, e.g., by flood retention. The null hypothesis of the test is that no change in variance (in time intervals) occurs. The robust measure for variance of a sample $X_1, \ldots, X_n$, Gini's Mean Difference, is defined as (Gerstenberger and Vogel 2016)

$$g_{k,n} = \frac{2}{(n-k+1)(n-k)} \sum_{k \leq i < j \leq n} |X_i - X_j|.$$

Gini's Mean Difference estimates the absolute difference between all data points and normalises their sum. It has many similarities to the classical variance but is superior in the case of non-normal distributions (Yitzhaki 2003). Moreover, it is a robust measure of variability, similar to the interquartile range IQR.

The test statistic for a change-point test based on Gini's Mean Difference is then defined as

$$G = \frac{1}{2\sigma\sqrt{n}} \max_{1 \le k \le n-1} |g_{1,k} - g_{k+1,n}|$$

with σ being the sample standard deviation of the sample. The asymptotic distribution to derive the critical value considering the significance level is the supremum of a Brownian Bridge. The Brownian Bridge is a stochastic process, most commonly known as its probability distribution function is related to the Brownian Motion under the condition of an identical starting and end value equal to zero, comparable to the two pylons of a bridge. The supremum of a Brownian Bridge is also known as Kolmogorov-distribution, e.g., used for the Kolmogorov-Smirnov-test, and available in many statistical computer programmes. The test is applied with a 5% significance level.

Based on the test statistic, a possible change-point in variance can be estimated by taking the point in time where the test statistics reaches its maximum, a common technique to derive an estimation of the change-point for change-point tests:

$$\hat{k} = \arg \max_{1 \le k \le n-1} |g_{1,k} - g_{k+1,n}|.$$

In Fig. 8.1, the application of Gini's Mean Difference test to detect significant changes in the variance of the time difference between consecutive flood events of a given type is illustrated for the R1-floods recorded at the Amberg/Vils gauge in Bavaria, Germany. The upper panel shows the flood peaks. Visually, no change can be detected for the flood peaks and indeed the application of the Wilcoxon-test does not give any significant change-point in mean. However, when having a look at the time difference in days between the consecutive flood events, a change becomes obvious. Gini's Mean Difference test detects a significant change in the variance of the time differences around 1975. From this time on, R1-floods occur more frequently. Yet, the mean of the flood peaks before and after the change-point almost stays the same (22.7 m^3/s before and 19.7 m^3/s after the change-point), while the skewness increases (2.1 before and 2.4 after the change-point).

The design of the test statistic only allows to test on one change-point in the time series. However, there might occur several change-points with either the same tendency (several increases, or several decreases of frequency) or with contrary tendencies (consecutive increase and decrease). Both phenomenons are of interest, as they might give notice on the type of change: a clear tendency towards increase or decrease or a cyclical behaviour of the occurrence of certain flood types, similar to sequences of flood-rich and flood-poor periods. Therefore, if a change-point is detected as significant, then the time series is split at this point and the test is applied to both partial series before and after the change-point, if they contain at least 10 observations. Of course, this affects the power of the test and hence the Bonferroni-rule is applied, i.e., the significance level is divided by the number of applications of the test per time series.

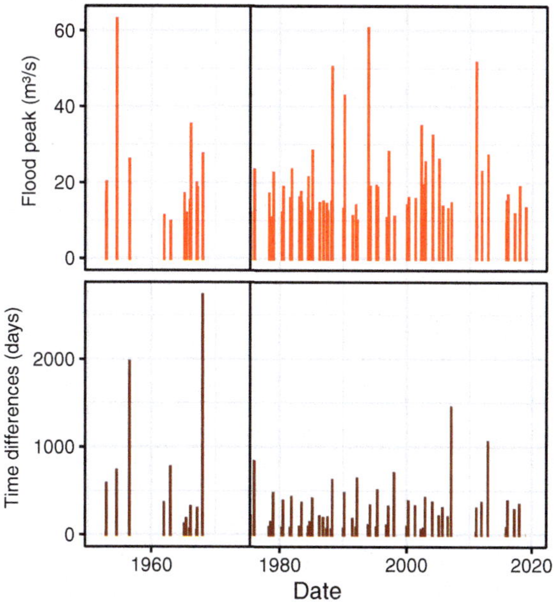

Fig. 8.1 Flood peaks and difference in days between two consecutive flood events of flood type R1 for the Amberg/Vils gauge in Bavaria, Germany. The black line indicates the detected significant change-point

8.1.2 Significant Changes in the Frequency of Flood Types in Bavaria, Germany

The change-point test is applied to the catchments in Bavaria in Germany (Sect. A.1). This region, especially the parts in the alpine foreland, is known to be affected by several heavy rainfall events in the recent years and is expected to faced by an increased frequency of heavy rainfall events over the next decades (Estrella and Menzel 2013; Poschlod 2022). Moreover, given the many catchments with rather high elevation in the uplands and the Alps, snow-induced floods played a major role in the flood regime of the recent century. Discharge data are available up to the year 2019 with different starting points of record begin, yet at least 30 years of records. The results of the test applied to each flood type separately show clear spatial patterns of significant changes of frequency of certain flood types in this region, when considering the most significant change-point (Fig. 8.2). The heavy-rain floods (R1) show a general increase in the alpine foreland, the Alps and the southern part of the south-western uplands, mainly since the years 1965 to 1975. In the northern parts of the uplands, however, heavy-rain floods tend to decrease in frequency since 2000. Changes in the frequency of long-duration-rainfall floods (R2) can mainly be observed for the alpine foreland and the Alps, with an increase for gauges close to the river Danube in the same period as for the R1-floods and a decrease for gauges close to the Alps in the recent years. R3-flood events do not change in frequency of occurrence. For snowmelt-induced floods (S1 and S2), a general tendency to a decreased frequency can be observed, mainly for those catchments with a high

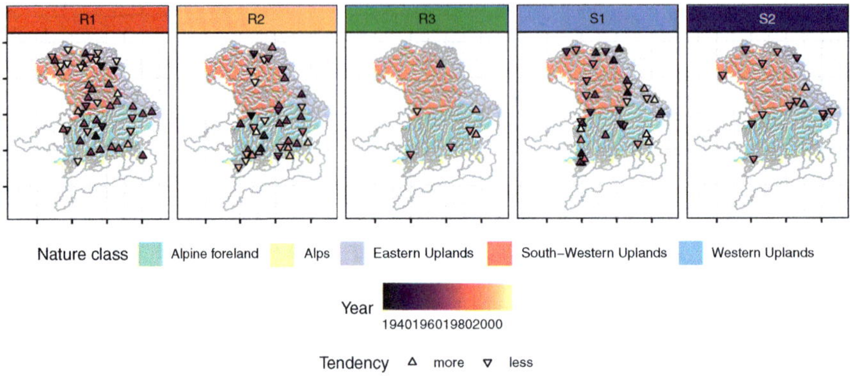

Fig. 8.2 Year of detected most significant change-point and tendency of frequency of occurrence of floods, differentiated by flood type, for the catchments in Bavaria, Germany

elevation. The time of change varies, starting for some catchments in the 1950s and for some in the 1980s. This is probably due to the different starting points of the records. If not only the most significant change-points are considered but all detected change-points, the picture does not change much. Instead, most change-points confirm the tendency of the main change-point, i.e., there is a consistent tendency to more or less frequent flood events of a given type.

One could think that the increase of events of one flood type automatically leads to a decrease of another flood type, as there can be only a limited number of floods per year. However, when comparing flood types R1 and R2, i.e., those flood types with the most detected change-points and contrary tendencies, this assumption cannot be confirmed (Fig. 8.3).

The results of the variance test therefore confirm the assumption of increasing frequency of heavy-rain floods in this region, while at the same time snow-induced floods decrease. This is line with several studies on changing floods for this region, as was pointed out before. Alternating flood-rich and flood-poor periods for the same catchment are not visible, though the occurrence of single flood-rich and flood-periods in this region is in line with the detected change-points (Fischer et al. 2022). The changes in the frequency of flood types have a crucial impact on water management in this region. With a decreasing number of snowmelt-induced floods, in spring probably the increase of water storage will be reduced, leading to a reduced amount of available water. At the same time, in summer many heavy-rain or flash floods will occur, which might lead to disastrous damages. As stated before, such floods will occur with increased frequency especially in the alpine foreland and the south-western uplands (Table 8.1). Interestingly, not necessarily the small catchments are affected by these changes, but catchments of more than 300 km^2 in size.

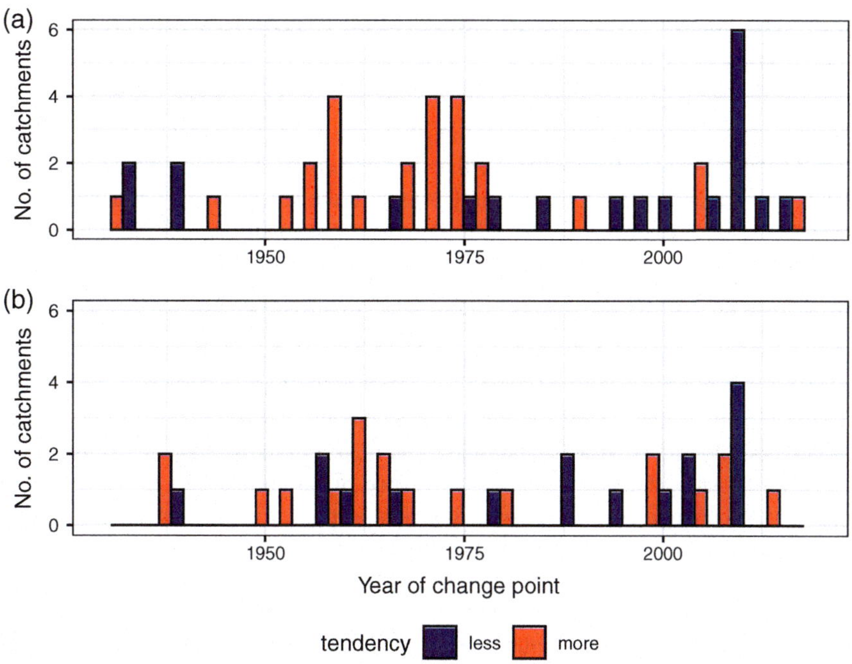

Fig. 8.3 Number of catchments with a significant change-point in frequency and the tendency of change per year for **a** flood type R1 and **b** flood type R2

Table 8.1 Overview of the catchments with a significant increase in frequency of heavy-rain floods, differentiated by catchment characteristics. Each catchment is only counted once, even if multiple change-points with an increase are detected

Region	No. of catchments in class	Thereof significant	Thereof significant (%)
Alps	25	3	12
Alpine foreland	48	12	25
Eastern uplands	32	2	5.90
South-western uplands	49	9	17.30
Catchment size			
<300 km^2	51	3	5.90
300–1000 km^2	58	10	17.20
1000–10000 km^2	38	8	21.10
>10000 km^2	16	5	31.25

8.2 Impact of Changes in Flood Type Frequency on Flood Statistics

In the previous section, significant change-points in the frequency of certain flood types were assessed. It was shown that heavy-rainfall floods tend to occur more frequently in several parts of the study region in the recent years, while snowmelt-induced floods occur less frequently. If winter precipitation does not change (or increase) by global warming, snowmelt-influenced events will become less frequent, but the number of rainfall-induced floods may increase. Such a change might also have an effect on the distribution of these flood types, as especially the skewness of the sample might change. Such changes in the distribution will be investigated in this section. It is of interest to evaluate how the joint distribution of the flood types, the TMPS model, is affected by changes of single flood types and how the AMS model, which does not differ between the types, can represent these changes.

In a first step, the mean relative difference of flood quantiles before and after the detected change-point is derived. For this purpose, for each catchment where a significant change in frequency for any flood type was detected, the most significant change-point (the one with maximum test statistics) of each flood type is taken and the sample is split at this point. Then, the distribution before and after that change-point is estimated and flood quantiles of different return periods are derived. This way, the distribution before any changes and the current state can be compared. Here, the return periods are chosen as $T = 10$, $T = 100$ and $T = 500$ to be able to compare changes in the lower and upper range of practically relevant return periods. From Table 8.2 and Fig. 8.4 it can be seen that the flood quantiles for heavy-rainfall floods (R1) strongly decrease for a return period of 10 years after the detected change-points for most catchments. This might be caused by more frequent small events of this type. However, for large return periods, for most catchments where a significant change-point is detected, an increase is estimated. This is an indicator for a changing skewness, probably caused by several large heavy-rainfall floods or an increase of the numbers of small floods from this type. Similarly, also R2 floods decrease for small return periods and increase for large return periods, though the change in

Table 8.2 Mean relative difference of flood quantiles for different return periods T under consideration of the most recent detected change point

Distribution	$T = 10$ (%)	$T = 100$ (%)	$T = 500$ (%)
AMS	0.3	22.2	50.2
TMPS	–2.9	–2.5	2.6
R1	–184	–0.6	9.1
R2	–34.5	2.3	0.7
R3	0.1	0.0	0.0
S1	–1.1	–0.7	–0.2
S2	–1.3	0.6	0.7

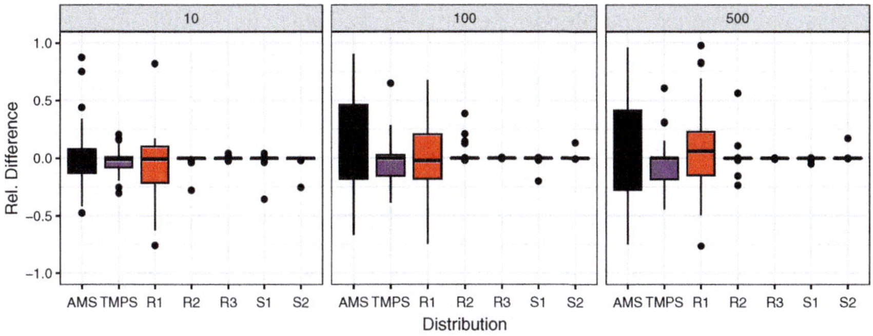

Fig. 8.4 Boxplots of the relative differences between flood quantiles of return period $T = 10$, $T = 100$ and $T = 500$ years before and after detected change-points for different distributions for all catchments in the region of Bavaria, Germany

the quantiles is smaller and mainly for return periods from 100 years on. Contrarily, snow-impacted floods show a general decrease or only small increase for large return periods. These results confirm the assumption of changing flood type frequencies. But how does this affect the joint distribution? When having a look at the TMPS model, this question can be answered: the overall changes are relatively small, with a decrease of the flood quantiles for small return periods and an increase for large return periods. Moreover, the standard deviation of the relative changes is also small (0.104 for $T = 10$, 0.179 for $T = 100$, and 0.281 for $T = 500$) and the boxplots show a clear tendency. Hence, it can be assumed that the impact on changes in frequency of single floods on the TMPS model is comparably small. This has two reasons: on the one hand, changes in frequency with contrary tendency balance each other when combined in the joint model. Therefore, for example, an increase of R1 flood quantiles and a decrease of S2 flood quantiles might lead to an overall small change in the TMPS model in the lower quantile range. On the other hand is the impact of a flood type on the TMPS model controlled by its relative frequency itself. Hence, an increase of frequency gives a flood type more weight in the TMPS model but it does not necessarily dominate all other flood types if the skewness is increased by many small flood peaks. When considering the AMS model and hence no distinction into the flood types, the picture changes. To evaluate the impact of the change in frequency of flood types on the AMS model, the most recent detected change-point of all flood types is taken and the sample is split into one subsample before the change-point and one after the change-point. Then, the GEV distribution is fit to each subsample. For the AMS model, the changes are much higher than for the TMPS model or in fact for any of the flood types. On first sight, this result might be unexpected. However, this phenomenon can be easily explained: in the AMS, all flood types are considered jointly without any distinction. Hence, changes in one flood type might affect the whole sample and the whole distribution. If R1 flood events increase in frequency, probably the skewness of the whole sample increases due to a few extra flood events of this type. Especially given the high skewness of R1 flood events with many small events and some extreme floods, a change in frequency has a high impact on the skewness of the whole sample, as only the extreme events will be considered in the AMS, while additional small floods will be replaced by larger floods of different flood types. This affects location, scale and skewness of the AMS model leading to increasing flood quantiles for higher return periods. Therefore, the AMS model is much more affected by changes of single flood types as the TMPS model, where the changes are taken into account type-specifically. However, it is hard to fully explain the changes when considering a whole region at once, having catchments in different geographical areas and of different catchment sizes. Therefore, the effect of changes will be discussed in more detail by two examples in the following.

The first example is that of the Amberg catchment already shown in Fig. 8.1, which is located in Bavaria, Germany. It has a catchment size of 438 km^2 and a mean elevation of 454 m a.s.l. The gauge is located at the Vils river, a tributary of the Danube river. In the period from 1952 to 2019, a total number of 151 flood events was identified, 73 or 48% thereof of flood type R1 (for the methodology see Sect. 6.2). For this flood type, a significant change-point in frequency was detected in the year 1970 with a tendency of an increasing frequency of R1-floods. For all remaining flood types, no significant change-points were detected. Thus, the sample of R1-flood events was split into two subsamples, one with events before 1975 (19 events) and one with events from 1976 on (54 events). Though the number of flood events before the change-point is not large, the absence of flood events also provides a lot of information. For each subsample, the type-specific distribution was fit using the methodology of Chap. 5. The samples of the remaining flood types were not split, as the non-existence of a change-point implies stationarity. Finally, the TMPS model was applied. Additionally, also the AMS model was applied, to the sample of all flood events before 1976 and to the sample from 1976 on. The results are given in Fig. 8.5.

The results mirror precisely the assumptions made based on the previous results: the change of frequency of R1 flood events leads to an increased skewness (increase from 2.1 to 2.4) and to a heavier right tail. The other moments are affected, too: the mean decreases from 22.8 to 19.7 and the standard deviation decreases from 14.1 to 10.4. Yet, the change in skewness has the greatest effect on large flood quantiles. This is inherited by the TMPS model, which shows an increase in the flood quantiles after the change-point from a return period of 30 years on. The remaining flood types play a less important role, as their frequencies were moderately reduced. These are also considered for the TMPS model and hence only small decreases of their flood quantiles can be observed. The AMS model is most affected by more frequent R1 events that dominate the range of small floods up to return periods of about five years. While in Fig. 8.5 it can bee seen that the distribution has a linear increase (in log-space) before the change-point in 1976 and thus has only a light tail, this changes to a heavy right tail after the change-point. The effect of change in frequency of one single flood type hence has a large impact on the AMS model. Contrary to the TMPS model, the AMS model does not weight the impact of each flood type and hence a change in the sample characteristics like location, scale and skewness affects the whole model. What do these results imply for practical purposes? First, it has to

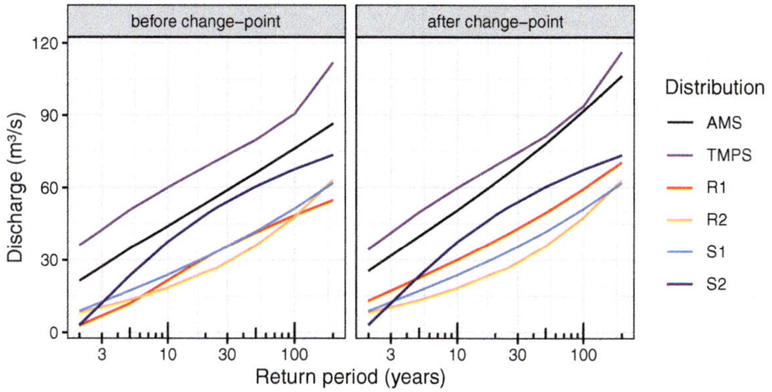

Fig. 8.5 Distribution functions of the flood types, the TMPS and the AMS models before and after the detected change-point for R1 floods in 1976 for the Amberg/Vils gauge

be noted that only a change in frequency was considered. This does not exclude a change in magnitude, though no evidence was found for this (using the Wilcoxon-test). However, even a change in frequency leads to significant increases in the flood quantile estimation, as the sample moments are affected by this change (see above). The flood quantile of return period 100 years of R1 floods increases from 49 m^3/s to 60 m^3/s (increase of 22%). The flood quantiles of the TMPS model do not change that much, as they weight the impact of each flood type, though with different weights before and after the change-point. For example, R1-floods have a proportion of 34% on all floods before the change-point and of 48% after the change-point. The flood quantile of return period 100 years of the TMPS-model increases from 91 m^3/s to 94 m^3/s (increase of 3%). The same quantile of the AMS model increases from 76 m^3/s to 92 m^3/s for the flood quantile of return period 100 years (increase of 21%). Thus, changes in frequency can strongly affect the estimation of the design floods if no type-specific distributions are considered.

While in the previous example, the change in frequency of one flood type clearly impacts the estimation of design floods, the following example demonstrates a contrary effect: the balancing of changes in frequency with reverse tendency of different flood types. For this purpose, the example of the Leucherhof catchment is used. The Leucherhof gauge is located at the Baunach river, a tributary of the Main river. It has a catchment size of 381 km^2 and a mean elevation of 351 m a.s.l., being located in the south-western uplands in Bavaria, Germany. For the period of 1930 to 2019, 200 flood events have been identified, 108 or 54% thereof of type R1, 24 or 12% of type R2, 11 or 5.5% of type R3, 36 or 18% of type S1 and 21 or 10.5% of flood type S2. Three significant change-points for a change in frequency were detected: one in 1975 with increasing tendency for flood type R1, one in 1983 for flood type S1 with decreasing tendency and one in 1985 for flood type S2 also with decreasing tendency. Thus, once again the impression is confirmed that heavy-rainfall floods increase in frequency over time while snow-related floods decrease in frequency. To be able to

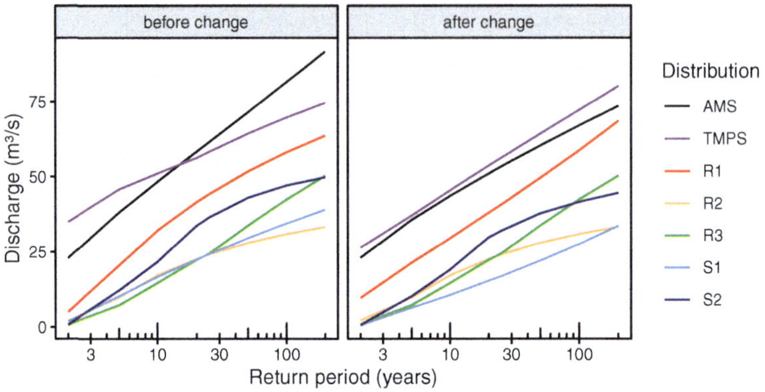

Fig. 8.6 Distribution functions of the flood types, the TMPS and the AMS models before and after the detected change-points for the Leucherhof/Baunach gauge

compare the distributions of the flood types before and after the change-points, each sample of a flood type with a change-point is split at the time of change. This allows a comparison of the flood quantiles before any change occurs to the current state where all changes occurred. The results are given in Fig. 8.6.

Compared to the previous example, the effects of the change-points are somewhat different. Again, the distribution of the R1-flood events has a slightly heavier tail after the change-point. The flood quantile of return period 100 years of R1 floods only changes by 0.6 m^3/s (1%). The same holds true for the snow-induced flood events, types S1 and S2, where also only small changes can be observed but with decreasing flood quantiles from a return period of 10 years and higher. When combining these changes in the TMPS model, again no great change of flood quantiles can be observed. A slight increase of flood quantiles can be observed from a return period of 50 years on, because the TMPSmodel captured the heavier tail of the R1-floods. Yet, the flood quantile of return period 100 years only increases by 3% (from 70 to 72 m^3/s). This demonstrates how the TMPS model considers reversed tendencies of changes of flood types and how these are balanced in the model. The AMS model, once again, is not able to make these distinctions. The decrease in frequency of snow-impacted floods and the increase of frequency of heavy-rainfall floods leads to a decrease of the combined variance from 0.98 to 0.63 and a lighter tail of the AMS model. As a result, the flood quantile of return period 100 years decreases by 18%.

The two examples provided in this section demonstrate the different effects of changes in frequency of flood types and how the different models handle these. While the TMPS model weights the impact of each flood type on the overall model by considering the relative frequency, the AMS model is not able to make such a distinction. Changes in frequency thus mostly have a greater impact on the AMS model. If reversed tendencies of change occur for different flood types, in the TMPS model these changes level each other out. The impact of change can vary between the return periods, as different flood types play a different role for certain ranges of return periods. For the AMS model, however, this leads to a change of skewness. The tendency and magnitude of changes in frequency thereby depends much on the location of the catchment. This makes it difficult to make general statements about regional changes.

The change-point test based on Gini's mean difference can be obtained by using the R code (R Core Team 2022) at the end of this chapter.

Program Code

```
GMD<-function(timediffs){
n<-length(timediffs)
parsum<-numeric(n-1)
mat<-matrix(NA, n,n)
for (i in 1:(n-1)){
for(j in (i+1):n ){
mat[i,j]<-abs(timediffs[i]-timediffs[j])
}
parsum[i]<-sum(mat[i,(i+1):n],na.rm=TRUE)
}
2/(n*(n-1))*sum(parsum)
}

# critical values are obtained from the supremum of a
# Brownian Bridge, here with significance level of 5%:

krit<-1.3582

GMDFunkt<-function(timediffs,List){

n<-length(timediffs)

gm<-numeric(n-1)

for(k in 2:(n-2)){
gm[k]<-abs(GMD(timediffs[1:k])-GMD(timediffs[(k+1):n]))
}

khat2<-which.max(gm)

# Test-statistic #
```

```
for(i in 1:(n-1)){
gm[i]<-gm[i]*i
}

sigma<-sd(timediffs,na.rm=TRUE)

TS<-max(gm)/(sqrt(n)*2*sigma)

if (TS>krit){
return (paste("Significant change-point at", khat2))
} else{return("No significant change-point")}

}
```

In this chapter, it was shown how the frequency of certain flood types can change over time. By application of a robust change-point test for the variance it was shown that a general tendency of increasing frequency of heavy-rainfall floods can be observed for Bavaria, Germany, while snow-induced floods tend to decrease in frequency. These changes have a crucial impact on flood frequency analyses. The TMPS model is able to capture the changes of single flood types and weights them according to their relative frequencies. This leads to mostly smaller changes of the flood quantiles over time compared to the AMS model, which is not able to make a distinction between flood types.

References

Blöschl G et al (2017) Changing climate shifts timing of European floods. Science 357(6351):588–590

Blöschl G, Hall J, Viglione A, Perdigao RAP, Parajka J, Merz B, Lun D, Arheimer B, Aronica GT, Bilibashi A, Bohac M, Bonacci O, Borga M, Canjevac I, Castellarin A, Chirico GB, Claps P, Frolova N, Ganora D, Gorbachova L, Gül A, Hannaford J, Harrigan S, Kireeva M, Kiss A, Kjeldsen TR, Kohnova S, Koskela JJ, Ledvinka O, Macdonald N, Mavrova-Guirguinova M, Mediero L, Merz R, Molnar P, Montanari A, Murphy C, Osuch M, Ovcharuk V, Radevski I, Salinas JL, Sauquet E, Sraj M, Szolgay J, Volpi E, Wilson D, Zaimi K, Zivkovic N (2019) Changing climate both increases and decreases European river floods. Nature 573(7772):108–111

Butler EE (2020) Heavy rain, come today. Nat Clim Chang 10:805–806

Burt T, Boardman J, Foster I, Howden N (2016) More rain, less soil: long-term changes in rainfall intensity with climate change. Earth Surf Proc Land 41(4):563–566

Estrella N, Menzel A (2013) Recent and future climate extremes arising from changes to the bivariate distribution of temperature and precipitation in Bavaria. Germany Int J Climatol 33:1687–1695

Fischer S, Lun D, Schumann AH, Blöschl G (2022) Detecting flood-type-specific flood-rich and flood-poor periods in peaks-over-threshold series with application to Bavaria (Germany). Stoch Environ Res Risk Assess

Fischer S, Schumann A, Bühler P (2019) Timescale-based flood typing to estimate temporal changes in flood frequencies. Hydrol Sci J 64:457–470

Gerstenberger C, Vogel D (2015) On the efficiency of Gini's mean difference. Stat Methods Appl 24:569–596

IPCC (2021) Climate change 2021: the physical science basis. Contribution of working group i to the sixth assessment report of the intergovernmental panel on climate change [Masson-Delmotte V, Zhai P, Pirani A, Connors SL, Pean C, Berger S, Caud N, Chen Y, Goldfarb L, Gomis MI, Huang M, Leitzell K, Lonnoy E, Matthews JBR, Maycock TK, Waterfield T, Yelekci O, Yu R, Zhou B (eds.)]. Cambridge University Press, Cambridge. In press

Kimuli JB, Di B, Zhang R, Wu S, Li J, Yin W (2021) A multisource trend analysis of floods in Asia-Pacific 1990–2018: implications for climate change in sustainable development goals. Int J Disaster Risk Reduct 59:102237

Kundzewicz ZW, Radziejewski M, Pinskwar I (2006) Precipitation extremes in the changing climate of Europe. Climate Res 31(1):51–58

Lun D, Fischer S, Viglione A, Blöschl G (2020) Detecting flood-rich and flood-poor periods in annual peak discharges across Europe. Water Resourc Res 56:e2019WR026575

Madsen H, Lawrence D, Lang M, Martinkova M, Kjeldsen TR (2014) Review of trend analysis and climate change projections of extreme precipitation and floods in Europe. J Hydrol 519:3634–3650

Poschlod B (2022) Attributing heavy rainfall event in Berchtesgadener Land to recent climate change - Further rainfall intensification projected for the future. Weather Clim Extremes 38

R Core Team (2022) R: a language and environment for statistical computing. R Foundation for Statistical Computing, Vienna, Austria. https://www.R-project.org/

Szolgayova E, Laaha G, Blöschl G, Bucher C (2014) Factors influencing long range dependence in streamflow of European rivers. Hydrol Process 28(4):1573–1586

Yitzhaki S (2003) Gini's mean difference: a superior measure of variability for non-normal distributions. Metron LXI, pp 285–316

Chapter 9
Type-Specific Hydrograph Generation and Flood Scenarios

Abstract Beside peak and volume, the shape of the hydrograph is an important characteristic of a flood event. As the timescale is a dimension for the spread of the hydrograph, it is logical to also use flood types thus derived to differentiate between typical hydrographs and to consider these differences in a type-specific hydrograph. The model of Chap. 5 is thus further extended to such a type-specific generation of flood hydrographs. Different probability density functions are applied and the challenges of generating type-specific hydrographs are outlined. The resulting flood hydrographs are then assigned with a given probability, such that flood scenarios based on ensembles of hydrographs are obtained. These scenarios can improve the planning of flood retention facilities due to in-depth knowledge on the shape of the flood hydrograph.

Flood frequency analyses aim to deliver design floods for practical purposes like the construction of a dam or the design of flood protection measures. The classical FFA uses distribution functions that are fit to records of floods to estimate quantiles for given return periods. Such procedures were demonstrated in Chap. 2 and in a more sophisticated way using flood types in Chap. 5. However, these procedures are limited to the consideration of one flood characteristic at a time, usually the flood peak. A flood event, though, is more complex and knowing the flood peak does not provide any information on the event itself. For practical purposes, knowledge on the whole flood event is important, since, e.g., flood protection measures should withstand not only the flood peak but have to hold back a significant part of discharge of the event. Storage of reservoirs has to be operated accordingly such that the flood peak is reduced and the storage content does not exceed critical limits causing damage to the building. A first step, thus, to better capture the whole flood event is to not only estimate the design flood peak, which is required due to governmental guidelines for flood protection, but to derive the corresponding flood volume, too. This can be realised by application of multivariate statistics. Alternatively, also the (mean) flood timescale can be used to obtain consistent pairs of peak and volume, as it provides a direct relation between both characteristics. Yet again, deriving flood peak and volume for a design flood only leaves again open one important factor:

S. Fischer and A. H. Schumann, *Type-Based Flood Statistics*, Water Science and Technology Library 124, https://doi.org/10.1007/978-3-031-32711-7_9

the shape of the hydrograph. As was demonstrated before in this book, the same flood peak (and also the same volume) can be caused by flood events with vastly different hydrograph shapes. Some approaches therefore try to estimate the design flood hydrograph (DFH). This can be done by deterministic approaches (Goswami 2020) or by using statistical distributions (Yue et al. 2002; Pramanik et al. 2010; Brunner et al. 2017). The latter is described in more detail in the following section for the special case of different flood types, proposed by Fischer and Schumann (2023).

9.1 Estimation of Type-Specific Design Flood Hydrographs by Probability Density Functions

From the previous chapters, it is already known how to derive the design flood peak for a given return period: one makes use of a distribution fitted to the sample of floods, either AMS, POT or type-specific. For practical purposes, normally the joint design flood of all flood types is required, i.e., one peak value obtained by either AMS, POT or TMPS model for each return period. Standard return periods are $T = 100$ or $T = 1000$ years, which are most relevant for design purposes such as the estimation of floodplains or the dimensioning of infrastructure. For exemplary purposes, in the following $T = 100$ will be used to construct the DFHs, though any other return period will work as well.

It was pointed out before, how the flood hydrographs may differ between the flood types. Especially the volume of a flood can take very different manifestations. Hence, it is hydrologically meaningful to take into account the different flood-generating processes that lead to such different flood events. Again, this can be done by considering flood types, as introduced in Sect. 6.2. At this step, the benefit of the proposed flood typology comes into play: the type-specific relationship between direct volume and direct peak Q_{dP}, the flood timescale. Since each flood type has a specific flood timescale which was used to differ between the flood types, this now can be used to estimate the direct volume from the direct peak. For each flood type j, the mean flood timescale is used, $\bar{TQ}_j$:

$$Vol_{dir} = Q_{dP} \cdot \bar{TQ}_j. \tag{9.1}$$

Originally, these flood timescales were only used to differ between the rainfall-driven floods. However, they also can be used to describe the peak-volume relation for the snow-induced flood events, but with higher uncertainty, as will be shown later on. By application of the type-specific mean flood timescale, different volumes for a given flood peak can be derived, depending on the flood type. For example, heavy-rain floods will probably have much smaller volumes than long-duration-rainfall floods, when considering the same flood peak.

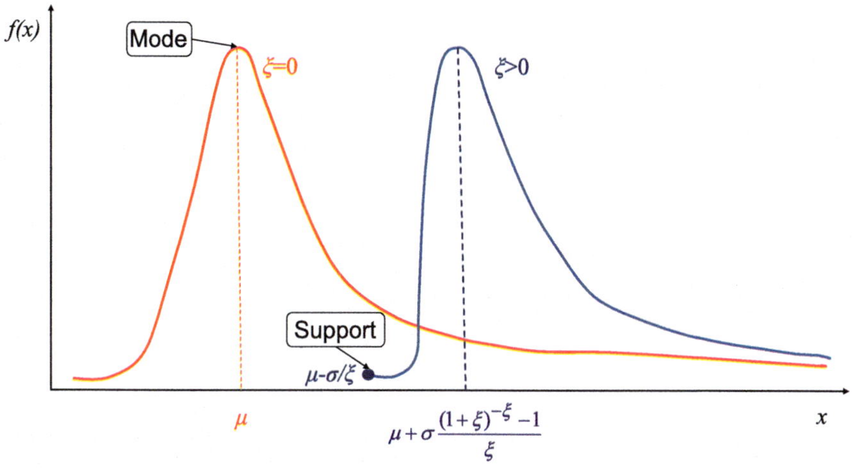

Fig. 9.1 Example of how the probability density of the GEV distribution can be used to describe a flood hydrograph. The larger the shape parameter of the GEV distribution, the steeper becomes the rising limb and the longer becomes the recession

Having the design flood peak and the type-specific flood volumes for a given return period, it remains to define the type-specific DFH. This is done statistically by making use of statistical distributions.

The advantage of using statistical distributions, more precisely probability density functions, for deriving DFHs lies in their ability to easily be adaptable to any hydrograph shape: the shape of pdfs for most distribution and parameter combinations already is similar to single-peaked hydrographs. By varying the parameters of the distribution, various shapes of the hydrograph (steep rise, long recession, even symmetric shape) can be derived. An excellent example in this case is the GEV distribution introduced in Chap. 2. The shape parameter of the GEV distribution controls if the pdf is symmetric ($\xi = 0$), left-skewed ($\xi < 0$) or right-skewed ($\xi > 0$). A right-skewed pdf has steep rising part and a long recession, thus embodying the typical shape of a flood wave. The larger the shape parameter, the steeper becomes the rising limb and the longer the recession (Fig. 9.1). With the scale and the location parameters, the position of the flood peak in time and the duration of the event can be controlled: the mode controls the point in time where the probability density exhibits its maximum, hence the peak. Moreover, the parameters can be used to define the support of the density function, thus the beginning of the flood event (in the case of $\xi > 0$). Similarly, other distributions are also suitable to model the flood hydrograph. Another advantage of using pdfs for modelling the DFH is that a pdf is always non-negative, which is required for DFHs, and that the integral of the pdf over the entire definition space is equal to 1. This makes it easily scalable to any desired flood volume. The temporal resolution can be chosen arbitrarily, though it is recommended to use at least hourly resolution to obtain smooth hydrographs also for small catchments.

Several different probability distributions are available that might be suitable to describe the shape of the flood hydrograph. For example, Brunner et al. (2017) considered nine different distributions as suitable for estimating the DFH, namely Normal, Lognormal, Frechet, Weibull, Logistic, Gamma, inverse Gamma, Cauchy and Beta distributions. The higher the number of parameters of a distribution, the higher the flexibility but also the complexity of the fitting procedure increases. However, all of these distributions are limited to the case of one flood peak of the DFH only. In Sect. 6.2, it was shown that especially floods driven by long-duration rainfall and/or several successive rainfall events very often have multiple flood peaks. Indeed, Fischer and Schumann (2023) found that in mean 30–40% of the R3 and S2 floods have multiple flood peaks and that these events contribute considerably to the largest flood peaks of the respective type. Most of these events have two flood peaks. In such a case, the standard unimodal distributions are not sufficient. For reservoirs, such flood events are more critical, as the first part of the flood wave may fill the flood storage and the second peak could result in an overtopping of the dam.

To take into account this phenomenon, mixture distributions can be considered. The number of components of the mixture model then defines the number of peaks considered for the DFH. Due to the complexity and increasing number of parameters, it is not recommended to apply mixture distributions with more than two components. Therefore, also only multi-peaked floods with two flood peaks are considered, which represents the majority of all observed multi-peaked flood event. The probability density function of a two-component additive mixture distribution (and therefore allowing for two flood peaks) is given by

$$f_{Mix}(x) = \alpha f_1(x|\theta_1) + (1-\alpha) f_2(x-\beta|\theta_2), \tag{9.2}$$

where f_1 and f_2 are two density functions of standard distributions with different parameter sets θ_1 and θ_2 and α is the weighting parameter, controlling the impact of each density distribution on the mixed density function and therefore defining the timing of the main peak. The parameter β is used to shift the support of the second density function such that the beginning of the second partial event is controlled. With β, thus, the beginning of the second rainfall event can be expressed. In Fig. 9.2, an example of a two-component mixture probability density function is given that consists of two LogNormal distributions. The Lognormal distribution has probability density function

$$f_{LN}(x) = \frac{1}{\sqrt{2\pi}\sigma x} \exp\left(\frac{-(\log x - \mu)^2}{2\sigma^2}\right), x > 0, \tag{9.3}$$

where μ and σ are the mean and standard deviation of the logarithm of the rv X, and is known to be right skewed, which is favourably for hydrographs since most hydrographs are assumed to have a steep rising limb and a long recession similar to the density of a right skewed distribution. It is therefore used, e.g., by Brunner et al. (2017) to model the DFH. With the mean or location parameter μ and the scale

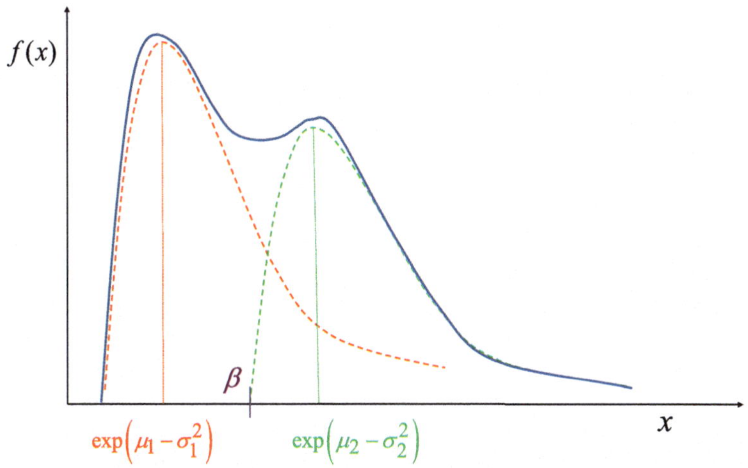

Fig. 9.2 Example of the mixture distribution of two LogNormal distributions with parameter sets $\theta_1 = (\mu_1, \sigma_1)$ and $\theta_2 = (\mu_2, \sigma_2)$ and of how the distribution parameters control the shape of the mixing density function

parameter σ, the timing of the peak of the corresponding hydrograph as well as the duration of the event can be controlled, since the mode of the LogNormal distribution is given by $\exp(\mu + \sigma^2)$. The support is always positive.

The complexity of the statistical distributions with often more than two parameters requires several flood characteristics to make an estimation of the DFH efficient, regardless of unimodal or bimodal hydrographs. The derived design flood peak and the type-specific flood volume alone are not sufficient to fit a probability density function for estimating the DFH, since the same combination of flood peak and volume might lead to different shapes of the hydrograph. Additional parameters describing the type-specific hydrographs therefore are required to obtain DFHs. Such parameters could be the rise time of the hydrograph, i.e., the time span between beginning and peak of the event, and the recession time, i.e., the time span between peak and end of the event (Fischer and Schumann 2023). Both parameters do not depend on the flood peak magnitude. Hence, from all observed hydrographs of a given flood type, the rise and recession times can be considered to define a parameter space. Since the typology usually is not selective, the most extreme cases of rise and recession time should not be included, as they might well represent also another flood type. It is therefore recommended to only consider the range between the 10% and 90% quantile denoted with QR (quantile range) for each rise and recession time.

With the flood event parameters 'design flood peak' and 'type-specific volume', as well as the QR of rise and recession time and the probability density function, the DFH can be estimated for each flood type. What remains is the selection of a suitable distribution from the many pdfs available. This can be done by classical statistical measures such as the root mean squared error, Spearman's ρ (or any other suitable dependence measure) or even classical distribution tests like the Kolmogorov-

Smirnov test between the normalised observed hydrographs (i.e., the hydrographs re-scaled to [0,1] by differentiation by the volume and subtraction of the flood peak) and the pdfs. Then, the DFHs can be derived. The following procedure on how to derive type-specific DFHs was proposed by Fischer and Schumann (2023).

To generate flood hydrographs with a certain flood peak and type (which specifies the volume of floods by the type-specific flood timescale), rise and recession time are varied within the QR to obtain an ensemble of DFHs. To fit the pdf of the type-specific distribution to these parameters, an optimisation procedure is applied that considers the minimisation of two criteria: the squared difference between the flood peak and the squared difference between the flood volume. For this purpose, the pdf which is defined in [0,1] is adapted to the peak and volume for the given return period by multiplication respectively re-scaling. The optimisation is repeated 100 times with randomly drawn starting parameters for rise and recession time in the range of QR. The consideration of a range for rise and recession time leads to several optimal parameters sets (equifinality) that represent various hydrograph shapes and hence an ensemble of possible DFHs. This ensemble includes best- and worst-case scenarios for different issues of flood design by considering the whole space of possible hydrograph shapes. The proposed approach also allows for extreme combinations of, e.g., short rise and very long recession times. Yet, these combinations are still in the range of possible durations for the respective flood type. Moreover, they are only considered as starting values and still have to fulfil the optimisation criteria for peak and volume, such that unrealistic combinations are automatically excluded.

It has to be considered that with the procedure introduced here, the direct (quick) flow is modelled. This means that the resulting DFH does not include the baseflow. The baseflow is quite variable between catchments and seasons and thus should be excluded when considering the flood event. However, for practical purposes, the total flow is required since flood protection measures have to be designed against the total flood event including baseflow. In this case, the baseflow has to be added afterwards to the DFH. A possibility to add the baseflow would be to estimate the type-specific mean baseflow for the range of the design flood peak and add it as constant to the DFH. For this purpose, for each flood type all flood events with a peak in a range of $\pm 25\%$ of the design flood peak are considered and the mean baseflow of these events is taken. For large design flood peaks beyond the observed range, the baseflow has to be extrapolated. This estimation has the advantage of considering the different flood-generating processes that are related to different states of baseflow. Of course, baseflow could also be estimated by other approaches, such as digital filters or recession. For an overview of different baseflow estimation methods, the study of Mei and Anagnostou (2015) is recommended.

Another information valuable to practitioners might be the probability of a flood type leading to the design flood peak. For example, if it is known that mostly heavy-rainfall flood events lead to the design flood with return period $T = 100$ years, the flood protection systems can be adjusted to this case: in summer, when heavy rainfall floods are most common, the storages should be empty to retain as much discharge as possible. On the other hand, if mostly snowmelt floods result in a design flood of a given return period, than the storages in spring should be empty but can be filled during summer to provide additional water in dry seasons. This probability can be derived empirically from the data, again by selecting a suitable range around the design flood peak and estimating the frequency of each flood type in this range. If this probability is assigned to the type-specific hydrographs, flood scenarios are derived that are of high practical relevance.

The whole procedure to derive DFH introduced in this chapter, starting with the event separation over flood classification, the TMPS model and the DFH, is summarised in Fig. 9.3. It has to be noted that in theory any flood type classification and any statistical model to derive the flood peak can be used. The proposed methods, however, offer a high consistency. The distributions proposed for modelling the respective type-specific DFHs proved to be the best choice for catchments in Central Europe but might be replaced for other regions of the world.

Examples of type-specific DFHs of catchments in Bavaria, Germany, for a design flood with return period $T = 100$ years are given in Fig. 9.4. For these catchments, the LogNormal distribution was selected to model the DFHs of flood types R1, R2 and S1, which are all characterised by a steep rise of the hydrograph (Fig. 9.5). For flood type S2, snowmelt-floods, the symmetric Cauchy distribution and the two-component Cauchy-mixture distribution were selected as best-fitting. It represents the smooth beginning and end of the snowmelt period best while also taking into account the considerable proportion of multi-peaked floods events for this type. Flood type R3 associated with several successive rainfall events required the use of a two-component mixture model using two LogNormal distributions due to several peaks, which is typical for this flood type, additional to the classical LogNormal distribution for the unimodal flood hydrographs. The optimisation was repeated 100 times with rise and recession times varied in QR. The shaded lines represent these realisations and thus uncertainty bands based on an ensemble of DFHs. The probability of each flood type to result in a design flood with return period $T = 100$ (top number of each panel) varies between the catchments, though a general tendency to heavy-rainfall floods becomes obvious. For these examples, large floods are particularly probable in the summer months. Snowmelt-induced floods, instead, are rarely relevant for these ranges of flood peaks, though rain-on-snow events might play a role for the Achsheim catchment. Of course, these patterns change with the catchments and especially the region they are located. Factors as climate, elevation and catchment state play a crucial role in the determination of the most relevant flood type.

The resulting hydrographs can be evaluated regarding their goodness of fit using the mean flood timescale and comparing it with the spread of the mean simulated hydrograph. This way, the ability of the optimisation to capture the shape of the hydrograph can be assessed. For example, for the catchments in Bavaria, Germany, the mean relative error between the type-specific flood timescale and the spread of the simulated hydrograph lies around 10% with only few outliers. This indicates an overall good performance of the procedure. Of course, such an evaluation is limited in its conclusion, as only one aspect of the hydrograph is considered. Moreover, only unimodal flood hydrographs can be evaluated as the flood timescale is not clearly related to the spread of multimodal hydrographs. How these issues can be overcome is shown in Sect. 9.2.

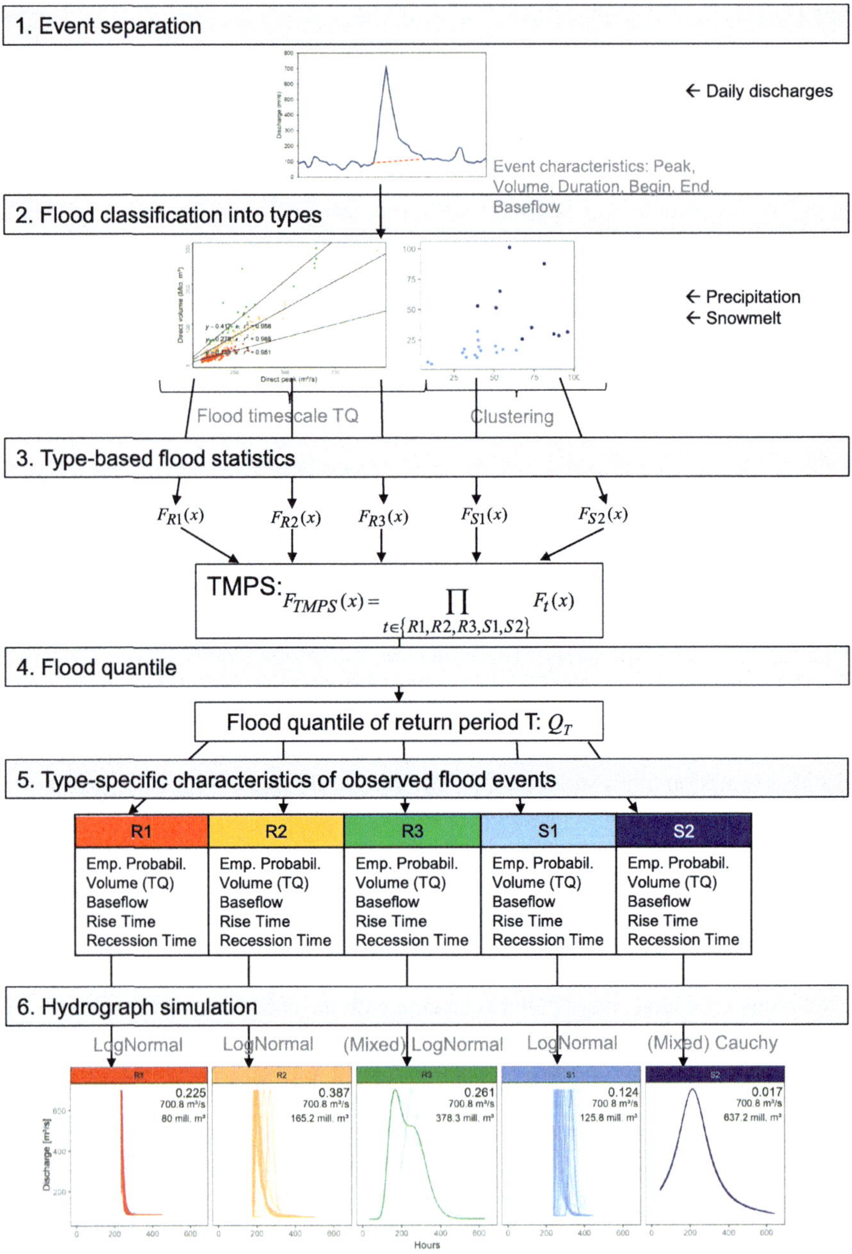

Fig. 9.3 Schematic overview of the steps leading to type-specific design flood hydrographs (modified from Fischer and Schumann 2023)

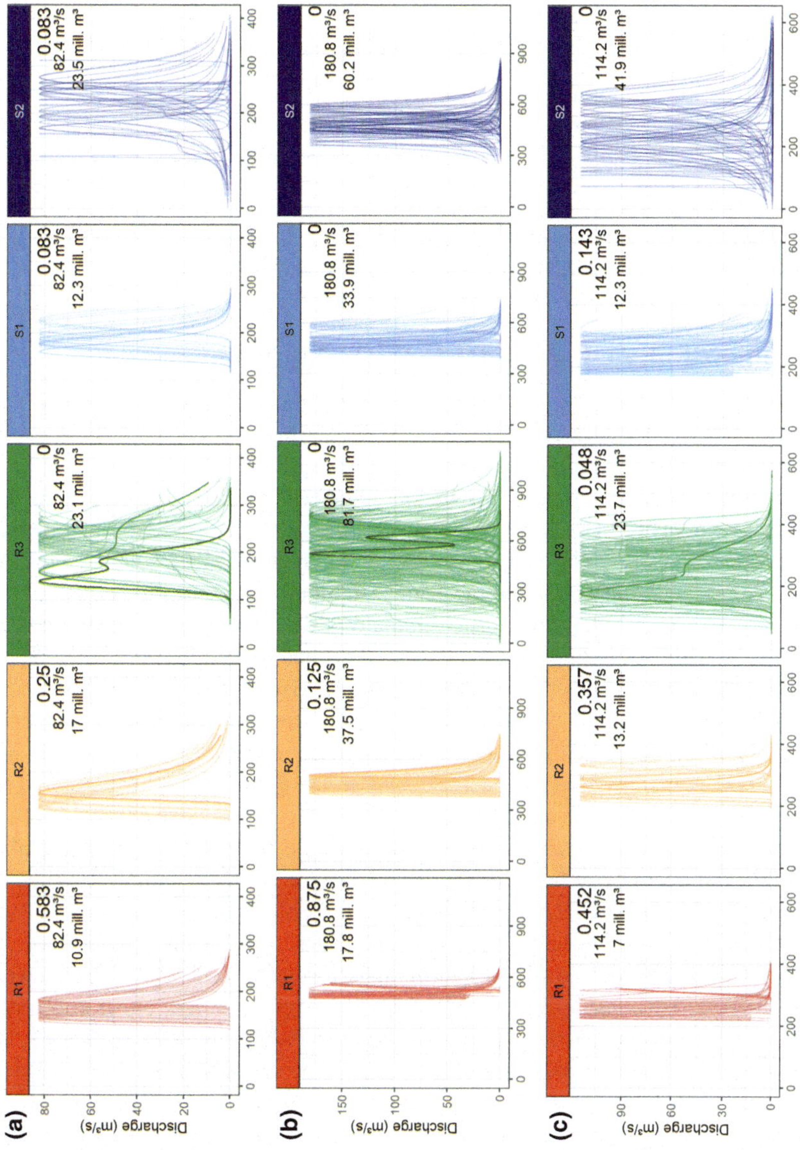

Fig. 9.4 Ensemble of synthetic design flood hydrographs for catchments **a** Achsheim (359 km^2), **b** Garmisch (249 km^2) and **c** Oberach (96 km^2). Thick lines represent the hydrographs with mean parameters

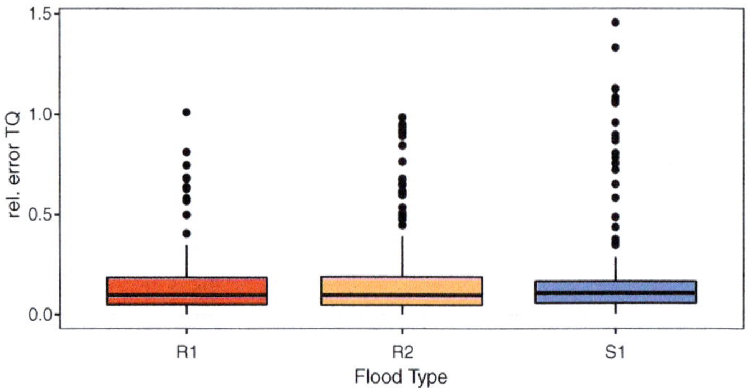

Fig. 9.5 Relative error between the observed mean flood timescale and the simulated spread of the hydrographs for the three flood types R1, R2 and S1 with unimodal hydrographs

9.2 Plausibility Check of DFH Ensemble Members

When simulating flood hydrographs, always the questions arises of the goodness-of-fit. For deterministic modelling of observed discharge, this is an easy task as the simulated discharge can be compared to the observed one. Similarly, stochastic models that aim to model discharge can be validated for periods of observed discharge. Standard goodness-of-fit measures in these cases would be the RMSE, squared deviation or maybe classical measures like the Kling-Gupta efficiency or the Nash-Sutcliffe efficiency. However, when aiming to simulate design flood hydrographs, the task of somehow validating them is not that easy. This has several reasons. First, design hydrographs are constructed for a flood peak of a given return period and the corresponding volume. Such events usually are not included in the observation series, especially when the return periods are large. Hence, one cannot compare the simulated flood events to observed ones. Second, in the optimisation of the DFHs, the flood peak and volume as well as additional parameters like the rise or recession time are included, the latter chosen from the observed range of all flood events. An evaluation of the goodness-of-fit based on these criteria hence is not meaningful. While the validation of DFHs may not be reasonable, the check of plausibility certainly is. With the procedure by Fischer and Schumann (2023) described in the previous section, one obtains an ensemble of flood hydrographs for each flood type. That is, when being interested in considering the DFH for a given return period, one is confronted with several flood hydrographs for each flood type, each with different shape of the hydrograph. While the probability of each flood type scenario helps to estimate the occurrence probability of each flood type and hence also of the type-specific volume, it remains unclear which of the many ensemble member of each flood type to consider. For practical purposes, it is recommended to consider the whole ensemble of DFHs to include best- and worst-case scenarios in the estimation, hydrographs with steep or slow rising limb and long or short recession. However, more information can be obtained when considering the ensemble. Though each of the ensemble members is a possible realisation of a hydrograph of the specific flood type, not all of them have the same probability. Keep in mind that flood event characteristics like the rise and recession time were estimated based on observed flood events and then combined in the optimisation. Not all of these times have the same probability and certainly not all of their combinations have. When comparing the simulated hydrographs with the observed ones, one can conclude on the plausibility of each ensemble member by having a look on how often the respective combination of rise and recession time occurred in the observed sample. More frequent combinations have a higher plausibility and should be considered as more frequent in practical applications than combinations that occur only rarely in reality.

To offer a more objective way of a plausibility ranking, Fischer and Schumann (2023) propose an approach based on multi-criterial decision making: the Technique for Order Preference by Similarity to Ideal Solution (TOPSIS, Hwang and Yoon (1981)). For this purpose, different performance criteria are chosen, $C^1, \ldots, C^J$. Then, for each simulated hydrograph S_k, $k = 1, \ldots, K$, and each observed hydrograph O_i, $i = 1, \ldots, n$, of a flood type, the decision matrix is build. For example, for three different performance criteria C^1, C^2 and C^3, the decision matrix would look like

	O_1			O_2			$\cdots$	O_n		
S_1	$c^1_{1,1}$	$c^2_{1,1}$	$c^3_{1,1}$	$c^1_{1,2}$	$c^2_{1,2}$	$c^3_{1,2}$	$\cdots$	$c^1_{1,n}$	$c^2_{1,n}$	$c^3_{1,n}$
S_1	$c^1_{2,1}$	$c^2_{2,1}$	$c^3_{2,1}$	$c^1_{2,2}$	$c^2_{2,2}$	$c^3_{2,2}$	$\cdots$	$c^1_{2,n}$	$c^2_{2,n}$	$c^3_{2,n}$
$\vdots$					$\vdots$					
S_K	$c^1_{K,1}$	$c^2_{K,1}$	$c^3_{K,1}$	$c^1_{K,2}$	$c^2_{K,2}$	$c^3_{K,2}$	$\cdots$	$c^1_{K,n}$	$c^2_{K,n}$	$c^3_{K,n}$

where $c^j_{k,i}$ is the criterion evaluated between S_k and O_j. Each of the matrix entries $c^j_{k,i}$ is multiplied with a weighting factor:

$$v^j_{ki} = w_i \cdot c^j_{k,i}.$$

If all criteria are considered as equal in their importance, all weights should be chosen equally with $w_i = \frac{1}{J \cdot n}$. However, when having certain practical applications in mind, the weights could be chosen accordingly. When considering the difference in rise time between observed and simulated hydrographs as criterion and aiming to consider the simulated hydrograph for the estimation of flood retention ponds, this criterion should have a high weight. Similarly, when considering the possibility of dyke breaches, more weight should be given to criteria related to the recession time or the duration. Fischer and Schumann (2023) recommend the use of three criteria: one minus the relative difference between observed and simulated rise time, one minus recession time respectively one minus skewness. Skewness in this case is defined as the difference between the mean day of the event and the day of the peak. Translating this to pdfs, this is analogous to compare mean and mode of the distribution, a typical measure for skewness. Due to the definition, all three criteria should be maximised.

The TOPSIS approach then estimates two measures, one for the deviation of each alternative to the best possible value, denoted with Utopia, and one for the deviation of each alternative to the worst possible value, denoted with Nadir:

$$D_k^+ = \sqrt{\sum_{j=1}^{3}\sum_{i=1}^{n}\left(v_{ki}^j - \max_{1\le i\le n, 1\le j\le 3}(v_{ki}^j)\right)^2}$$

$$D_k^- = \sqrt{\sum_{j=1}^{3}\sum_{i=1}^{n}\left(v_{ki}^j - \min_{1\le i\le n, 1\le j\le 3}(v_{ki}^j)\right)^2}.$$

A ranking of the simulated hydrographs of a given flood type then can be obtained by considering

$$\bar{C}_k = \frac{D_k^-}{D_k^+ + D_k^-},$$

which is largest for the best-ranked alternative, as then the distance to Nadir is maximised and the distance to Utopia is minimised.

With the proposed framework, an ensemble of type-specific flood hydrographs for a given design flood peak is obtained. These can be used to test the performance of the reservoir. For example, the whole ensemble of the hydrographs can be used in a simulation to find out when the reservoir fails. The probability of failure then can be combined with the plausibility derived from the MCDM. An a priori definition of critical hydrographs is not possible, as the failure of the reservoir depends mainly on the construction and operation of it.

Probability density functions and mixture distributions are statistical functions that can be directly used to estimate synthetic design hydrographs. It is important to consider the flood-generating processes, e.g., by applying flood types, since the hydrograph shape depends much on them. For a given return period, the DFH can be derived for each flood type and by estimating the probability of each flood type, flood scenarios can be derived. Using varying starting parameters for hydrograph characteristics like rise and recession time, ensembles of DFHs can be obtained that represent best- and worst-case scenarios. In combination with MCDM, the most plausible hydrograph shapes can be determined.

References

Brunner MI, Viviroli D, Sikorska AE, Vannier O, Favre AC, Seibert J (2017) Flood type specific construction of synthetic design hydrographs. Water Resourc Res 53(2):1390–1406

Fischer S, Schumann A (2023) Generation of type-specific design flood hydrographs. Hydrol Sci J

Goswami M (2020) Generating design flood hydrographs by parameterizing the characteristic flood hydrograph at a site using only flow data. Hydrol Sci J 1–19

Hwang CL, Yoon K (1981) Multiple attribute decision making - methods and applications. A state-of-the-art survey. CRC Press, Boca Raton . CRC Press, Boca Raton

Mei Y, Anagnostou EN (2015) A hydrograph separation method based on information from rainfall and runoff records. J Hydrol 523:636–649

Pramanik N, Panda RK, Sen D (2010) Development of design flood hydrographs using probability density functions. Hydrol Process Int J 24(4):415–428

Yue S, Ouarda TB, Bobée B, Legendre P, Bruneau P (2002) Approach for describing statistical properties of flood hydrograph. J Hydrol Eng 7(2):147–153

Part V
The Spatial Dimension of the Flood Typology

Classical flood statistics focus at one gauge or station at a time. However, floods are not limited to one point in space. Instead, floods are complex spatial phenomena, where spatial processes such as precipitation or snowmelt play a role as well as the interaction between catchments. An example is flood superposition, where flood waves of main river and tributary occur synchronously and overlap downstream, thus probably increasing the flood peak and the flood volume. This spatial interaction becomes more relevant the larger the catchment is. Especially large river basins are often affected by flood superposition. Moreover, widespread floods, i.e., floods that occur in several catchments simultaneously, can cause severe damages to large parts of infrastructure and economy. It is therefore important to consider floods as a spatial phenomenon and to include the spatial interaction of floods in flood frequency analyses of large catchments. Again, flood types help to better understand the flood-generating processes and how these change within a large catchment. In this part, spatial statistical concepts are introduced that take into account the impact of tributaries and flood types. They make use of classical statistical multivariate models, namely copulas and Vine-copulas. Similar to the univariate flood frequency analyses, the aim is to obtain flood scenarios, i.e., design flood hydrographs that are obtained by different combinations of flood types and interactions of catchments and that are assigned with a probability. A statistical model chain is introduced that provides such design floods for a downstream gauge dependent on the upstream gauges for different meteorological scenarios.

Chapter 10
Spatial Analyses of Floods

Abstract When considering several catchments in a river basin jointly, the spatial dependence between them has to be considered, too. However, the dependence may vary with the flood-generating conditions. Spatial flood types can help to better understand the interplay between catchments and thus to improve spatial statistics. In this chapter, a methodology is proposed to define the focus point of flood in a large basin under consideration of the tributaries and to incorporate this information in a spatial model.

While for small to mesoscale catchments, the hydrograph alone can be used to characterise the hydrological processes that led to the flood event, for large catchments the runoff generation is more complex. If a catchment belongs to the macroscale class, not only spatially heterogeneous meteorological and hydrological processes have to be considered, but also the interaction of sub-catchments. Especially the flood superposition plays a crucial role in this case: the simultaneous occurrence of floods in a tributary and the main river can lead to an increase of the flood peaks at the downstream gauge if both flood waves occur at the same time. If the temporal difference between both flood waves is small, the peak as well as the volume is increased. If the flood peaks arrive temporally shifted at the outlet gauge, solely the flood volume is increased. Often, only the impact of a tributary generates an extreme event downstream, as it was, e.g., the case for the 1993 flood event at the lower Rhine, where the impact of the Moselle river tributary increased the flood to its multiple. The investigation of the role of flood superposition hence takes a more and more prominent role in flood investigation in the recent years. Many studies for central Europe have investigated how flood superposition may alter the flood event downstream, positively or negatively (Bloeschl et al. 2013; Skublics et al. 2016). But it is not always the natural river structure that leads to unfavourable flood superposition. Often, anthropogenic impacts cause or worsen flood superposition in a basin. A famous example is the Rhine river basin. Here, flood protection systems in the upper Rhine led to a temporal shift of the flood events and hence a superposition of the main river Rhine and its tributary, the Neckar river, such that flood events downstream were increased. A detailed analyses of this phenomenon is given by

S. Fischer and A. H. Schumann, *Type-Based Flood Statistics*, Water Science and Technology Library 124, https://doi.org/10.1007/978-3-031-32711-7_10

Vorogushyn and Merz (2013). Still, not all flood events are alike. Depending on the shape of the hydrograph, the probability of unfavourable flood superpositions might differ significantly. In a study by Fischer et al. (2021), it was shown that heavy rainfall floods lead less frequently to a flood superposition than long-duration rainfall floods, but if they occur simultaneously, the peak is increased much more than for other flood types.

Hydrograph-based flood types help to characterise flood superposition in a river basin. For simultaneous flood events, the change of a flood type from upstream to downstream gauge might indicate a flood superposition. For example, using the flood typology from the previous Sect. 6.2, if two R1 flood events occur at the upstream gauge and the tributary gauge and an R2 flood event occurs at the downstream gauge after confluence, this is an indicator that both floods from upstream gauge and tributary gauge arrived at the confluence with a time shift. This way, the usually small volume of the two R1 floods is increased and the duration prolonged, leading to an increase of the flood timescale and hence a different rainfall-driven flood type. If, however, at the downstream gauge a flood of type R1 occurs, too, it can be assumed that both upstream flood events occurred simultaneously at the confluence, leading to an increase of the flood peak but not of the duration of the downstream event. An example of such events is given in Fig. 10.1. Three gauges in the Danube river basin are considered: the Passau-Ingling gauge located at the Inn river close to the confluence with the Danube, the Hofkirchen gauge at the river Danube, upstream of the confluence with the Inn, and the Achleiten gauge, which is located downstream of the confluence of Danube and Inn (see also Sect. A.2 for more information on the Danube basin). The two flood events given in the figure are among the most extreme events that ever occurred in the upper Danube. However, they are quite different regarding their flood hydrographs. In 2013, the two flood waves from the upper Danube and the Inn arrived almost simultaneously with one day difference at the confluence. Both flood events are characterised by a large flood peak but only comparably small volume (flood type R1). Due to the simultaneous occurrence,

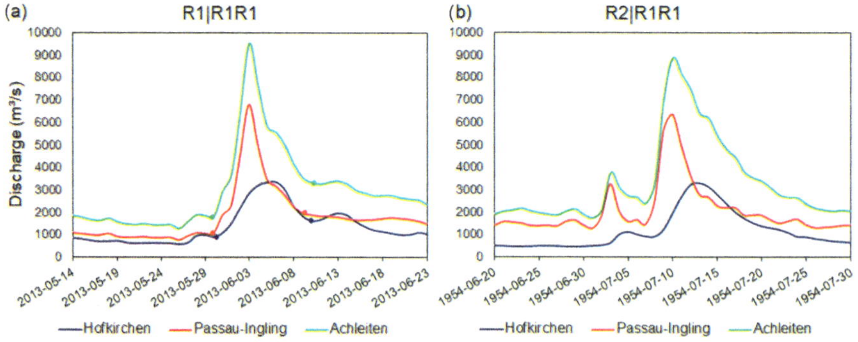

Fig. 10.1 Hydrographs of two flood events in the Danube basin with **a** simultaneous occurrence and **b** shifted occurrence of the flood peaks of the upstream gauges (modified from Fischer and Schumann 2021)

the downstream flood event at the gauge after confluence is not increased much in duration. However, the peak increases much in Achleiten due to the nearly perfect superposition of both peaks at the point of confluence. Contrary, in 1954, the two flood hydrographs occurred shifted in time, with the peak of the Inn flood event reaching the confluence when the flood event in the upper Danube has almost ended. This shifted occurrence resulted in a prolongation of the downstream event in Achleiten with a much longer recession. The flood type changes from R1 for the upstream gauges to R2 for the gauge after confluence. The change of the flood type thus can be used to characterise the flood superposition and how a flood event changes downstream. One could argue that the flood-generating processes remain the same, however, if heavy rainfall occurs in both tributaries, this should be considered as a spatial rainfall for the whole basin. Additionally, the change of the flood type may also indicate the change of the flood-generating processes. If the flood type changes from snowmelt-induced in the upstream parts to rainfall-driven in the downstream parts, this indicates that for the whole catchment, rainfall was the main driver, while in the higher elevations, snowmelt was a main component in flood generation. Examples are given in the first section of this chapter.

Besides the hydrological effects of superposition, also the statistical effects have to be considered. For example, the dependence structure between the floods of main river and tributary depends on the interplay between both. The dependence differs if a flood event origins mostly from the main river compared to a mixture event raised from both, tributary and main river. To capture the impact of the tributary on a flood event, a new flood typology is proposed that is based on the flood volume proportions. It will form the basis for the spatial flood statistics of large river basins and is introduced in the second section of this chapter.

10.1 Using Flood Types for the Analyses of Flood Superposition

In this study, the flood typology introduced in Sect. 6.2 is used to analyse the impact of superposition on a flood at an outlet gauge. For this, the flood types at the main river are compared to those in the tributaries.

Once again, the example of the upper Danube is considered with the Achleiten gauge after confluence of Inn and Danube rivers taken from Fischer and Schumann (2021). For the Achleiten gauge, distinct patterns in the flood type combinations occur, allowing to make inference on the flood wave superposition (Table 10.1). While R1 flood events in Achleiten are mostly caused by two R1 events in the tributary and the main river, an R2 flood event most probably occurs due to an R1 event in the Inn tributary and an R2 flood event at Hofkirchen. In the first case, a large temporal overlay of both tributary and main river can be assumed. This leads to an increase of the peak but small increase in duration, as it was the case for the 2013 flood (see above). Interestingly, also a flood event with a longer duration in

Table 10.1 Empirical probability of the combination of event types of tributary, main river and outlet gauge for the three most frequent event types at the Achleiten/Danube gauge after confluence (modified from Fischer and Schumann 2021)

Type Achleiten	Type Passau	Type Hofkirchen	Relative frequency
R1	R1	R1	0.705
R1	R1	R2	0.016
R1	R1	R3	0.016
R1	R1	S1	0.016
R1	R2	R1	0.213
R1	R2	R2	0.016
R1	R2	S1	0.016
R2	R1	R1	0.359
R2	R1	R2	0.051
R2	R2	R1	0.487
R2	R2	R2	0.103
R3	R2	R1	0.500
R3	R2	R2	0.500

the tributary can cause a short flood event at the outlet gauge, if the temporal delay between tributary and main river is small. All remaining combinations leading to R1 flood events are negligible since their probability of occurrence is small. Obviously, a temporally similar occurrence, which would be required to generate an R1 flood event at the outlet gauge, is not probable in these cases. Here, one should consider that the flood types are always related to the catchment. An R1-flood in a large catchment could remain an R1-flood downstream even if overlaid with an R2-flood from a tributary as the increase in catchment area is significant for the specification of flood events.

R2 flood events at the outlet gauge most often occur due to a temporal delay in the occurrence of shorter flood events in tributary and/or main river. For example, more than one third of all R2 flood events in Achleiten is caused by two R1 flood events in tributary and main river. Only the temporal delay between the occurrence of the flood peaks at the outlet gauge can lead to this change in flood type. Most interestingly, R3 events in Achleiten are not caused by R3 events in any of the tributaries. Instead, they are generated by R2 events in Passau-Ingling and R1 or R2 events in Hofkirchen. Here, the temporally delayed superposition of the two flood events leads to a flood event with higher volume but smaller peaks in Achleiten. However, the only small number of events of type R3 can lead to some uncertainty in this case.

As a second example, the Rhine river basin is considered. Details on this basin are provided in the Appendix in Sect. A.3. A special case for this basin is the flood type R0 considered additionally for the confluence gauge. This flood type refers to flood events with small flood timescale (high peaks and small volumes) in a period of high baseflow (Sect. 6.2.3).

Table 10.2 Empirical probability of the combination of event types of tributary, main river and outlet gauge for the three most frequent event types at the Andernach/Rhine gauge after confluence

Type Andernach	Type Cochem	Type Kaub	Relative frequency
R0	R1	R1	0.700
R0	R1	R2	0.033
R0	R2	R1	0.217
R0	R2	R2	0.050
R1	R1	R1	0.731
R1	R1	R2	0.025
R1	R2	R1	0.192
R1	R2	R2	0.051
R2	R1	R1	0.043
R2	R1	R2	0.106
R2	R2	R1	0.128
R2	R2	R2	0.638
R2	R3	R2	0.085
R3	R2	R2	0.400
R3	R3	R3	0.600
S1	R2	S1	0.230
S1	S1	S1	0.462
S1	S1	S2	0.308
S2	R1	S2	0.286
S2	R2	S2	0.114
S2	S2	R1	0.171
S2	S2	S2	0.429

Similarly distinct patterns as for the Danube can be detected for the Rhine basin at the Andernach gauge, which acts as the outlet gauge in this study (Table 10.2), though for the Rhine basin also the impact of snowmelt has to be considered. Interestingly, combinations for event types R0 and R1 are very similar, both originating mostly from an R1 flood in the main river (Kaub gauge) and an R1 or R2 flood in the tributary Moselle (Cochem gauge). Again, the temporal delay of the occurrence of the peaks at the confluence has to be small. In general, it can be observed that the main river in most cases has the same event type as the outlet gauge. This means, the event type is not altered much by the tributary. For the rainfall-induced floods, it hence can be concluded that the temporal delay of peaks of Moselle tributary and main river is small. For snowmelt floods, the situation is different. For the Rhine river basin, which is impacted much by the Alps in the south, often snowmelt floods in the upper Rhine occur together with rainfall-induced floods of the Moselle tributary.

The results show how a flood event in large river basins with several tributaries can be altered due to superposition of the flood waves. This, of course, has a direct impact on the magnitude of the flood peak and hence on the extremity of a flood event. Hence, it is of importance that this factor is also considered in flood statistics for large basins. Flood types can help to better understand the changes of flood events that take place in a large basin. Though of course a hydraulic consideration should be preferred, this is not always possible as most hydraulic models require knowledge on which and how much flood protection systems failed and how the inundation of the floodplains has influenced the individual flood events. This is hardly possible for each flood event and especially critical for extreme floods, where many of these structures may have been damaged. Flood types, instead, allow for a simple-to-use comparison of flood events in a basin and an assessment of flood event alteration.

10.2 A Spatial Flood Typology

The flood types considered so far focus on the meteorological causes of the flood events and how these have an impact on the flood hydrograph. However, when considering a gauge after a confluence in a large river basin, the sole consideration of the flood-generating processes may not be sufficient. Instead, it is important to know which of the tributaries contributes most to the flood event downstream, as this has an impact on the spatial dependence structure which is required for the spatial modelling of floods. In classical spatial statistics in hydrology, especially when using copulas to consider simultaneous flood events, differences in the spatial dependence according to the focus point of an event are usually not considered. However, when aiming at understanding the flood-generating processes, it is important to consider these differences. A possibility to classify these spatial differences is a spatial flood typology. Such a typology was proposed by Fischer and Schumann (2021), referred to as volume-based flood typology. It has the aim to differentiate floods by comparing the spatial distribution of runoff among the tributaries and the main river and thus implicitly taking into account the spatial distribution of rainfall or snowmelt. This typology focuses on the volumes and their generation while considering the impact of the tributaries. It classifies flood events as originating mainly from the main river, one or more tributaries or as being a mixture type with similar contributions from both main river and a tributary. Hence, it is applied to $(n_T + 1)$-tuples consisting of the gauge after confluence, the upstream main river gauge and one or more tributary gauges, where n_T denotes the number of upstream and tributary gauges. The most common application would be a triple consisting of gauge after confluence, upstream main river gauge and one tributary gauge, as for example for the previous example of the Danube. All flood events at the gauge after confluence are considered and the relation between the direct volume of these and flood volumes of the corresponding flood events at the tributary and upstream main river gauges is calculated and divided by the relation between the catchment sizes for normalisation purposes:

$$V_{rel;i}(j) = \frac{V_i(j)}{V_C(j)} \cdot \frac{AE_C}{AE_i}, \quad i = 1, \ldots, n_T, \; j = 1, \ldots, n \tag{10.1}$$

with $V_k(j)$ being the direct volume of the flood event j at gauge k and C denotes the outlet gauge. $V_{rel;i}(j)$, normalized to [0, 1] by a min-max-scaling, are clustered by application of the kMeans clustering algorithm into $2^{n_T} + 1$ volume-based flood types $T_1, \ldots, T_{2^{n_T}+1}$, one for each tributary respectively the main river and one mixture type, for which the proportion of each tributary is similar. This general clustering approach makes this typology applicable to any basin without defining any thresholds and hence analysing the flood regime first. The number of volume-based flood types can be reduced or increased by hydrological knowledge on the basin under consideration, for example, by identifying improbable combinations for the mixture types. If a flood event at the outlet gauge was not separated for all tributaries or the upstream main river, it can be assumed that this event does not belong to the corresponding tributary-type at this side nor is a mixture type.

As an illustrative example, the volume-based flood typology is applied to the example of the Danube-Inn-confluence (see Sect. A.2 for more details) and hence a triple of gauges, aiming to define three volume-based flood types. The kMeans algorithm applied to $Vol_{rel;\text{Inn}}$ and $Vol_{rel;\text{Upper Danube}}$, delivered the three clusters shown in Fig. 10.2 for Achleiten. Based on this, the three volume-based flood types are defined as Inn (Cluster 1), Upper Danube (Cluster 2) and Mixtype (Cluster 3). Examples of the flood-generating precipitation and the respective volume-based flood types of the flood events for this basin are given in Fig. 10.3. Clearly, the different focus points of the flood-generating rainfall events can be seen. For the volume-based type Upper Danube, the large amounts of rainfall fall down close to Alps in the right-hand tributaries of the Danube river in the German part. For the example of the volume-based type Inn, a distinct rainfall cluster can be seen for the lower Inn catchment. For the Mixtype, the rainfall is spatially more uniformly distributed with no clear focus point. Hence, by considering the flood volumes, the spatial distribution of the flood-generating rainfall can be considered implicitly. This extends the available information on the flood-generating processes.

The volume-based flood types do not only describe the spatial distribution of the flood events but the also may help to differ between small and large floods. According to the large catchment area of the Danube, most of the flood events with

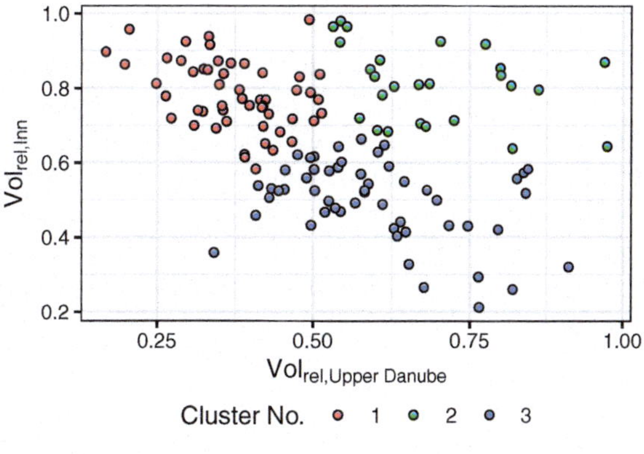

Fig. 10.2 Results of the kMeans clustering of volume-based flood types for Achleiten gauge

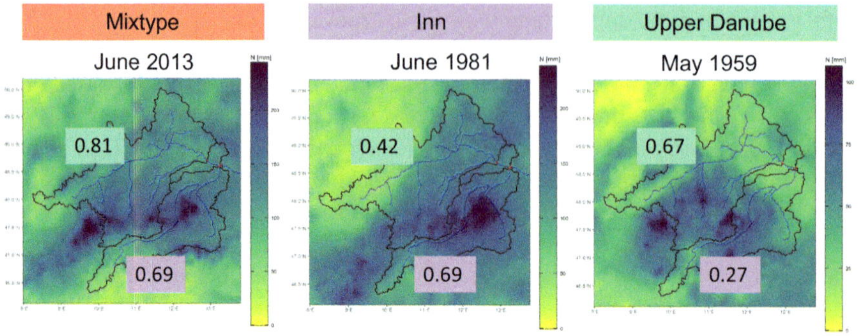

Fig. 10.3 Spatial precipitation of three flood events in the Upper Danube and their respective volume-based flood types. Numbers indicate the Vol_{rel} values for the Inn tributary respectively the Upper Danube

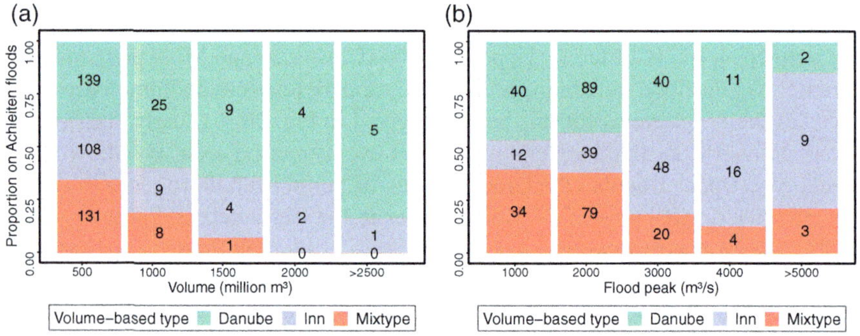

Fig. 10.4 Proportion of each volume-based type on the different classes of **a** volume and **b** peak magnitudes for the Achleiten/Danube gauge. The numbers indicate the number of events per class. (modified from: Fischer and Schumann 2021)

large volumes are generated by a flood event that mostly originates in the upper Danube (Fig. 10.4). On the contrary, floods originating mostly from the Inn basin lead to flood events with comparably smaller volumes at the gauge after confluence. This is meaningful, since the Inn basin is characterized by a much faster runoff-reaction than the Danube. Here, the flood volumes are in general smaller but the peaks are higher. The largest flood peaks in Achleiten, though, are mostly caused by the Mixtype, while medium flood peaks often originate from the Inn river. This implies that in case of extreme flood peaks, mostly both, tributary and main river, are affected. Considering this in flood protection, this means that one has to prepare for high flood peaks when both, tributary and main river, will likely experience large floods.

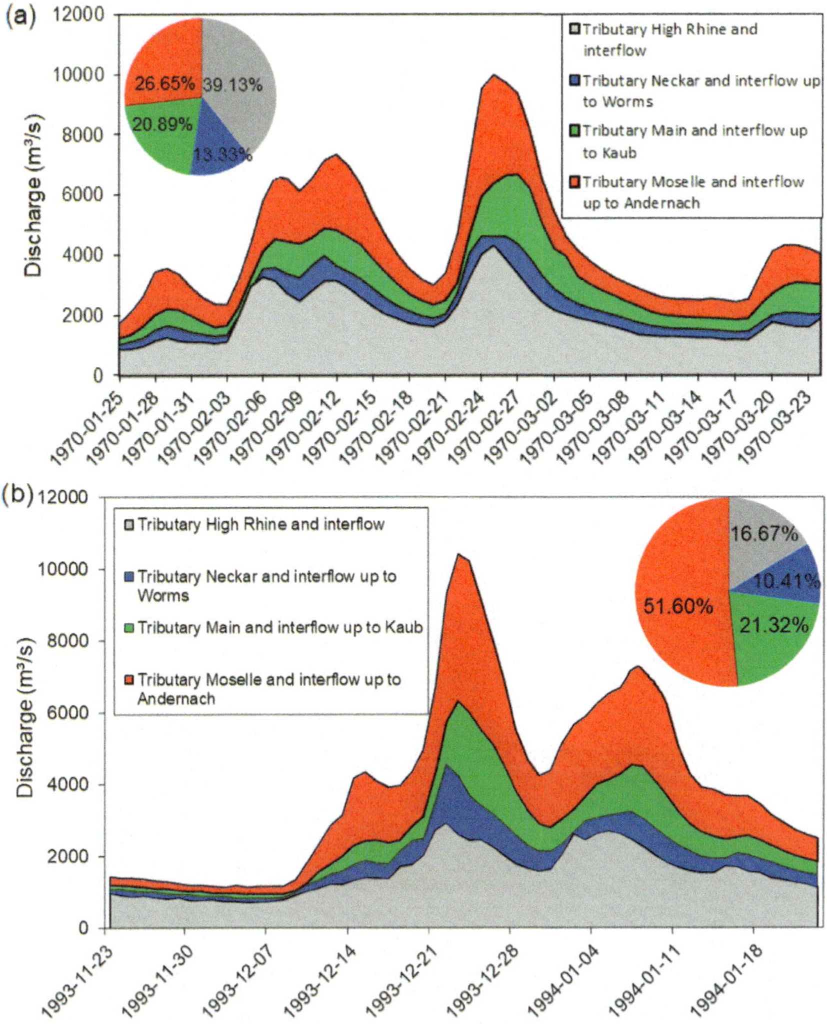

Fig. 10.5 Two flood events in the Rhine basin with the proportion of each tributary on the downstream event highlighted by the hydrographs as well as illustrated in a pie chart (relative to the catchment size) for volume-based flood types **a** "Alpine" and **b** "Moselle"

As a second example, the more complex Rhine basin is considered (see Sect. A.3 for more information). Due to the more complex structure and the several tributaries, a triple of gauges is not sufficient to describe the spatial dependence. Instead, the triple at the confluence of Moselle and Rhine is considered, consisting of the Andernach, Cochem and Kaub gauges, as well as two additional upstream segments. One, representing the upper Rhine mainly impacted by the Alps, is described by the Maxau gauge, the other, Worms, describes the intersection from Upper Rhine

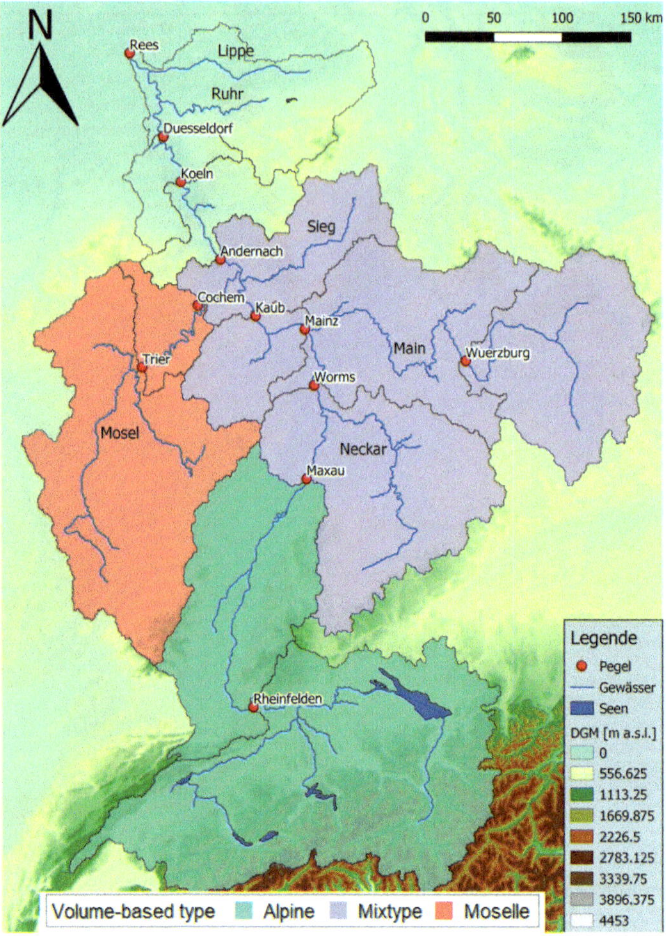

Fig. 10.6 Map of the Rhine basin where the regions corresponding to the volume-based flood types are highlighted

downstream to the confluence of Rhine and Neckar river. Taking into account these overlaps, i.e., the relative difference of the flood quantities between gauges, the effects of the tributaries without suitable gauges can be considered summarily. With this approach, four relative flood volumes are obtained that can be compared to the volume in Andernach. The volume-based flood typology has to be extended for such as case. This is done by multiplying the relation of the flood volumes with the relation between the catchment sizes. Then, these Vol_{rel} are classified into three volume-based flood types: Alpine, Moselle and Mixtype (Fig. 10.6). Of course, also a more differentiated classification is possible by, e.g., taking into account the tributaries Neckar and Main separately. However, since the data basis is limited and the impact of the Moselle river is of main interest for the flood generation downstream, it

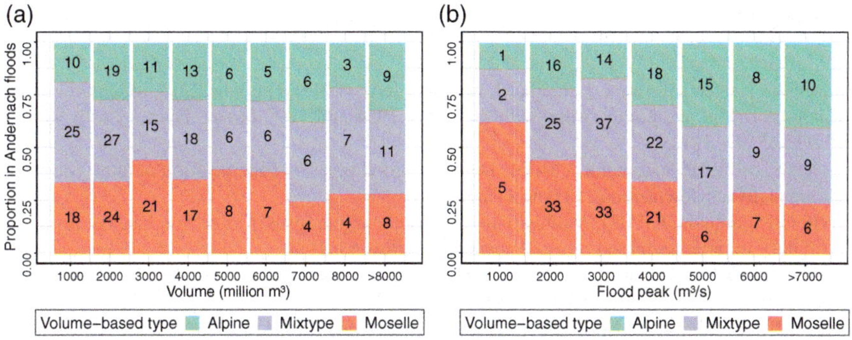

Fig. 10.7 Proportion of each volume-based flood type on **a** flood volume and **b** peaks at Andernach/Rhine gauge. Numbers indicate the number of observed events per class (modified from Fischer and Schumann 2021)

was decided to merge these two volume-based types in the Mixtype. By application of the kMeans clustering the classification into he three volume-based flood types is obtained. Two examples of flood events and their corresponding volume-based flood types are given in Fig. 10.5. For the example of the flood event with volume-based flood type "Alpine", it can be seen that a large proportion of the flood volume has its origins in the upper Rhine. A similar proportion is only reached by the Moselle tributary. However, if the flood event has the volume-based flood type "Moselle", as it was the case for the disastrous 1993 flood event, more than 40% of the downstream volume is generated in the Moselle tributary, while the other tributaries only play a minor role.

Comparing these three volume-based flood types for the Andernach gauge, the proportion of each type shifts for increasing flood peaks (Fig. 10.7). While small-peaked flood events are mostly caused by flood events with main impact in the Moselle river, large-peaked events occur for each of the volume-based flood types with equal probability.

The consideration of focus points of floods and thus implicitly of rainfall and snowmelt in a basin and hence the spatial dependency between main river and tributaries can be described by a spatial flood typology. With the proposed typology, it is possible to measure the impact of tributaries on an outlet gauge and therefore the flood superposition. The applied typology delivered significant differences in the generation of large peaks for the basins. Moreover, it will form the basis for the statistical simulation of floods in large basins in the next chapter.

References

Blöschl G, Nester T, Komma J, Parajka J, Perdigão RAP (2013) The June 2013 flood in the Upper Danube Basin, and comparisons with the 2002, 1954 and 1899 floods. Hydrol Earth Syst Sci 17(12):5197–5212

Fischer S, Bühler P, Schumann A (2021) Impact of flood types on superposition of flood waves and flood statistics downstream. J Hydrol Eng 26(7)

Fischer S, Schumann A (2021) Multivariate flood frequency analysis in large river basins considering tributary impacts and flood types. Water Resourc Res 57(8)

Skublics D, Blöschl G, Rutschmann P (2016) Effect of river training on flood retention of the Bavarian Danube. J Hydrol Hydromech 64:349–356

Vorogushyn S, Merz B (2013) Flood trends along the Rhine: The role of river training. Hydrol Earth Syst Sci 17(10):3871–3884

Chapter 11
Multisite Type-Specific Event Generator Under Consideration of Spatial Dependence

Abstract In this chapter, the considerations regarding the interplay of flood types and their spatial dependence from the previous chapter are transferred into the development of a multisite event generator. The spatial dependence of synchronous floods between the catchments is modelled with a Vine-Copula approach, using the dendritic structure of a river network. The general model structure is outlined and the challenge of considering the impact of dams as well as sparse gauging network is discussed. The procedure is illustrated by an example of the Rhine river basin with its several tributaries.

When considering floods in mesoscale catchments and how to model them, often the focus is laid on the meteorological and catchment processes that lead to such floods. Examples are many deterministic models or stochastic event generators. For macroscale catchments, however, it was shown before in Chap. 10 that a consideration of event flood types alone is not sufficient. Instead, also the spatial aspects of such large catchments and the interaction between tributaries have to be considered. Here, the most important aspect is flood superposition. Especially for large river basins with many tributaries, flood superposition can play a crucial role in the flood generation. When considering the impact of the tributaries, both, the tributary as well as the main river, are equally important for the generation of an event type downstream at the confluence (Fischer et al. 2021). But, the impact differs between flood types. Flood peaks of waves with shorter duration will less likely superimpose downstream compared to peaks of long-duration flood events. However, short duration flood events are often characterized by steep hydrographs with the larger flood peaks, and hence a superposition of peaks is more critical. Hence, for large river basins, the joint consideration of a hydrograph-based flood typology and the flood superposition may be beneficial. Some combinations are more probable than others and the flood regime of the main river is a mixture of these different cases. To assess the probability of extreme events, the probabilities of favourable and unfavourable coincidences and superpositions of flood waves from tributaries with floods in the main river have to be analysed. The complex network structure of large river basins as well as the dependence between flood events of the tributary and the main river due

S. Fischer and A. H. Schumann, *Type-Based Flood Statistics*, Water Science and Technology Library 124, https://doi.org/10.1007/978-3-031-32711-7_11

to macro weather situations requires multivariate statistical models. Often, copula models are used to describe these multivariate structures (Chen et al. 2012; Huang et al. 2018). However, in these approaches the flood type is mostly ignored. To overcome this problem, the spatial dependency as well as the local flood-generation have to be considered in the multivariate model. For example, one could combine the two different flood typologies proposed in Sect. 6.2 and Chap. 10: one for the characterisation of the meteorological drivers for each catchment and one for the classification of the spatial aspects determining the superposition of flood waves in a river basin. These two classifications then should form the foundation of a statistical simulation procedure, which allows the generation of simultaneous peak-volume pairs of flood events at the main river and the tributaries under consideration of the dependence structure, the flood type and the probabilities of flood scenarios. Such an approach was described by Fischer and Schumann (2021) and will be introduced in this chapter in detail.

11.1 Model Chain for the Multisite Stochastic Event Generator

The consistent generation of peak-volume pairs of flood events in a river basin requires a statistical model that does not only take into account the spatial dependence but also the river network and the flood generation. Such a model therefore consists of several steps which are linked closely. Here, a multisite stochastic model is described that makes use of the flood typology from Sect. 6.2 and the volume-based flood types from Chap. 10. The model chain is designed for a gauge after a confluence, denoted with C, with an arbitrary number of n_T tributary gauges. This means, there is the main river with a gauge located downstream below the confluence of the main river and n_T tributary rivers. For sake of simplicity, in the following the upstream gauge at the main river will be labelled as tributary, too. This way, a general formula for an arbitrary number of tributaries can be proposed. Of course, the most common case is an upstream gauge at the main river, a tributary gauge and a confluence gauge downstream. In this case, $n_T = 2$. Such triple pairs are most often considered when analysing flood superposition (Guse et al. 2020; Fischer et al. 2021).

A first and important step when aiming to analyse flood events spatially, is the ability to generate arbitrarily large synthetic samples of events with the same statistical distribution as the observed ones. This allows for a broader range of statistically possible event combinations and an analysis of the probabilities. The starting point is the gauge after confluence and the river basin is then modelled upstream recursively. Hence, the procedure for the statistical simulation of synthetic samples of flood events is developed with the aim to generate arbitrary large samples of pairs of peak and volume for the confluence gauge that have the same distribution as those observed. Again, the stratification into flood types and volume-based types is crucial since only then the flood-generating processes and the spatial dependencies can be captured correctly by such an approach. Moreover, the respective peak-volume pairs of the tributaries have to be generated consistently, i.e., in a way that they match the downstream event. This requires the consideration of the dependence structure between the tributaries and the confluence gauge. The simulation procedure described in Fischer and Schumann (2021) consists of seven elemental steps, which will be described in the following sketch. This model chain allows to generate a sample of N pairs of peak and volume, stratified by flood type and volume-based type.

1. Derivation of a vector of flood volumes for the tributaries $(V_1(j), \ldots, V_{n_T}(j))$ corresponding to the flood event j, $j = 1, \ldots, n$, at the confluence gauge and assessment of the corresponding flood types and volume-based types.
   Result: List of flood event volumes of the confluence gauge with respective volume-based flood type and corresponding flood volumes of the tributaries.
2. For each volume-based type at the confluence gauge: Fitting of a distribution to the time series of all flood volumes for each tributary separately and application of this marginal distribution for transformation of the flood volumes derived in step 1 into exceedance probabilities in [0,1]. Fitting of a n_T-dimensional copula to the transformed pairs of volumes of the tributaries.
   Result: Dependence structure between the volumes of the tributaries for each volume-based type.
3. For each volume-based type at the confluence gauge: Generation of a sample of random pairs using the copula derived in step 2. Choice of the sample size according to the proportion of the respective volume-based type, p_{T_i}, i.e., $N_{T_i} = N \cdot p_{T_i}$, $i = 1, \ldots, n_T$. Transformation of these pairs back to $\mathbb{R}^+$ by using the respective marginal distribution (quantile function) for each tributary.
   Result: Simulated tuples of volumes for the tributaries with respective volume-based type.
4. Calculation of the volume of each event for the gauge after confluence by the sum of the volumes of the tributaries for each pair of volumes.
   Result: Simulated volumes for the gauge after confluence and corresponding volumes for the tributaries with assigned volume-based type.

[Optional: Bias-correction for distinct volume-based types if a systematic bias can be observed. For this, the median error between the observed and simulated quantiles of the volumes, q_{obs} and q_{sim}, can be used.]

5. For each volume-based type and each 10%-quantile range of the observed volumes at the tributaries: Estimation of the probability of occurrence of each event type R1, R2, R3, S1 and S2 for each tributary and transmission of this probability to the simulated volumes with respective quantile ranges and volume-based type of the simulated volume pairs.
   Result: Event types of the simulated volumes for the tributaries.
6. For each 10%-quantile range of the observed volumes and each volume-based type at the gauge after confluence: Estimation of the probability of occurrence of each event type R1, R2, R3, S1 and S2 at the confluence gauge conditional on the event types of the tributaries and transmission of this probability to the

volumes with respective quantile ranges and volume-based type of the simulated volume pairs.
Result: Event types of the simulated volumes for the confluence gauge under consideration of the flood type of the tributaries.

7. For each event type R1, R2, R3, S1 and S2: Estimation of the flood peak corresponding to each simulated flood volume of the confluence gauge and the tributary gauges by application of the flood timescale (Chap. 9).
Result: Consistently simulated pairs of peak and volume with corresponding event type and volume-based type for the confluence gauge as well as the respective flood events for the tributaries.

It has to be noted that of course also the gauge after confluence could be included into step 2 and 3 and hence step 4 would be omitted. However, this way a consistent estimation of the sum cannot be guaranteed. Instead, the sum of the tributaries can be larger than the simulated volume of the confluence gauge. This is unfavourable for applications later on. Moreover, by the method described here, a regionalisation and hence application to an ungauged confluence is possible. For this purpose, the estimated probabilities of occurrence for the event types simply have to be estimated regionally or by a nearby located gauge. The same holds true for step 7, where a regionalisation is easily possible as shown by Fischer et al. (2019) and later on in Part VI of this book.

An overview of the procedure is given in Fig. 11.2.

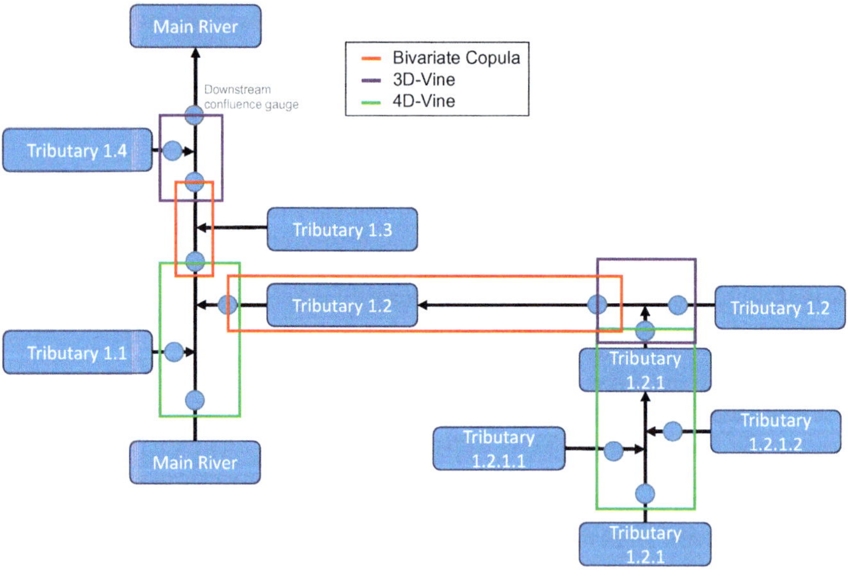

Fig. 11.1 General scheme for extending the multisite stochastic event generator to upstream gauges by using (Vine) copulas (from: Fischer and Schumann 2021)

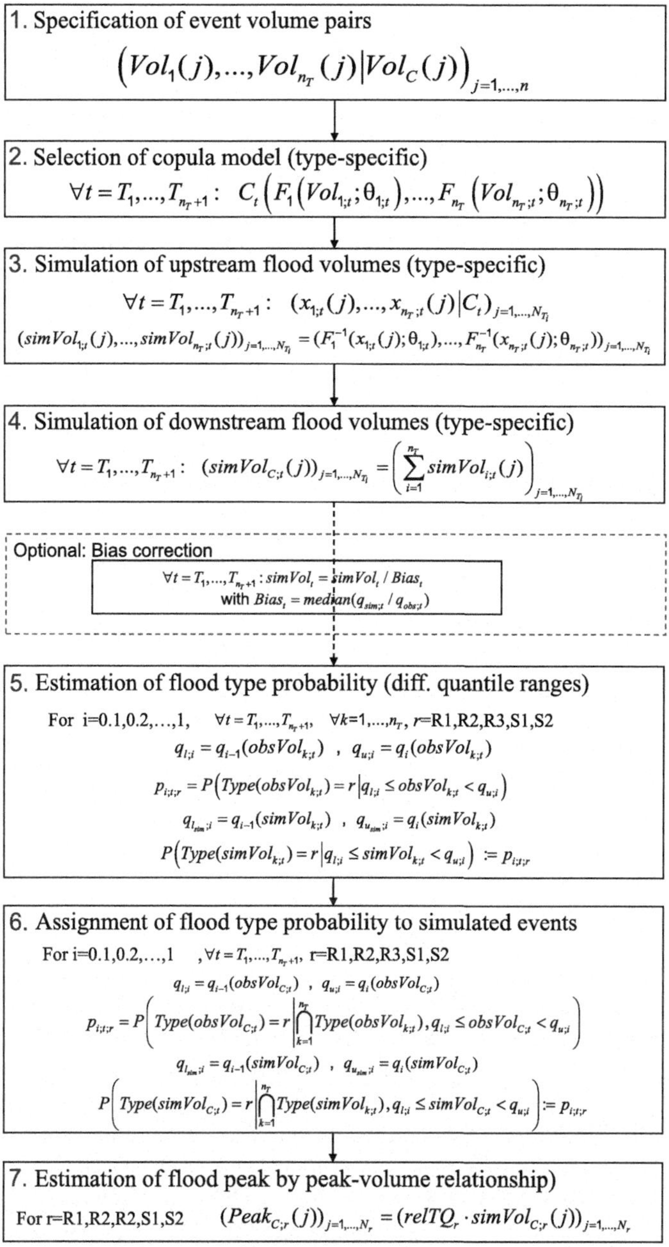

Fig. 11.2 Schematic overview of the statistical spatial simulation (from: Fischer and Schumann 2021)

The model chain can be extended to further gauges located upstream. In this case, another Vine-Copula model has to be fitted under consideration of the flood volumes of the downstream gauges. This way, the model chain is recursively extended to upstream gauges (Fig. 11.1) by application of the conditional copula probability. More precisely, a copula model is fit for the upstream gauge and the consecutive tributary. Then, the simulated peak-volume pairs from the tributary gauge are used and by application of the conditional probability, the upstream volumes are derived, for each volume-based flood type separately. In theory, any arbitrarily complex river network could be considered by this extension. However, data availability limits such approaches. It has been recommended by Fischer and Schumann (2021) to not apply the model using copulas with dimension larger than three and no more than level-one-tributaries, i.e., tributaries that belong directly to the main river.

11.2 Performance Measures for Flood Event Generation

A comparison of the generated samples to the observed one regarding goodness of fit is difficult. Since the aim is to model the distribution of flood peak and volume and not the exact events as best as possible, standard methods for comparing distribution functions regarding goodness of fit, e.g., those introduced in Sect. 2.4, offer a possibility of comparison. Statistical tests can provide information, if the distribution of the generated sample is suitable for the observed sample. However, they cannot provide information on the goodness of fit. The same holds true for the information criteria, which can only be used to compare several distributions. However, visual comparison of observed and generated sample is possible, e.g., by a QQ-plot. A challenge when using the QQ-plot in this case is the sample uncertainty of the observed peak respectively volume samples. The mostly short observation periods make an estimation especially of large quantiles difficult. One option to handle this uncertainty is to consider the sample uncertainty also for the generated samples. The advantage of a stochastic event generator is that arbitrary large samples can be generated. For example, $n = 1000$ values for the flood volume of each flood type and tributary could be generated. Instead of comparing all the flood volumes or peaks of a flood type with the much smaller number of observed flood peaks of the respective flood type, sub-samples should be drawn with the same sample size as the observed flood peaks. This can be repeated several time, e.g., 1000 times, due to the large number of generated flood peaks, similar to a bootstrap approach. With these sub-sample, uncertainty bands can be defined for the QQ-plot. More precisely, for each data point corresponding to a quantile of the QQ-plot, the 90% and 10% quantiles of all sub-samples are estimated. These form the uncertainty limits. The median of each quantile of all sub-samples is used to display the standard dots in the QQ-plot.

The QQ-plots allow for the evaluation of the distribution of the simulated flood volumes or flood peaks compared to the distribution of the observed ones.

However, this visual inspection does not deliver an objective measure of the goodness of the proposed generator. For practical purposes, one is interested in measures that provide information on the relative deviation of the estimated flood quantiles. Again, a direct comparison is difficult. But by comparing the corresponding quantiles, the generated distribution can be compared to the empirical distribution. A measure for this is the Trimmed Normalised Mean Squared Error (TNRMSE) defined as

$$TNRMSE = \sqrt{\frac{1}{n-[n\delta]} \sum_{1=1+[n\delta/2]}^{n-[n\delta/2]} \left(\frac{Q_{i;obs} - Q_{i;sim}}{Q_{i;obs}} \right)^2}. \tag{11.1}$$

The trimming factor δ omits that the largest and smallest quantiles are considered in this measure, as these would have a large impact on the results. Usually, δ is chosen as 5% to only omit the most extreme values. $Q_{i;obs}$ and $Q_{i;sim}$ are the i-th quantile of the observed respectively generated sample.

11.3 Multisite Stochastic Flood Event Generation for the Rhine Basin

Using again the example of the Rhine river basin introduced in Chaps. 6 and 10 and for which details are provided in Sect. A.3, the procedure is illustrated. For the Rhine river basin with target gauge Andernach, three main volume-based flood types were identified for the tributaries: Moselle, Alpine and Mixtype floods (Chap. 10). Additionally, six flood types have been identified: R1, R2, R3, S1 and S2 as described in Sect. 6.2 and one additional flood type R0, which is related to floods associated with high baseflow (Sect. 6.2.3). R0 floods are a sub-type of R1 floods with a similar peak-volume relation but they occur during a phase of high baseflow. Fischer and Schumann (2021) suggest to separate such floods by using a threshold for the baseflow. More details are given in Sect. 6.2.3.

For this river basin, first the triple of confluence gauge Andernach, tributary gauge Cochem (belonging to the Moselle river) and tributary gauge Kaub (located upstream of the confluence of Rhine and Moselle rivers in the Rhine river) is considered. The steps described for the model chain are applied as follows:

1. For each flood event observed at Andernach gauge, the corresponding flood events for Cochem and Kaub are selected. Corresponding events are defined as having a flood peak at a time of +/– three days compared to the flood peak in Andernach. The time difference can be adjusted according to the distance between the gauges.
2. For each of the three volume-based flood types "Moselle", "Alpine" and "Mixtype", a copula is fitted to the volume pairs of synchronous floods of Cochem and Kaub gauges. As marginal distributions, the GEV distribution is used for Cochem gauge and the PearsonIII distribution is used for Kaub gauge (selected by Anderson-Darling test and AIC/BIC). Bivariate copulas are chosen as Tawn Type 2 (Moselle, Kendall's $\tau = 0.30$), t-Copula (Alpine, $\tau = 0.56$) and BB8 copula (Mixtype, $\tau = 0.48$) by application of the AIC/BIC criteria.

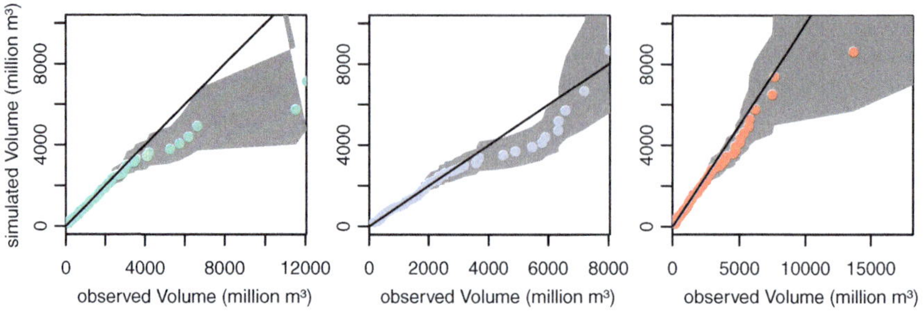

Fig. 11.3 QQ-plots and 95% uncertainty bands (based on 1000 repetitions, grey) for the simulated and observed flood volumes at Andernach gauge with volume-based flood type "Moselle" (left), "Alpine" (centre) and "Mixtype" (right)

3. A sample of $N = 100000$ pairs of flood volumes for Cochem and Kaub is generated and transformed by application of the marginal distributions, with 35556 pairs of volume-based flood type "Moselle", 26032 pairs of volume-based type "Alpine" and 38412 pairs of the "Mixtype". The proportion of each volume-based flood type is estimated from the observed sample of flood events.
4. The flood volume of each simulated event for Andernach gauge is estimated by the sum of the simulated volumes of Kaub and Cochem gauges.

The simulated flood volumes can be evaluated regarding their goodness-of-fit by application of QQ-plots (Fig. 11.3). Since the aim is to reproduce the statistical distribution of the observed flood volumes, QQ-plots with confidence bands are particularly useful as they compare the observed and simulated quantiles, as described above. The results reveal that the flood volumes up to a quantile of about 95% can be captured well by the stochastic event generator for all volume-based flood types. However, the extreme flood events are difficult to simulate, especially for the Moselle type. It remains to be investigated how this is transferred to the flood peaks, which is done in the following steps.

6. For the observed flood volumes of Kaub and Cochem gauges, 10% quantile ranges are defined respectively. Within each of these ranges, the probability of occurrence of each flood type R0, R1, R2, R3, S1, and S2 is estimated. According to these probabilities and with respect to the quantile ranges, for the simulated flood volumes in Kaub respectively Cochem, the flood types are assigned.
7. For the observed flood volumes of Andernach gauge, 10% quantile ranges are defined respectively. Within each of these ranges, the probability of occurrence of each flood type R0, R1, R2, R3, S1, and S2 is estimated given the respective flood type at Kaub and Cochem gauges. According to these probabilities and with respect to the quantile ranges, for the simulated flood volumes in Andernach, the flood types are assigned.
8. By application of the type-specific flood timescale, i.e., the peak-volume relationship, the flood peak corresponding to each simulated volume is estimated for each flood type of Andernach, Cochem and Kaub gauges (see Chap. 9).

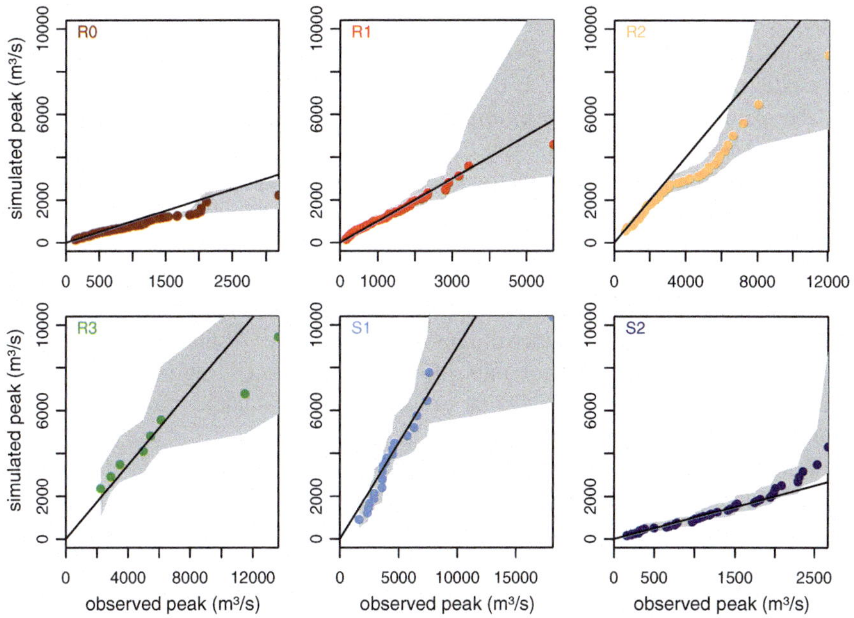

Fig. 11.4 QQ-plots and 95% uncertainty bands (based on 1000 repetitions, grey) for the simulated and observed flood peaks at Andernach gauge with flood type R0, R1, R2, R3, S1 and S2 (from top left to bottom right)

The tendency of underestimation of extreme floods is transferred from the simulated volume to the simulated peaks, though mainly for flood types R2 and R3, the rainfall-driven floods with longer duration. For the majority of flood events, the stochastic event generator delivers reliable estimation of flood peak and volume for each flood type and volume-based flood type (Fig. 11.4).

In a next step, the respective flood peaks and volumes for the upstream gauges Worms and Maxau have to be simulated. First, the Worms gauge is considered, as this is the gauge closest to the already simulated Kaub gauge. The steps 1–7 are repeated, though for a bivariate case. That is, first a copula is fitted to the pairs of observed synchronous flood events of Kaub and Worms for each volume-based flood type. The copulas Tawn Type 2 (Moselle, $\tau = 0.63$, and Alpine, $\tau = 0.73$) and survival BB1 (Mixtype, $\tau = 0.59$) under usage of the PearsonIII distribution as marginal distribution for the flood volumes in Worms are selected. Using the conditional probability of the fitted copulas, for each simulated flood volume in Kaub, the corresponding flood volume in Worms is estimated, stratified by volume-based type. Similar to steps 5 and 6, the flood type is assigned to the simulated flood volumes in Worms. Finally, the peak is estimated by the type-specific flood timescale. Having the simulated flood volumes in Worms, the model chain can be extended to the Maxau gauge, being the one located upstream of Worms. Again a bivariate approach is used considering the flood volumes in Worms, where the Tawn

Type 2 copula is chosen for all three volume-based flood types under usage of the PearsonIII distributions as the marginal distribution of the flood volumes in Maxau (Moselle: $\tau = 0.35$, Alpine: $\tau = 0.71$, Mixtype: $\tau = 0.46$).

The example of the Rhine basin emphasises clearly how dependency of flood events is transferred in space. The dependency between the tributaries and main river varies within the basin with the flood type and the dominating meteorological drivers. As a result, also the dependency structure, expressed by the copula model, varies, mainly between the volume-based flood types. The impact of tributaries and intermediate areas becomes obvious if we compare the dependencies between two consecutive gauges at the main river in terms of Kendall's τ. For example, the dependency between the gauges for the volume-based flood type "Alpine" clearly decreases downstream. The differentiation into different runoff-generating processes and the consideration of the impact of tributaries thus provides valuable information for the multivariate modelling.

Besides the graphical comparison, the quality of the model should also be evaluated by quantitative performance measures, such as the TNRMSE (Eq. 11.1). The results show that the simulation error for the distribution of flood peaks at the confluence gauge Andernach is overall small for the flood types R0, R1 and R2 (Table 11.1). For R3, an error estimation is not meaningful for all gauges due to the very small data basis. For the snow-impacted flood types S1 and S2, the errors are slightly larger. This might have two reasons: the data samples for these types are much smaller than for the rainfall-driven flood types causing model uncertainty. Moreover, the peak-volume relation was developed especially for rainfall-induced flood types. An extension to snow-impacted floods is not straightforward. However, the results are still within an acceptable range.

For the Cochem gauge, representing the Moselle tributary, similarly small errors as for Achleiten are obtained. Only the flood type S1 results in higher errors of 40%. For the Kaub gauge, errors are larger but still mostly around 30%. In general, errors increase with the size of the intermediate area to the gauge after confluence. This means that the error is forwarded during the recursive procedure and increases with every new simulation step. However, having in mind the few data that have been used to simulate volumes of gauges located upstream far away from the confluence gauge, errors are still in an acceptable range, at least for the flood types R1, R2 and S1, the most important flood types for extreme flood events in the Rhine basin. For other basins, where the samples might be more uniformly distributed regarding the frequency of flood types, the errors might decrease due to a better representation of all flood types. However, this application also emphasises the limitation of the model: too many gauges located upstream and with large growth of the intermediate area will result in higher uncertainties at a target gauge further downstream due to error propagation.

It has to be noted that the multisite stochastic event generator delivers pairs of peak and volume of direct flow, i.e., without baseflow. For many large basins, as for example the Rhine river basin, baseflow is an important component and defines a large proportion on the total flow. Fischer and Schumann (2021) point out that the partition of the baseflow depends much on the flood type and of course varies between

Table 11.1 TNRMSE between observed and simulated flood peak quantiles differentiated by flood type for the gauges in the Rhine basin. TNRMSE could not be calculated for R3 due to insufficient data base for Andernach and Cochem gauges. (modified from Fischer and Schumann 2021)

Gauge	R0 (%)	R1 (%)	R2 (%)	R3 (%)	S1 (%)	S2 (%)
Andernach	11.57	9.49	12.88	–	23.79	13.42
Cochem	–	13.78	10.49	–	39.75	8.9
Kaub	–	29.32	20.25	35.77	30.82	28.3
Worms	–	20.80	23.66	56.73	44.5	26.8
Maxau	–	54.04	29.65	70.94	62.12	36.60

the seasons. If the simulated peak-volume pairs should represent the total runoff, as required, e.g., for the estimation of design floods in flood statistics, the baseflow component has to be added to the simulation results afterwards. This can be done by estimating baseflow in a type-specific way as proposed by Fischer et al. (2019) or by using classical baseflow estimators such as linear filters (Mei and Anagnostou 2015), similar to the generation of hydrographs in Chap. 9.

11.4 The Impact of Dams and the Gauging Network

For the development of the model chain in Sect. 11.1 it was assumed that all processes are stationary. This means that neither the distribution of the flood volumes nor the spatial dependency change in time or magnitude. For many large basins, this is reasonable assumption and the flood event generator will work well in these cases. However, there might exist basins where these assumptions are violated. One human impact on the flood regime is given by retention facilities such as dams and reservoirs. Dams alter flood events by reducing the flood peak and volume and often release the flood event with a time-shift. However, dams to not alter all flood events alike. The degree of alteration differs by the size of the event (in terms of volume and peak). Flood events with small volumes are altered much, while flood events with large volumes are reduced in volume only until the storage capacity of the reservoir is reached. The most critical case of flood protection by reservoirs is given if the spillway of a dam is activated. Depending on the volume of the rising limb of the hydrograph and the flood retention capacity, the peak of the flood is more or less affected in such cases. Due to this moderation of the flood event, the spatial dependency in a catchment is altered, too. These impacts cannot be considered by a classical copula since hydrological coherences are changed artificially by the dams. This leads to overestimation of the flood volume especially in the medium quantile ranges, where these events are most affected by flood retention. The assumption of stationarity, thus, is not valid in this case. At the same time, the considered models for spatial dependency assume (implicitly) that the runoff of tributary and main river adds up to the runoff at the downstream gauge. This is a valid assumption when the measuring gauges are located close to the confluence. However, if the increase of the drainage area to the confluence becomes more significant, the spatial dependency as well as the assumption on the changes of flood types might not be straight-forward. Both aspects will be discussed in this section by examples of river basins for which such phenomena occur.

11.4.1 The Impact of Dams: Elbe River Basin

A great example for a basin where the impact of dams changes the spatial dependency of flood events is the Elbe river basin in Czech and Germany, the second-longest river in Germany. The Elbe river basin has a total catchment size of 148269 km^2 and drains parts of Austria, Poland, Czech Republic and Germany. Its head lies in the Giant Mountains in Poland and after a length of 1094 km it enters the North Sea. The flood regime of the German part of the Elbe river is characterized by several factors, including precipitation, snowmelt, and the river's topography. It is susceptible to flooding due to its low gradient and extensive floodplains. Heavy rainfall in the basin's upper reaches can cause significant increases in water levels downstream, leading to flooding. Snowmelt can also contribute to the river's flow and increase the risk of flooding. The Elbe basin was affected by many large flood events in the past. Well-known floods in the German parts are those in the years 1954 and 1958 but also lately in 2002 and 2013. In former time, the Elbe River was very vulnerable to floods caused by snowmelt, but such events became rarer nowadays. Instead, often heavy rainfall in the upper parts of the basin are caused by a specific weather pattern which results from a low pressure trough position in Central Europe (a so-called Vb weather pattern). In the 1950s and 1960s, especially in Czech Republic, several reservoir systems have been installed. Nowadays, the total storage volume of reservoirs in the Elbe river basin exceeds 4000 million m^3. The flood storage capacity is 573 million m^3. Most of this storage volume was built between the years 1955 and 1988. This has a large impact on the flood events downstream. How to possibly handle such non-stationarity in the data is proposed in the following by using the example of the Elbe river basin. For this study, only a small part (28%) of the Elbe basin is considered, starting close to the border of Germany and Czech Republic and ending at the Barby gauge. It was chosen according to the alteration of landscape downstream: the Barby gauge is representative to characterise the flood generation processes of the Elbe River, since downstream of Barby the flood regime changes into a low land regime with very long flood events and flat hydrographs. Hence, flood waves observed in Barby can be transferred to gauges downstream. This is why for this case study the Barby gauge (catchment size 94060 km^2) was chosen as the end point of this study for the Elbe basin. The Dresden gauge (catchment size 53096 km^2) was chosen as the starting point. It is located closely to the German-Czech border. Besides the Saale River tributary represented by the Calbe-Grizehne gauge (23719 km^2), the Mulde River tributary is of special interest as it has a pronounced flood regime (see the example of Zwickau gauge in Chap. 4), for which the Golzern

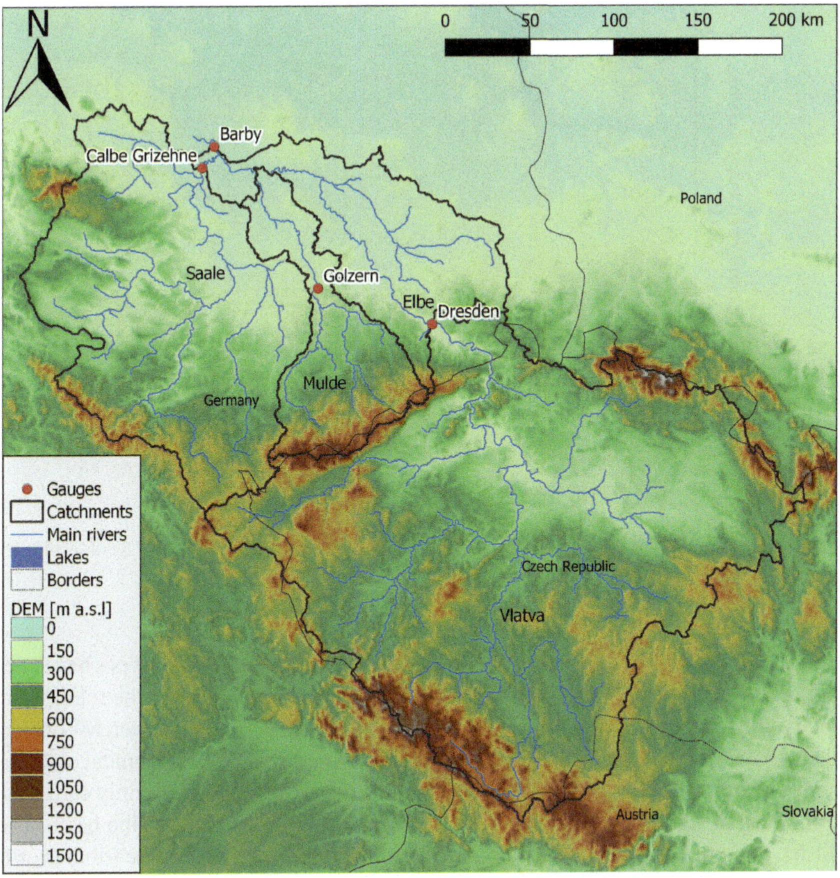

Fig. 11.5 The Elbe river basin until Barby gauge after confluence

gauge is considered. With only 5432 km^2 catchment size, the Mulde basin belongs to the mesoscale catchments. The longest joint observation period available for these four gauges is 1910–2013, the basis of the study (Fig. 11.5).

For the volume-based flood types of the Elbe basin, three different flood types are considered: "Upper Elbe", "Tributary" and "Mixtype". The volume-based flood type "Upper Elbe" characterizes flood events mainly originating in the upper part of the Elbe, strongly impacted by the Vltava River in the Czech Republic. The "Tributary" type includes flood events that are strongly affected by runoff which originates from the Mulde and/or Saale tributaries. Of course, also two different volume-based flood types, one for each tributary, could be considered here. However, since the main interest lies in the characterisation of the flood regime at the Barby gauge by assessing the impact of tributaries on the river reach between the Dresden and Barby gauges, this typology is sufficient here. Moreover, it will keep the multisite statistical model, which will be used later on for the simulation, simple, having to

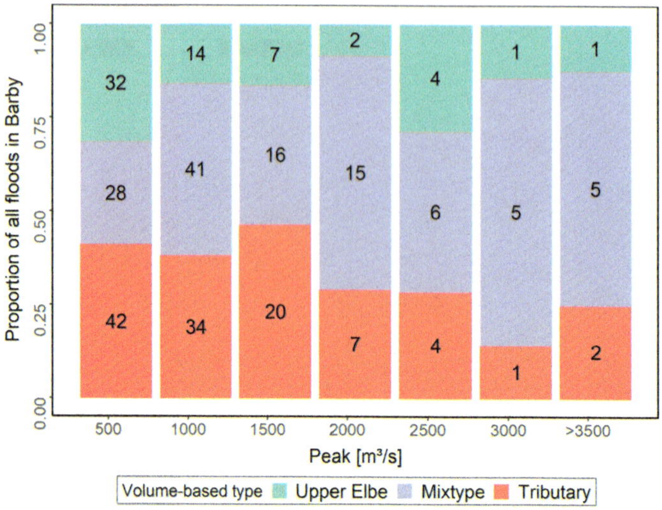

Fig. 11.6 Proportion of each volume-based flood type on the peaks at Barby/Elbe gauge. Numbers indicate the number of observed events per class

consider only 3D-vines instead of 4D-vines. The volume-based flood types have been determined as before using the kMeans cluster algorithm applied to the relations of the volumes multiplied with the relation of the catchment sizes. However, to consider the tributaries Mulde and Saale jointly, their volumes as well as catchment sizes are summed up for each event. The flood events with largest peaks are mainly caused by a mixture of floods from both tributaries with events in the Upper Elbe basin (Fig. 11.6). For smaller peaks, often separate floods mainly occurring in the tributaries or the upper Elbe are responsible.

The impact of the dams becomes evident mainly by analysing the spatial dependency of the flood events. However, the volume-based flood types are not affected alike by dams, which is due to the different spatial focal points of the types. For example, the volume-based flood type Upper Elbe only shows very small dependencies between the flood volumes of the tributaries and the volumes at the Dresden gauge. With a correlation of $\tau = 0.18$, no suitable dependence model can be found. Instead, the bivariate consideration leads to serious overestimation of the right tail, where the small correlation coefficient leads to largest variance.

For the remaining two volume-based flood types, "Tributary" and "Mixtype", the copula-based bivariate modelling works well. However, it turns out that the dependence structure changes with increasing quantile. For example, for the Mixtype and flood volumes below the 60%-quantile of all volumes of this type, Kendall's τ is only –0.23 while it increases for flood volumes above the 60%-quantile to 0.63. By using one copula model for the whole range of all quantiles, then an overestimation of the right tail occurs. The consideration of such a threshold takes into account the different impact of the dams for small and large flood volumes: for small volumes, most of the flood event is stored by the dams and does not have a large impact on the downstream gauge. For large flood volumes, the storage is not sufficient and some discharge is released during the event which has an impact on the downstream flood.

Both cases cannot be captured by the model chain, as only stationary behaviour can be described. However, extensions are possible, e.g., the definition of a threshold from which on the copula model changes. Yet, this requires detailed knowledge on the reservoir capacity and its operation. Moreover, aspects like seasonal differences in the storage have to be considered as well as the development of operations over time, which are typical for reservoirs with multiple uses. A general recommendation on how to incorporate reservoirs in the model therefore cannot be given.

11.4.2 The Impact of the Gauging Network: Inn River Basin

The Rhine river basin considered in the previous section is an example for a complex basin that has multiple impact factors for flood generation, such as the Alps, the many tributaries and the large spatial extension. A basin with quite different challenges is the Inn basin. The Inn is a tributary of the Danube river. Fischer and Schumann (2021) demonstrated how important the Inn is for flood events in the Danube in southern Germany. Moreover, the Inn was topic of many flood studies, especially when analysing extreme floods in this region (Bloeschl et al. 2013; Schroeter et al. 2015). It has an interesting river network structure, as the main river, the Inn, has its origin in the Alps (Fig. 11.7). Rather close to the confluence of the Inn with the main river Danube, there is the confluence of the Inn with its main tributary, the Salzach. The Salzach river is known to have a fast-reacting catchment which is potentially relevant for the generation of extreme floods. Besides these two main tributaries, there is also a comparably small tributary, the Alz river, where flood events are considerably reduced due to the impact of the lake Chiemsee. In this example, the Passau-Ingling gauge is chosen as gauge after confluence, which is located closest to the mouth of the river Inn into the Danube river and hence represents the whole river basin, while the two corresponding tributary gauges are Burghausen (Salzach) and Eschelbach (Inn) (Table 11.2).

Following this structure and under the consideration of the available measuring gauges, the following three volume-based flood types can be defined: Inn, Salzach and Mixtype. These volume-based type differ significantly in their role in flood generation (Fig. 11.8). The Inn type is only rarely responsible for extreme floods, neither in the peak nor the volume. Types Salzach and the Mixtype, instead, are both equally responsible for the largest observed flood peaks and volumes.

The spatial multisite stochastic event generator is applied to the Inn basin using the structure given in Fig. 11.9. While the confluence is modelled by the three gauges Passau-Ingling, Burghausen and Eschelbach with a 3D-Vine-copula, the upper Inn is modelled recursively by three 2D-copulas of two consecutive gauges each: Eschelbach - Wasserburg, Wasserburg - Rosenheim o.d.M. and Rosenheim o.d.M. - Oberaudorf. Since for the Alz no representative gauge is available, this tributary is

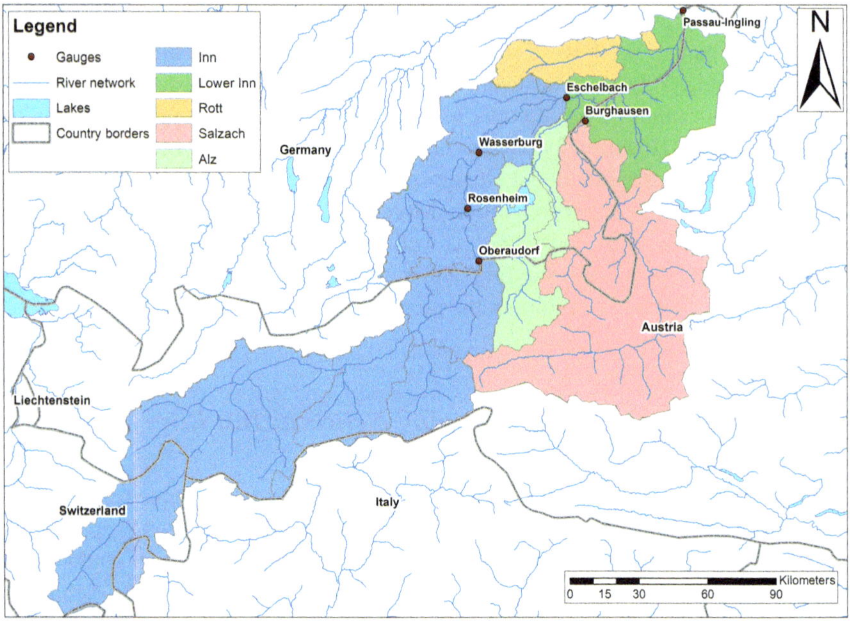

Fig. 11.7 The Inn river basin with selected gauges that fulfil the criteria for the application of flood types

Table 11.2 Catchment characteristics of the catchments considered for the Inn river basin

Catchment	River	Area (km^2)	Mean elevation (m a.s.l.)	River kilometre
Passau-Ingling	Inn	26040	1259	3.1
Burghausen	Salzach	6655.1	1268	–
Eschelbach	Inn	13320.5	1551	87.76
Wasserburg	Inn	11960.4	1673	158.67
Rosenheim o.d.M.	Inn	10153.5	1844	185.5
Oberaudorf	Inn	9713.2	1902	211

considered implicitly in the modelling of the gauge after confluence, Passau-Ingling. The respective copulas that are selected as best-fitting are listed in Table 11.3. As marginal distribution, the GEV distribution is selected as best-fitting for all volume-based flood types. The dependence between the gauges is not as high as it was for the Rhine basin. One reason could be that the conditions of the flood generation and here especially the orographic impacts determine a greater variability of spatial dependencies of floods between sub-basins. With regard to this complexity, the density of the hydrometric observation network seems to be insufficient. Between the confluence of Salzach and Inn and the gauge after confluence, Passau-Ingling, not only a large

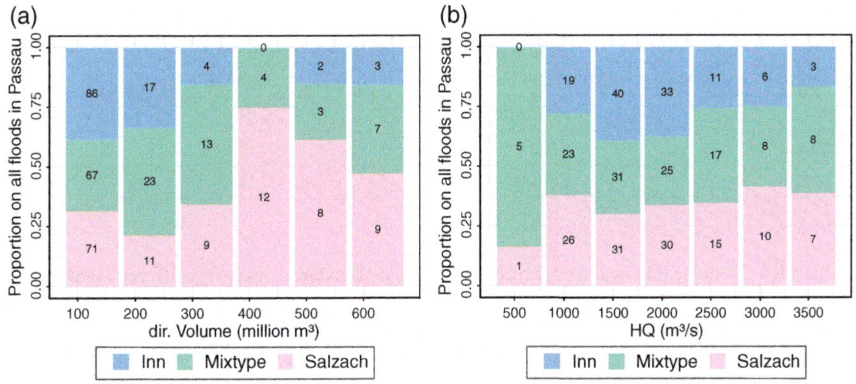

Fig. 11.8 Proportion of each volume-based type on the different ranges of observed **a** flood peak and **b** flood volume for the Inn river basin. Numbers represent the number of observed events in the respective range

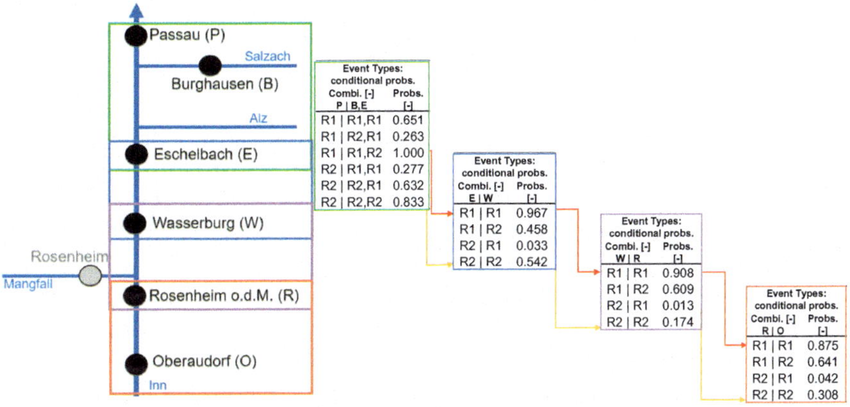

Fig. 11.9 Spatial structure of the event generator for the Inn river basin and conditional probabilities for the flood type combinations. Arrows demonstrate the path of the most probable combinations of flood types upstream for an R1 flood event (red) or an R2 flood event (yellow) in Passau

intermediate catchment is located but also the confluence with the Rott tributary. This means that runoff from this significant intermediate catchment cannot be considered explicitly in the dependence model. This was investigated further by having a look at the TNRMSE values for the stochastic event generator (Table 11.4). While the general impression of the error distribution is confirmed (smaller errors for downstream gauges and rainfall-driven floods), there is one very striking aspect of the results: The error for heavy-rainfall floods for the Passau-Ingling gauge after confluence is significantly higher than the errors of the upstream gauges. With 24.6% it is more than twice as high as the error at the two tributary gauges, Eschelbach and Burghausen. Moreover, this phenomenon only occurs for the flood type R1. This supports the

Table 11.3 Estimated copulas and Kendall's τ for the volumes of the Burghausen and Eschelbach tributary gauges and the Passau-Ingling gauge after confluence differentiated by volume-based type

Volume-based type	Copula	Kendall's tau
Inn	t-Copula	0.33
Salzach	Joe	0.28
Mixtype	Gumbel	0.50

Table 11.4 TNRMSE between observed and simulated flood peak quantiles differentiated by flood type for all catchments considered in the Inn river basin. TNRMSE cannot be calculated for flood type R3 and S2 for all gauges due to insufficient data

Gauge	R1 (%)	R2 (%)	R3 (%)	S1 (%)	S2
Passau	24.6	8.9	10	18.7	–
Eschelbach	11.8	8.9	–	30.5	–
Burghausen	10.5	8.4	8.4	8.4	–
Wasserburg	17.9	18.1	–	36.3	–
Rosenheim	16.4	19.2	–	50	–

assumption that an important part of the R1 events, resulting from uphill rainfall in this alpine region, is ignored when considering only the two tributary gauges Eschelbach and Burghausen for the flood generation in Passau. However, this is mostly limited to the heavy-rainfall floods, which are spatially less extended and hence may only affect the intermediate catchment where no observations are available. This example demonstrates that the structure of the gauging network in relationship to the heterogeneity of the flood generation has a crucial impact on the performance of the stochastic event generator. Significant gaps in the observation networks increase the uncertainty of the event generator for specific flood types (here the R1-floods). Yet, such an analysis can also provide valuable information, as it can be used to specify the deficits of hydrometry in river basins.

For large catchments and complex river networks, the classical consideration of flood-generating processes by using flood types has to be extended by a spatial consideration of floods including tributaries and flood superposition. By combining flood types and volume-based types in a multivariate copula model, consistent flood events can be estimated for a whole basin at the same time. The dendritic structure of the river network thereby is taken explicitly into account. Such a stochastic event generator forms the basis for the definition of spatial design flood scenarios. Challenges occur when dams have an impact on the spatial dependency structure or if the gauging network has significant gaps.

References

Blöschl G, Nester T, Komma J, Parajka J, Perdigão RAP (2013) The June 2013 flood in the Upper Danube Basin, and comparisons with the 2002, 1954 and 1899 floods. Hydrol Earth Syst Sci 17(12):5197–5212

Chen L, Singh VP, Shenglian G, Hao Z, Li T (2012) Flood coincidence risk analysis using multivariate copula functions. J Hydrol Eng 17:742–755

Fischer S, Bühler P, Schumann A (2021) Impact of flood types on superposition of flood waves and flood statistics downstream. J Hydrol Eng 26(7)

Fischer S, Schumann A (2021) Multivariate flood frequency analysis in large river basins considering tributary impacts and flood types. Water Resourc Res 57(8)

Fischer S, Schumann A, Bühler P (2019) Timescale-based flood typing to estimate temporal changes in flood frequencies. Hydrol Sci J 64:457–470

Guse B, Merz B, Wietzke L, Ullrich S, Viglione A, Vorogushyn S (2020) The role of flood wave superposition in the severity of large floods. Hydrol Earth Syst Sci 24:1633–1648

Huang K, Chen L, Zhou J, Zhang J, Singh VP (2018) Flood hydrograph coincidence analysis for mainstream and its tributaries. J Hydrol 565:341–353

Mei Y, Anagnostou EN (2015) A hydrograph separation method based on information from rainfall and runoff records. J Hydrol 523:636–649

Schröter K, Kunz M, Elmer F, Mühr B, Merz B (2015) What made the June 2013 flood in Germany an exceptional event? A hydrometeorological evaluation. Hydrol Earth Syst Sci 19(1):309–327

Chapter 12
Estimation of Multivariate Flood Probability Scenarios

Abstract It was demonstrated in the previous chapters of this part how important the consideration of spatial dependency and of the different flood-generating mechanisms in the catchments is for the generation of consistent flood events in large river basins. This idea is further extended in this chapter, where a methodology to obtain type-specific flood scenarios based on the multisite spatial model in Chap. 11 is proposed. Such scenarios are most relevant for many practical applications for water management. The assignment of probabilities to each scenario allows for an evaluation of the most critical scenarios. The proposed model is the natural extension of the type-specific hydrograph scenarios given in Chap. 9 but in this case the different combinations of flood types of the upstream catchments are considered to allow for a spatially consistent scenario evaluation of the flood events.

The high practical relevance of design floods was already pointed out in Chap. 9. To obtain such design floods, usually univariate flood frequency analyses is applied. Such approaches may be sufficient when considering catchments of the mesoscale with limited complexity of the river network. However, the larger the catchment, the more complex are the flood-generating processes and the impact of tributaries and flood wave superposition. Hence, simple univariate flood frequency analyses are insufficient for the planning of flood protection measures for large catchments. For a design flood downstream, the flood-generating processes and thus the flood types (Sect. 6.2) as well as the return periods upstream have to be specified. Again, the volume-based types (Chap. 10) can be used to define the spatial dependency. However, such an approach only delivers design flood peaks. It was emphasized in Chap. 9 that design floods should include more than a flood peak but rather a whole design flood hydrograph. In the case of a river basin, such a DFH scenario does not only consist of the hydrograph at the downstream gauge, but also hydrographs associated with the respective return period for the upstream gauges are required. The many options for possible combinations of flood types and volume-based types upstream make such multisite DFHs more complex. The multisite stochastic event generator introduced in Chap. 11 can be used to estimate such complex but consistent scenarios.

S. Fischer and A. H. Schumann, *Type-Based Flood Statistics*, Water Science and Technology Library 124, https://doi.org/10.1007/978-3-031-32711-7_12

12.1 Design Flood Peak Estimation Under Consideration of Flood Types and Tributaries

Design flood peaks are flood peaks that occur with a given exceedance probability respectively are characterised by a given return period. As introduced already for the univariate case, for this purpose, statistical distributions are fit and design flood peaks are estimated via quantiles. In the multisite case, i.e., for large catchments, besides the univariate distribution of the flood peaks at the gauge under consideration, also the upstream gauges have to be considered. This leads to conditional probabilities, where the downstream design flood peak for a given return period is estimated given the upstream flood peaks. Additionally, the considered information can be extended by incorporating flood types.

For the estimation of return periods conditional on the flood type and the tributary, the dependence structure between the tributary, the upstream main river and the confluence gauge has to be estimated. The aim is to obtain a return period respectively a design flood peak for a flood of a given flood type at the gauge after confluence given the return periods and the flood types at the tributaries and the main river upstream. This allows a more flexible estimation of return periods that takes into account the flood type and the confluence of floods. Since a large amount of data is required for the estimation of this multisite model, the statistical simulation proposed in Chap. 11 is used to generate a sufficiently large data sample. Such a multisite approach is, e.g., proposed by Fischer and Schumann (2021).

Consider once again a gauge after a confluence, denoted with C, with an arbitrary number of n_T tributary gauges. Then, for each combination of flood types, a Vine-Copula approach is applied (Sect. 3.3). Here, a $(n_T + 1)$D-Vine is used. In the following, the methodology will be described for the special case $n_T = 2$, since for this case still easy-to-handle analytical formula exist and a triple of gauges is the most common case. Theoretically, the principle is applicable for arbitrary n_T. To estimate the probability of a flood event at the confluence gauge, Eq. 3.6 is applied with W being the flood peaks of the confluence gauge, U being the flood peaks at the main river gauge upstream of the confluence and V being the flood peaks at the tributary gauge, transformed to $[0, 1]$ by their margins. Marginal distributions are selected by the AIC/BIC criteria. The copulas and the respective parameters are estimated with a Maximum-Likelihood approach, where the best fitting model is selected by the AIC. As input for the conditional probability in Eq. 3.6, quantiles for given return periods and flood types are estimated for each tributary and the main river. This is done with the type-based statistics given in Chap. 5. The respective return periods for the confluence gauge are then derived by optimisation of the probability given in Eq. 3.6 and the application of the annual distribution of each flood type. This way, any combinations of flood types and return periods of the tributaries can be considered, e.g., an R1-flood with return period 25 years in the tributary and an R3-flood with return period 90 years in the main river. It is of interest, which of these combinations lead to the design flood peak with given return period at the gauge after confluence and how plausible this combination is. It is well-known that a downstream flood of a given return period can be caused under unfavourable conditions by a confluence of flood events with much smaller return periods. Indeed, there exist many possible combinations of flood events that might lead to the same return period downstream. Not all combinations might be meaningful from a hydrological perspective, since the hydrological processes in the tributary and main river are connected. For example, it is unlikely for a flood event with return period 200 years to occur in the tributary while in the main river only a flood event with 2-year return periods occurs. In this case, the flood-generating processes would be spatially extremely heterogeneous. This drawback of multivariate statistics is discussed in many hydrological applications, such as Volpi and Fiori (2012), Graeler et al. (2013) and was also mentioned in Chap. 3. To overcome this problem, critical ranges can be estimated, e.g., by using the empirical frequency or the density of the data (Chebana and Ouarda 2011; Volpi and Fiori 2012). Similarly, also not each of the flood type combinations has the same probability to result in the design flood peak of given return period.

When considering a higher number than a triple of gauges, the procedure of estimating design flood peaks has to be extended successively upstream. This means, once the most probable flood type is obtained for a scenario at a gauge, the most probable flood type for causing this flood has to be estimated for the upstream gauge using the conditional probability of the (bivariate) copula and so on. By estimating the empirical probability of each combination, the critical scenarios can be identified. For this, in the $\delta\%$-confidence-range of the estimated T-year return period flood at the gauge after confluence, the empirical probability of each combination is estimated under consideration of the flood event type. For these combinations, the probability of each volume-based flood type is then estimated. By application of the TMPS-model (Chap. 5), for the obtained return periods, design flood peaks are estimated.

The relevance of combinations might change with changing return periods. Therefore, it is of interest to consider several return periods for the design flood of the gauge after confluence. This way, e.g., it can be investigated which combinations are relevant for the right tail and thus for extreme flood events. Of course, these ranges are also those with highest uncertainty due to the data scarcity.

Once again, the large Rhine river basin is used as an example. This example comes from Fischer and Schumann (2021). The empirical frequencies of the flood type combinations for tributary, upstream and downstream gauges at the Rhine that lead to a flood event at the Andernach gauge after confluence under consideration of the flood types at the Kaub (main river, upstream) and Cochem (tributary) gauges reveal that the most frequent combinations are (R1|R1R1), (R2|R2R1), (R2|R2R2), (S2|S2S2) and (S2|R1S2). (R2|R2R1) in this case refers to an R2-flood in Andernach given an R2-flood in Cochem and an R1 flood in Kaub and so on. Obviously, for this basin, rainfall-driven as well as snowmelt-induced flood events may lead to flood events downstream. This is not surprising since the Alps in the head part of the basin have a considerable impact, especially in spring when snowmelt begins. On the other hand, the flatter parts of the basin may often be affected by rainfall. If the observed flood events in Andernach are compared regarding their flood type combinations, it becomes visible that the largest (direct) peaks have been generated by either the combination (R1|R1R1), a combination of two heavy rainfall floods which lead,

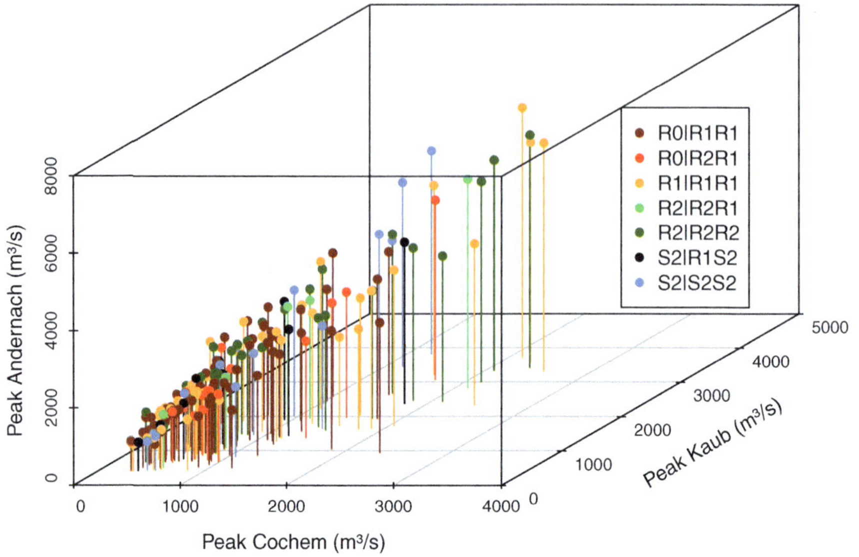

Fig. 12.1 Most frequent combinations of event types of the tributary and the main river gauge for the peaks at the Andernach gauge after confluence. Vertical lines ensure readability in the 3D-plot

due to superposition of peaks, to a type R1 flood at the gauge after confluence, or (R2|R2R2) (Fig. 12.1). However, also the combination of two snowmelt-floods led to large peaks, even though the peak of the tributary was comparably small.

This interdependency of flood peaks in the basin can be considered in the estimation of the design floods. When specifying a design flood with a return period of 100 years, as this is the most-frequently used return period for practical purposes, the required return periods of the upstream and tributary flood peaks vary much with the flood type. The most critical combination in this case is the combination of two heavy-rainfall floods (R1) in tributary and main river. Such flood events are associated with large peaks but comparably small volumes and a synchronous occurrence thus increases the flood peak downstream. Indeed, under these assumptions it would be sufficient to have a 55-year flood in the main river overlaid with a 25-year flood in the tributary to obtain a 100-year flood downstream (Fig. 12.2). Less critical combinations are obtained for snowmelt-floods (S2), where much larger return periods are required to obtain a 100-year flood downstream: for example, a 70-year flood at the main river and a 50-year flood at the Moselle.

In Table 12.1 from Fischer and Schumann (2021), all scenarios are summarised and probabilities for each combination are given. Similar to the flood types, also the volume-based flood types favour certain combinations. Hence, certain volume-based flood types have a higher probability to cause a 100-year flood than others. These types describe the spatial dependence and hence some spatial scenarios are more probable than others to result in extreme floods. To obtain similar results for the upstream gauges Worms and Maxau, the flood type in Kaub is taken and successively

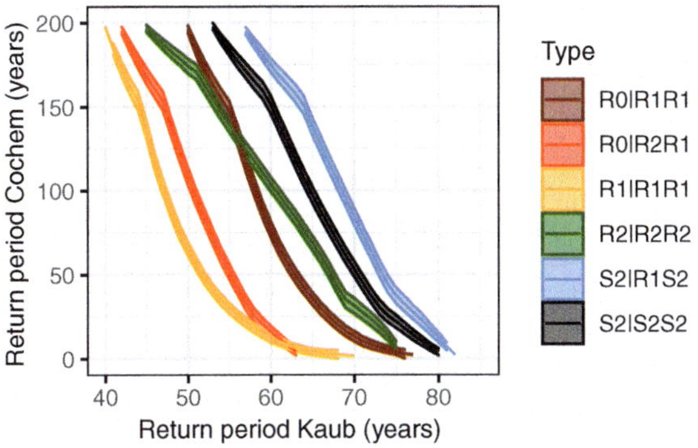

Fig. 12.2 Mean combination and 95% confidence intervals of return periods of peaks in the Moselle tributary (Cochem gauge) and upper Rhine (Kaub gauge) that can cause a peak with return period of 100 years at the Andernach gauge after confluence for different combinations of event types. Confidence bands are derived from optimisation. (modified from Fischer and Schumann 2021)

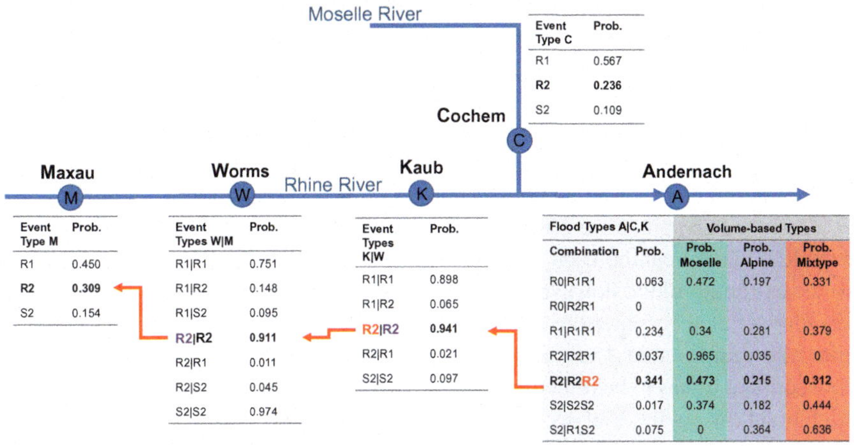

Fig. 12.3 Scheme on how to estimate the probability of each combination successively for upstream gauges in the multisite generator (modified from Fischer and Schumann 2021)

the most probable combination for this type given a flood in Worms and then the combination for the flood type in Worms given Maxau is estimated using the bivariate copula between two successive gauges upstream. This is illustrated in Fig. 12.3.

The most probable combination for a 100-year flood event in Andernach is the synchronous occurrence of two R2 type floods in the Moselle tributary and the main river Rhine that results in a flood of type R2 at the gauge after confluence. The flood type downstream does not change in this case, implying that both flood events occur nearly at the same time at the confluence and overlap much. For this combination, nevertheless large annualities of the flood event in the Moselle are required in combination with moderate annualities in the main river. In contrast,

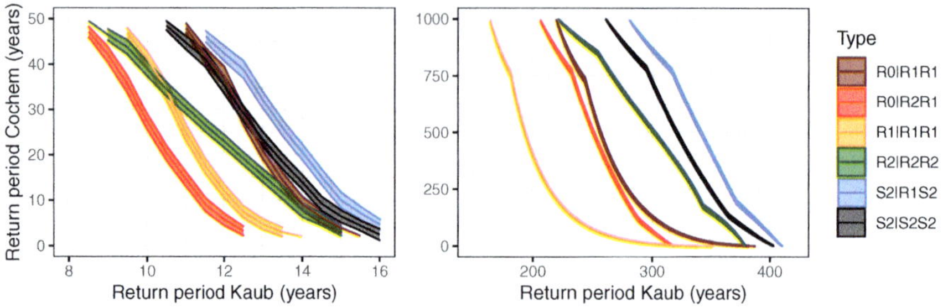

Fig. 12.4 Combination of return periods of peaks in the Moselle tributary (Cochem gauge) and upper Rhine (Kaub gauge) that can cause a peak with return period of 20 years (left) and 500 years (right) at the Andernach gauge after confluence for different combinations of event types (modified from Fischer and Schumann 2021)

much smaller annualities are required in the case of the synchronous occurrence of two heavy-rainfall floods (R1), the second most-probable combination to result in a 100-year flood in Andernach. Several other flood type combinations, instead, have a probability of zero to result in a 100-year flood. Regarding the spatial dependency, it becomes obvious that the extremity of the flood event alters within the basin. For some combinations, the upstream gauge has a larger return period than the downstream gauge, implying that the flood peak is diminished in this section. This mainly depends on the intermediate runoff and possible tributaries in these sections.

Analogous results for a 20- and 500-year flood event in Andernach are given in Fig. 12.4. These can be used to compare the relevance of combinations according to the extremity of the downstream event. The larger the downstream return period, the more critical becomes the combination (R1|R1R1). Instead, for small annualities like a 20-year flood in Andernach, the combinations (R0|R1R1) and (R0|R2R1) are most critical. It has to be remembered that such floods are associated with periods of high baseflow but themselves only have a small peak. Hence, they occur quite often in spring time when snowmelt in Alps starts or in summer time when the discharge of the Rhine river was fed by melting glaciers in the Alps. Hence, they are relevant for the small but frequent floods in the basin.

The multisite estimation of design flood peaks allows for a better understanding of the flood-generating processes and the spatial dependence in large catchments. Yet, it is limited to the flood peak only and does not provide information on the flood hydrograph. How to include such information will be discussed in the next section.

12.2 Multisite Scenarios for Design Floods

In Chap. 9, it was pointed out how the estimation of design flood hydrographs can improve the knowledge base for flood protection planning. Of course, this is not only true for the univariate case, i.e., the consideration of only one catchment at a time. Indeed, knowledge on the DFH may be even more important in large catchments, where flood events are complex and impacted by many factors such as flood superposition and combinations of different flood-generating processes. For such large catchments, not only the DFH of the downstream gauge is relevant, but also knowledge on the flood hydrographs of tributaries and upstream gauges might help to better prepare for a flood event. In this section, such DFH scenarios are introduced. Again, it is made use of the flood types and the volume-based flood types introduced before. These classifications are used to describe the flood-generating processes in the catchment as well as the spatial dependency and extension of the event. Each derived scenario is addressed with a probability, such that the most critical and probable scenarios could be selected for flood protection.

The general scenario remains the same: the gauge after confluence of tributary and main river is chosen as target gauge for which the scenarios for a given return period are derived. Combinations of flood types in the tributaries and main river are considered.

First, for each flood type at the gauge after confluence and for a given return period, type-specific return periods of floods in the tributaries and upstream in the main river are estimated using the conditional probability in Eq. 3.6. Second, the flood peaks corresponding to the return periods are estimated using the type-specific distributions introduced in Sect. 5.4. Third, the probability of each combination is estimated from the observed sample together with the probability of each volume-based flood type. This results in a table similar to Table 12.1. Fourth, the type-specific flood volume for each design flood peak is estimated by making use of the flood timescale, the relation between flood peak and volume. This results in pairs of flood peaks and volumes for each scenario. Finally, the corresponding flood hydrograph is estimated for each of the peak-volume pairs.

In this case, once again probability density functions could be used to obtain such hydrographs. However, the challenge when dealing with large catchments is to obtain consistent estimation of hydrographs. This is not straightforward when applying densities since many different shapes could result in an optimal fit to the flood peak and volume. Instead, Fischer and Schumann (2021) propose to rescale observed flood hydrographs of flood events. More precisely, for each flood type and volume-based flood type combination, flood events from the observed sample are chosen that were a result of the same combination. These flood events are then rescaled to match the flood peak and volume of the given return period. This way, an observed flood event, which is consistent in space in the catchment, is used as exemplary design flood hydrograph. Due to the catchment size it is sufficient to consider a daily resolution of discharge.

To illustrate the procedure, once again the example of the Rhine river basin from the previous sections is considered taken from Fischer and Schumann (2021). For example, for a 100-year flood in Andernach with flood peak 10600 m^3/s, different scenarios and the respective flood hydrographs at the different gauges can be compared. In Fig. 12.5, the two most probable combinations of flood types in tributary and main river that lead to a 100-year flood in Andernach are given. Since all three volume-based flood types have similar probability of occurrence, all three are

Table 12.1 Probability of occurrence of the most frequent combinations of flood types for the 100-year flood in Andernach and selected combinations of annualities

Flood type			Prob. of occurrence of combination	Probability of occurrence of volume-based type			Return period of flood peak		Worms		Maxau	
Andernach	Cochem	Kaub		"Moselle"	"Alpine"	"Mixtype"	Cochem	Kaub	Flood type	Return period peak	Flood type	Return period peak
R0	R1	R1	0.063	0.472	0.197	0.331	250	50	R1	77	R1	69
							85	60	R1	109	R2	9.3
							10	70	R1	115	S2	14
R0	R2	R1	0	–	–	–	280	40	R1	71	R1	66
							100	50	R1	77	R1	69
							20	60	R1	80	R1	71
R1	R1	R1	0.234	0.34	0.281	0.379	250	40	R1	27	R2	7.5
							60	50	R1	27	R2	7.5
							15	60	R1	27	R2	7.5
R2	R2	R1	0.037	0.965	0.035	0	70	50	R2	14	R2	9
							65	100	R1	130	R1	224
							50	150	R1	130	R1	247
R2	R2	R2	0.341	0.473	0.215	0.312	180	50	R2	49	R3	4.2
							100	60	R2	52	R2	5
							50	70	R2	240	S1	25
S2	S2	S2	0.017	0.374	0.182	0.444	300	50	S2	83	S2	10
							150	60	S2	141	S2	11
							75	70	S2	166	S2	11
S2	R1	S2	0.075	0	0.364	0.636	200	60	S2	98	S2	36
							90	70	S2	110	S2	39
							10	80	S2	120	S2	42

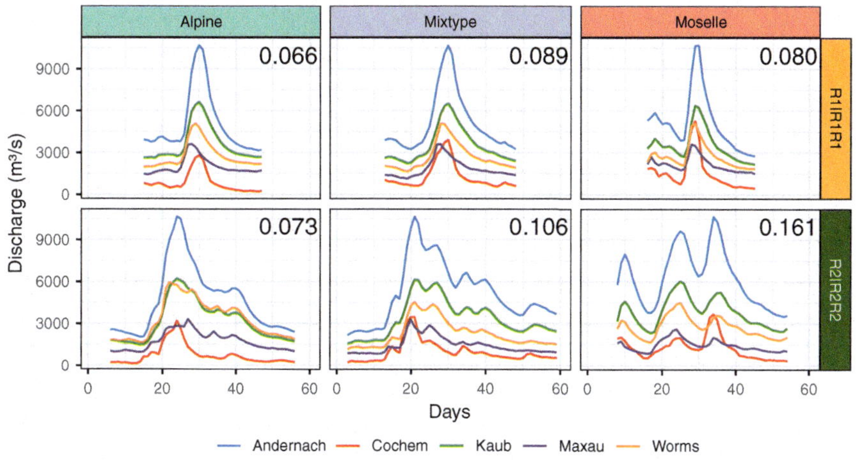

Fig. 12.5 Flood event scenarios for the most probable combinations that lead to a 100-year flood at the Andernach gauge after confluence, differentiated by the volume-based flood types. The numbers indicate the probability of the occurrence of each volume-based flood type for each combination. (modified from Fischer and Schumann 2021)

included in the scenario analysis. The flood hydrographs are obtained by selecting the flood events of given combination and volume-based flood type from all observed events that have smallest deviations from the estimated peaks and volumes for the 100-year flood given in Table 12.1 at all gauges and by rescaling the events to the estimated peaks and volumes.

Figure 12.5 emphasises the differences in superposition and between the volume-based flood types. It can be seen that R1 flood events are steeper than the R2 flood events and require an almost perfect superposition to generate such large flood events in Andernach. For the R2 flood events, the superposition is more shifted in time. Moreover, for the volume-based flood type "Moselle", it can be demonstrated that the Moselle flood event in Cochem exceeds those in the upper Rhine. For the "Alpine" events, large flood peaks can be observed for the upper Rhine only (Maxau and Worms gauges), indicating a flood generation in the alpine and Neckar area. For the "Mixtype", Moselle and upper Rhine contribute similarly to the flood event. This scenario analysis once again emphasizes the importance of differentiation between the flood types as well as the flood superposition.

Similarly, flood scenarios can also be derived for the Inn river basin introduced in the previous chapter (Fig. 11.9). It was pointed out before that large floods in this basin occurred mostly due to heavy rainfall events. Hence, it not surprising that more than 50% of the most frequent scenarios for the design flood with return period 100 years emerge from this flood type. As shown in Fig. 12.6, the most probable combinations are heavy-rainfall floods in all subcatchments with a focus on either the Inn river, where the Eschelbach and Wasserburg gauges at the Inn contribute with a comparably high flood peak, or a mixture between the main tributaries in the

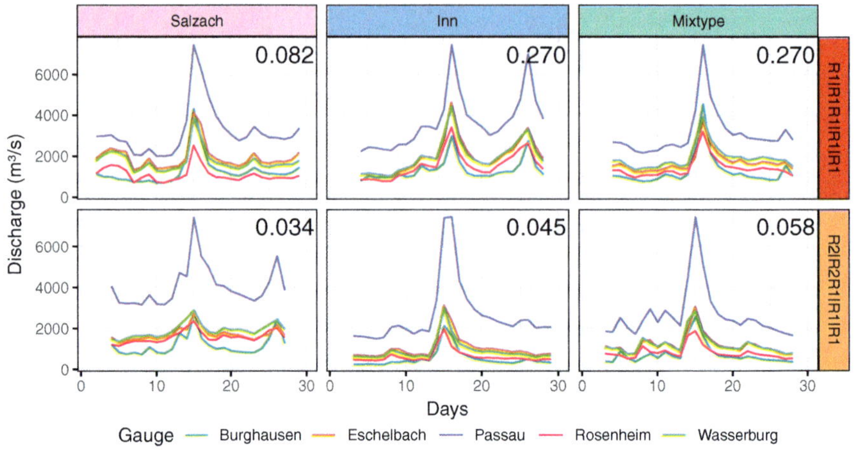

Fig. 12.6 Flood event scenarios for the most probable combinations that lead to a 100-year flood at the Passau-Ingling gauge after confluence, differentiated by the volume-based flood types. The numbers indicate the probability of the occurrence of each volume-based flood type for each combination

basin, where Burghausen (Salzach) and Eschelbach gauges contribute similarly to the downstream peak. The scenario for an R2 flood event downstream in Passau-Ingling also reveals an interesting pattern. In this case, it is most probable that heavy-rainfall floods occur in the Inn tributary, while in the Salzach tributary an R2 flood event occurs. Due to superposition, this then results in an R2 flood event downstream in Passau-Ingling. The probability for such a scenario is smaller than for an R1 flood in Passau-Ingling, but still should be considered in the planning of flood protection measures, as the higher flood volume compared to the R1 flood events might be challenging in this case.

The flood scenarios for a given return period, which were discussed here, extend classical design flood estimation. Besides the design flood peak also the flood volume and the shape of the design flood hydrograph are obtained. In this approach, design floods are not only obtained for a single gauge but consistently for a basin with several gauges. This allows a spatially consistent definition of design floods. The estimation of probabilities of each scenario helps to better understand which would be the most probable combination, while the remaining scenarios can be used for assessments of best- and worst-case scenarios. Flood protection systems can be planned according to these scenarios, i.e., upstream flood protection systems can be coordinated with downstream systems to reduce the flood as much as possible, taking into account the spatial focus point of the different flood scenarios.

Spatial flood scenarios help to better understand the flood-generating mechanisms in large river basins and the impacts of tributaries on flood events. They consider the probability of certain flood type combinations as well as the flood superposition for the resulting design flood event. The most probable flood hydrograph combinations for each scenario make the consistent planning of flood protection in large basins possible and offer the possibility to identify the most critical basin reactions, e.g., the most critical tributary. Moreover, best-and-worst-case scenarios can be defined.

References

Chebana F, Ouarda T (2011) Multivariate quantiles in hydrological frequency analysis. Environmetrics 22:63–78

Fischer S, Schumann A (2021) Multivariate flood frequency analysis in large river basins considering tributary impacts and flood types. Water Resourc Res 57(8)

Gräler B, van den Berg MJ, Vandenberghe S, Petroselli A, Grimaldi S, De Baets B, Verhoest NEC (2013) Multivariate return periods in hydrology: a critical and practical review focusing on synthetic design hydrograph estimation. Hydrol Earth Syst Sci 17(4):1281–1296

Volpi E, Fiori A (2012) Design event selection in bivariate hydrological frequency analysis. Hydrol Sci J 57:1506–1515

Part VI
Regionalisation of Flood-Type-Specific Distributions

Classical flood frequency analyses as introduced in the previous parts are based on a sample of observed floods. However, this requires a gauging network and in the best case long records of measurements. The installation and operation of such gauging networks is expensive and demanding, which is why usually only few gauges if any are installed in a river basin. Nevertheless, flood frequency analyses are required for ungauged parts of the basin, too. Regionalisation allows to transfer knowledge from observed catchments to ungauged catchments by taking into account hydrologically similar regions. This makes the application of flood frequency analyses possible. In this part, approaches to regionalise the statistical models of the previous parts are proposed.

The typically methodology of regionalisation of flood statistics is based on three main steps:

1. Catchments are compared to each other on the basis of selected attributes in order to identify hydrological similarities and to group similar catchments into homogeneous regions.
2. Based on the designated similar areas, a regionalisation approach is developed for each region using watershed attributes to determine the quantiles or statistical distribution functions.
3. For an unobserved catchment, the attributes are estimated and then used to determine to which of the previously formed groupings (homogeneous regions) this catchment can be assigned. The previously developed regionalisation procedure for this region is applied to estimate the flood statistics from the attributes of the ungauged catchment.

We used a different approach here. For each flood type, different characteristics relevant to flood generation were selected to define hydrologically similar units. The resulting units differ between flood types as well as between the parameters of their distribution functions. Here, a grouping of parameters into shape, scale and location parameters was used. The parameters sought for unobserved areas are then each given as the median values of the known parameters of all gauges within these groups.

The grouping, or more precisely the clustering of catchments, is the key element of this regionalisation. It is described in Chap. 13. The regionalisation of parameters based on this clustering is introduced in Chap. 14. Chapters 15 and 16 make use of the proposed regionalisation by extending the framework described in the previous parts to ungauged catchments. This allows for the generation of type-specific pairs of flood peak and volume and the corresponding flood hydrographs.

Chapter 13
Definition of Hydrologically Similar Regions

Abstract The definition of hydrologically similar regions is one of the key tasks in regionalisation. This chapter introduces an approach on how to include flood types in the definition of hydrologically similar regions. Flood types provide additional information on the flood-generating processes in a region and hence might enhance the understanding of the relevant catchment, precipitation and runoff attributes. For this purpose, the method of hierarchical clustering is proposed. Moreover, the optimal number of clusters has to be found. It is demonstrated on how type-specific characteristics can be considered as well as how the resulting groups differ between flood types.

13.1 Some Basic Remarks on Regionalisation

Flood frequency analyses, as any statistical analyses, are based on samples of observed events, e.g., flood peaks, volumes or whole flood waves. In hydrology, this requires a gauging network. For the most stream-gauging systems, discharge in streams is a derived result. The time-based water level, or stage, is measured continuously at each stream-gauging station. Rating curves, that are relationships between stages and discharges, are applied to transform these data into discharges. These curves are site-specific and temporally variable, as the cross-section is changing. Periodic discharge measurements are required to set up rating curves, to control them, and to update them. The discharge is measured indirectly by measuring flow velocity and cross-sectional area. Obviously, this is a costly process, but the quality of discharge data depends on these rating curves as well as on the registration of water levels. Yet, the installation and operation of such gauges requires financial means as well as qualified human resources. Hence, the number of gauges that are installed in a river basin usually is limited. Small catchments are often not equipped with gauges at all or gauges only provide short records of observation, as the gauging systems have only been recently installed or the measurements have been discontinued. Such short records usually do not provide sufficient information for flood frequency analyses. In many parts of the world, a systematic gauging network is very

S. Fischer and A. H. Schumann, *Type-Based Flood Statistics*, Water Science and Technology Library 124, https://doi.org/10.1007/978-3-031-32711-7_13

fragmentary. The missing information and insufficient data basis for many catchments makes regionalisation one of the main research topics in hydrology (Sivapalan et al. 2003). Regionalisation is defined as the transfer of information (parameters, variables) from gauged catchments to ungauged catchments (Bloeschl and Sivapalan 1995). For this purpose, the gauged catchments should be hydrologically similar to the ungauged catchment. Here, the aim is to transfer the results of flood frequency analyses to data sparse and ungauged regions. However, still many open questions exist and new challenges have to be faced when dealing with regionalisation. The biggest challenge lies in the definition of hydrologically similar regions, as these form the basis of the transfer of information in space. Incorrect assumptions and hence wrongly identified homogeneous regions have a large impact on the outcome of the regionalisation and might lead to serious errors in the result. Therefore, there exist various approaches which aim to identify hydrologically similar regions based on different assumptions. These approaches can be grouped into two main concepts of similarity: spatial proximity and hydrological similarity (Merz and Bloeschl 2005). Spatial proximity is based on the assumption that distance is correlated to similarity, i.e., catchments with small distance to each other have a more similar runoff regime than catchments that are located further away from each other. On the other hand, hydrological similarity is based on the assumption that similar catchment attributes lead to similar flood regimes. Approaches following this assumption try to group catchments according to statistical similarity or homogeneity. Examples are clustering approaches, principal component analysis or multiple regression, where the runoff (as flood quantile or other variable) is aimed to be explained by a combination of several catchment attributes (Nathan and McMahon 1990). All these approaches as well as both concepts of similarity have benefits and drawbacks, depending much on the region of consideration (climate, variability), the data availability but also the aim of regionalisation. Various regionalisation studies exist with quite different outcomes on the suitability of these approaches (Odry and Arnaud 2017).

Though regionalisation is always based on the definition of hydrologically similar regions and the transfer of information from observed catchments to an ungauged catchment within such a region, there are several ways to regionalise flood statistics: The type of distribution function can be specified and its parameters are regionalised or the flood quantiles themselves are regionalised. Another option is based on regional flood frequency curves, where the flood quantiles expressed by some statistical terms are proportional to the catchment area. The most famous approach in this context might be the index flood (Dalrymple 1960). It is based on the assumption that the statistical distribution of the floods within a homogenous region is similar except for a scaling parameter that reflects the specific local features.

What is usually not considered in conventional regionalisation are the flood-generating mechanisms. Few studies emphasised how flood-generating characteristics could improve regionalisation, as the catchment reactions and thus the homogeneous regions might vary between different types of floods (De Michele and Rosso 2002). First attempts to incorporate such information are for example the seasonality index (Burn 1997) which might implicitly characterise the flood-generating process of the largest flood in a year. To go one step further, once more flood types can be considered in regionalisation. Similar to the flood statistical approaches described in the previous parts, flood types can add information on the flood-generating processes in a catchment. Depending on the applied approach, different characteristics of the catchments will be relevant. Flood conditions are determined mostly by the precipitation regime, initial soil moisture, land use, and soil and slope conditions. If floods are differentiated into types according to the different meteorological drivers, the type-specific relationships between catchment characteristics and flood statistics should become more tight. When considering flood types as a proxy in the definition of homogeneous groups, more information on the catchments are gained and the definition of hydrologically similar catchments is improved. In this chapter, clustering approaches will be used to group catchments according to their catchment, precipitation and runoff attributes, but with distinction by flood types. The clustering algorithms that will be used are introduced in the following section. In the following chapters it is then demonstrated on how to use flood types in regionalisation.

13.2 Clustering Approaches

As already described in the previous section, one crucial step in regionalisation is the definition of hydrologically similar regions. There exist numerous approaches to do so. A mathematical, objective approach that does not require many assumptions is clustering. Clustering is defined as the identification of patterns in large data sets. Since it does not assume any pre-defined structure of the patterns, it is among the unsupervised approaches. Many different clustering approaches exist that differ in their way of defining the cluster as well as in the properties of the resulting clusters. Usually, all cluster algorithms are based on some kind of similarity measure, e.g., the distance between the data points. It is no easy task to find the best-performing clustering algorithm for the task at hand and no general recommendation is given here. However, two examples of how to define clusters are given in the following: hierarchical clustering and longitudinal clustering. Both approaches focus on different types of input data, the hierarchical clustering on simple input data consisting of one value of each variable, and the longitudinal clustering on curves as input data. These two clustering techniques will be used later on to define hydrologically similar regions.

13.2.1 Hierarchical Clustering

Hierarchical clustering refers to a clustering method that is build upon a tree-like representation of the clustered objects. Here, an agglomerative approach is presented, which merges the branches of the tree in each step according to highest similarity to form the clusters. As a linkage criterion, Ward's minimum variance criterion is applied. This criterion minimizes the total within-cluster variance by application of the squared Euclidean distance. More precisely, in each step, those clusters are merged to a new cluster that lead to smallest increase of the within-cluster variance. In the example in Fig. 13.1, in the first level of merging there are six clusters (Catchments A–D, Catchment E, Catchments F–G, Catchment H and Catchment I). In the next step, clusters are merged further to four clusters, then three and then two clusters. Note that the order of the catchments is chosen by the algorithm and that the input order does not have an impact on the outcome. For hierarchical clustering, it is mandatory that all attributes have values in [0, 1], which can be reached by simple transformation, e.g., min-max-scaling.

It remains open how the number of clusters to be selected can be determined optimally. By using again the cluster variance as a criterion, the optimal number of clusters can be defined. For this purpose, after each clustering step (level) the total sums of squared distances from the cluster centres (TSS) are calculated. The total sum of squares for a number of K clusters and n observations of p attributes is defined as

$$TSS = \sum_{k=1}^{K} \sum_{i \in S_k} \sum_{j=1}^{p} (x_{ij} - \bar{x}_{kj})^2, \tag{13.1}$$

where S_k is the set of all observations in cluster k, x_{ij} is the ith observation of the jth attribute and $\bar{x}_{kj}$ is the jth attribute of the cluster centre for the kth cluster. The choice of the optimal cluster number is again a trade-off between minimal variance and the smallest possible number of clusters. Especially for regionalisation, a high number of clusters with only few members is disadvantageous as the uncertainty of the transferred information is then also high. Fischer and Schumann (2021) suggest to plot the TSS against the number of clusters and use a t-test between the slope of two fitted regressions to determine the number of clusters which leads to a significant change in the reduction of the variance. This follows the idea of finding a change-point from which on the variance does no longer decrease significantly (Salvador and Chan 2004). In the given example, the optimal number of clusters was identified as four, consisting of the catchments A–D, two single-member cluster with catchment E respectively catchment I and one cluster with catchments F–H.

13.2.2 Clustering of Longitudinal Data

Often, attributes that are considered for regionalisation do not consist of a single value that can be used directly for regionalisation. Instead, they might consist of curves or functions of data. One example are hypsometric curves, which will be introduced in the next section. In such cases, conventional clustering approaches like the hierarchical clustering cannot applied. However, there exists clustering approaches that explicitly take into account shapes of curves for clustering. One example is the Clustering of Longitudinal Data (CLD) in combination with kMeans clustering. CLD focus on the shape of the curve, usually being applied to progress curves of

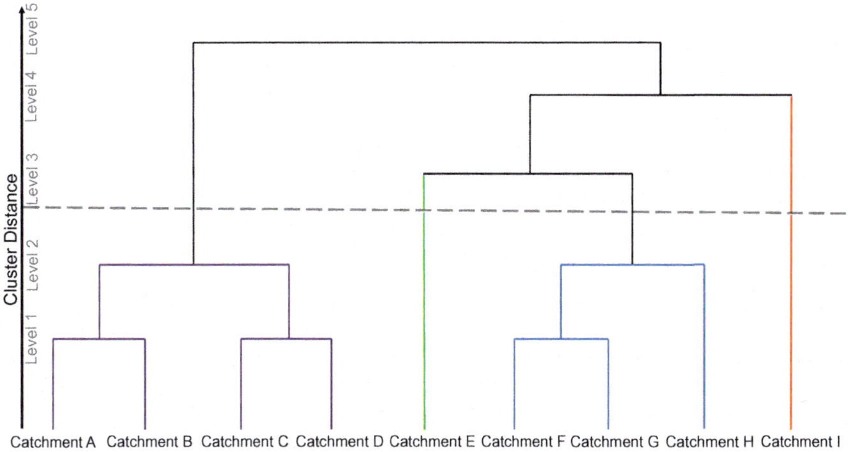

Fig. 13.1 Schematic overview of hierarchical clustering

measurements, instead of the measured value (Genolini et al. 2016). Fischer and Schumann (2021) state that

> CLD in combination with kMeans makes clustering of several samples with ordered indices possible by a modification of the classical kMeans with Euclidean distance, where classical clustering methods fail to classify these ordered samples according to their shape and not their values.

In this approach, first, the curves of attributes are separated into a pre-defined number of equidistant sections. This number depends on the resolution of the available data as well as (for hydrological regionalisation) the catchment size. Then, the curves are clustered according to their shapes. The partition obtained is evaluated regarding a goodness-of-fit criterion. One example is the Calinski-Harabasz criterion (Calinski and Harabasz 1974) (also Variance-Ratio-Criterion), which is defined as

$$CHC_k = \frac{SS_B}{SS_W}\frac{n-k}{k-1}, \tag{13.2}$$

where k is the number of clusters and n is the number of observations. SS_B is the overall between-cluster variance with

$$SS_B = \sum_{i=1}^{k} n_i \|m_i - m\|^2, \tag{13.3}$$

where n_i is the number of observations in cluster i, m_i is the cluster mean of cluster i and m is the overall mean of the data. $\|x\|$ denotes the Euclidean Norm. Similarly, SS_W is the within-cluster variance defined as

$$SS_W = \sum_{i=1}^{k} \sum_{x \in C_i} \|x - m_i\|^2, \tag{13.4}$$

where C_i denotes cluster i. This idea is similar to the Ward-criterion introduced for hierarchical clustering but the Calinski-Harabsz criterion is used as a performance criterion and not for merging. Since the algorithm might not always find the global but only the local optimum, it is recommended to repeat the clustering several times with different starting parameters (rerolling). The repetition should also take into account different numbers of clusters to evaluate which number is optimal.

Well-defined clusters have a large between-cluster variance (different clusters deviate much from each other) and a small within-cluster variance (values within a cluster do not deviate much). The optimal partition with the optimal number of clusters therefore has the highest Calinski-Harabasz criterion. In the example given in Fig. 13.2, this would mean that the curves are grouped into four clusters, as in the left panel the highest Calinski-Harabasz criterion is obtained for a number of four clusters in the first run (black dot). The four clusters clearly represent different types of curves. Those curves in cluster D for example have a concave shape, while those in cluster A have a constant slope.

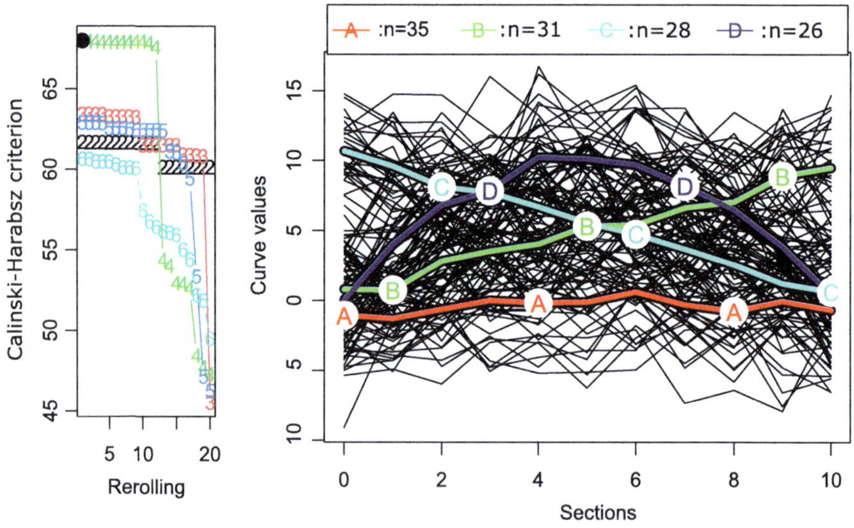

Fig. 13.2 Clustering of 120 curves using the CLD with kMeans clustering and the Calinski-Harabasz criterion for evaluation. The numbers in the left panel represent the number of clusters, while the y-axis gives the values of the Calinski-Harabasz criterion. The x-axis of the left panel describes the number of rerollings in the algorithm. Numbers on top of the right panel represent the numbers of curves assigned to each cluster, while the coloured curves are the mean curves of each cluster

13.3 Selection of Catchment, Precipitation and Runoff Attributes

When aiming to define hydrologically similar catchments, the first fundamental steps consist of making a decision on the similarity measures. Similarity can be defined in many ways and quite different characteristics can be used for the grouping. Distance-based measures are a simple way to define similarity (see Sect. 13.1), as no knowledge on the hydrological processes is required. At the same time, this is also the major drawback of these approaches. When ignoring the hydrological aspects of similarity, vastly different catchments might be grouped together just because they are located close together. One is also faced with the problem of how distances between catchments could be defined. The distances between gauges are an insufficient measure as well as the distances between two centres of gravity if the catchment shapes and areal sizes differ. In the worst case, grouped catchments might have totally different flood regimes and belong to different river basins. Thus, commonly, other measures of catchment similarity are considered in regionalisation. Two aspects have to be taken into account: Floods depend (1) on the spatial and temporal distribution of precipitation and (2) the reaction of the catchment on meteorological loads. Thus, characteristic values of the precipitation regime and catchment characteristics should be applied to define similarity measures. The use of runoff characteristics is not allowed here because the similarity measures must be determined in the application phase for ungauged catchments. As the proposed flood types (Chap. 6.2) were defined by consideration of different meteorological drivers, which initiate different runoff processes, the relationship between floods and catchment characteristics becomes more tight if the sample of observed floods is subdivided into types. More precisely, type-specific groups of hydrologically similar catchments are defined by making use of catchment attributes and precipitation attributes which differ between flood types.

13.3.1 Precipitation and Runoff Attributes

Precipitation and runoff characteristics have a direct impact on the flood regime in a catchment. In addition to the characteristics of flood-triggering rainfall events, the seasonal distribution of precipitation (rain or snow) and the total amount of precipitation in their relationships to soil retention capacity, evapotranspiration, and infiltration (i.e., to the soil moisture conditions of the watershed) are relevant here. An option to consider the climatic wetness state of watersheds in its seasonal variability are low flow characteristics. Of course, at was shown in the previous parts of this book, such characteristics depend much on the flood type. Fischer and Schumann (2021) propose to use type-specific versions of precipitation attributes. For example, they use the α-quantile of the D-day precipitation sum $Prec_D$ which should represent the flood-inducing sum of rainfall. To obtain this variable, the median of the type-specific duration D of precipitation is taken and the maximum precipitation sum for this duration of D days derived from all analysed flood events of this type is calculated. To avoid a bias caused by very long rain events, D was defined as the median number of days that led to 80% of the total amount of rainfall for each flood event. For example, for heavy-rainfall floods of type R1, D might only be one or two days and for flood type R2 around five days while for long-duration rainfall floods of type R3, D might be equal to ten or more days. Further possible precipitation attributes are the type-specific characteristics of rainfall intensities. To characterise the retention capacities of watersheds in the underground and the variability of the monthly wetness conditions, the weighted mean of the averages of the monthly minimums of daily discharges can be used (hereafter referred to as retention capacity). To estimate a single type-specific value, a weighted mean of these twelve long-term averages is estimated. The monthly weights are defined by the mean frequency of occurrence of the flood type in the respective month. The weights account for the seasonally different occurrence of the flood types. For example, flood type R1 might be the most dominant flood type in summer while S1 might dominate in winter.

All runoff attributes are considered in a scale-invariant way, i.e., they do not depend on the catchment size to make them comparable between the catchments. The weighted mean of averages of the monthly minimums of daily discharges is divided by the catchment area to make it scale independent. As mentioned before, a major drawback in the use of runoff attributes lies in the fact that they usually can only be derived for gauged catchments. For example, the estimation of meaningful water retention capacities requires the combination of observed runoff data with rain data for flood events. This is of course contradictory to the aim of regionalisation, where such attributes have to be available for ungauged catchments. Nevertheless, these attributes can also be derived for ungauged catchments from observed data at gauged neighbouring catchments (Fischer and Schumann 2021). As these only serve as input values for the clustering of homogeneous groups, regionalisation of such attributes can be simplified by basic assumptions. For rainfall characteristics, (Fischer and Schumann 2021) propose distance-based approaches, i.e., to use a simple k-nearest neighbour approach (with $k = 10$), though in its weighted version and type-specifically. More precisely,

$$Prec_D(y) = \frac{\sum_{i=1}^{k} \omega_i Prec_D(x_i)}{\sum_{i=1}^{k} \omega_i}, \tag{13.5}$$

where $\omega_i = \frac{1}{|AE(x_i) - AE(y)|}$ are the weights, y is the ungauged catchment, x_i, $i = 1, \ldots, 10$, are the nearest neighbour catchments, AE is the catchment size and $Prec_D$ is the α-quantile precipitation for D days (D type-specific and fixed). Fischer and Schumann (2021) demonstrated that this regionalisation resulted in quite a small bias of 10–20%, being only slightly higher for the snowmelt-induced floods. Similarly, also rainfall intensity and the retention capacity for each flood type are regionalised. This way, a set of regionalised type-specific attributes is derived for ungauged catchments from gauged neighbouring catchments.

13.3.2 Catchment Attributes

Catchment attributes are much easier to obtain than precipitation and runoff attributes, as they are usually also available for ungauged catchments, e.g., by GIS data. Various catchment attributes exist and may be used for regionalisation, the choice depending much on the study region. However, also the aim of the regionalisation might play a role when selecting catchment attributes for defining hydrologically similar catchments. For example, when aiming to regionalise the shape parameter of the distribution of floods, the catchment reaction is important and one has to consider attributes that can describe the potential of the catchment to generate extreme floods. Other parameters such as the location parameter might require more basic variables like the catchment size and elevation. Generally speaking, the choice should be done in a type-specific way.

Common choices of catchment attributes are catchment area or mean elevation to describe the main characteristics of the catchment. Additional attributes that describe the catchment reaction can be considered, too. Examples are the river network density, which indicates the permeability of the soil together with the slope conditions and the lateral flows at the surface, the soil type or the land use, which serve as proxy for catchment reaction. When dealing with flood types, especially one catchment characteristics should be used to detect coherences between catchments: the hypsometric curve (Fischer and Schumann 2020). It can be used to consider the temporal sequences of the ablation of the snow cover in a catchment, as it gives notice on the different elevation proportions and possible plateaus, where snowmelt occurs at the same time according to the air temperature, which depends mainly on elevation. For example, the exemplary hypsometric curve in Fig. 13.3 reveals that the corresponding catchment has a few parts with very high elevation, while the remaining parts slightly decline in elevation with a small plateau in the lowlands.

However, hypsometric curves do not provide one single value that can be used for defining hydrologically similar groups of catchments, which makes them not directly useable for regionalisation purposes. However, if they are abstracted, they can be considered nevertheless. For this purpose, only the shape of the hypsometric curve is considered, i.e., no information on elevation or catchment size (these can be considered separately in the regionalisation). The hypsomteric curves in a region are then clustered into different groups, each group representing a set of curves of similar shape. Fischer and Schumann (2021) propose to use Clustering of Longitudinal Data (CLD) in combination with kMeans clustering (Sect. 13.2.2). For this purpose, the hypsometric curves are separated into a pre-defined number of equidistant sections. For mesoscale catchments and 1x1 km resolution, 50 sections are sufficient. Then, the curves are clustered according to their shapes (see Fig. 13.4). The five resulting clusters clearly represent different shapes of hypsometric curves. Those curves in cluster D for example have steep declines in the upper parts of the catchments, while those in cluster A have a smooth shape of elevation. The membership of a catchment to any of these groups then can be used for regionalisation.

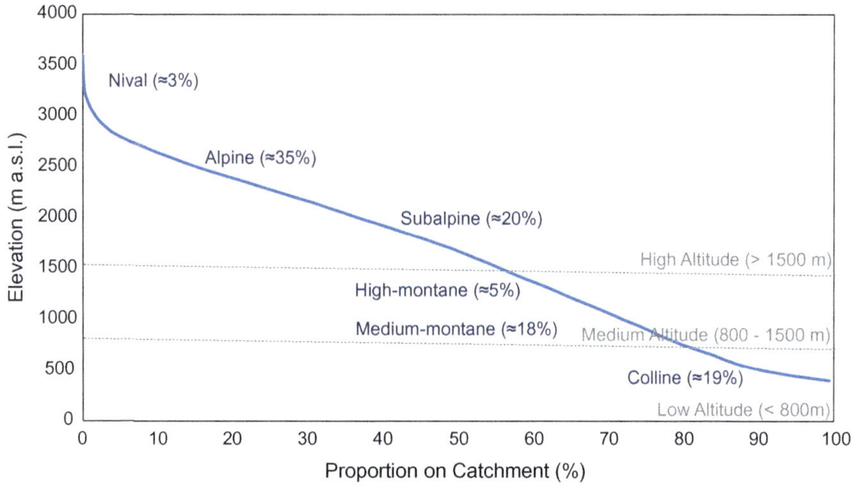

Fig. 13.3 Hypsometric curve and elevation levels for the Wasserburg/Inn gauge in Bavaria, Germany

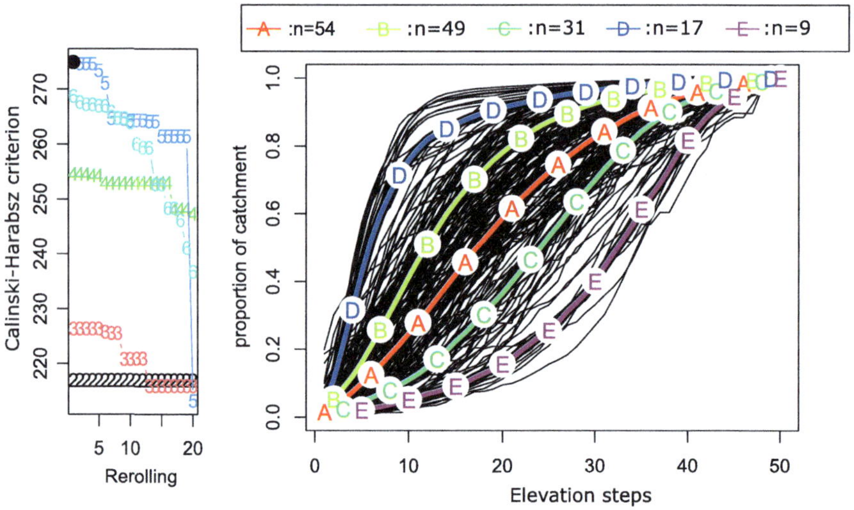

Fig. 13.4 Clustering of the hypsometric curves of 164 catchments in Bavaria, Germany, using the CLD with kMeans clustering and the Calinski-Harabasz criterion for evaluation. The numbers in the left panel represent the number of clusters, while the y-axis gives the values of the Calinski-Harabasz criterion. The x-axis of the left panel describes the number of rerollings in the algorithm. Numbers on top of the right panel represent the numbers of curves assigned to each cluster. (modified from: Fischer and Schumann 2021)

13.4 Definition of Homogeneous Clusters

With a variety of available runoff and catchment attributes at hand, groups of hydrologically similar catchments can now be formed. However, usually, not all attributes can be used for this approach at the same time and the addition of further attributes may not provide any new information. If attributes are strongly correlated with each other, a consideration of all of these attributes does not add much information to the definition of the homogeneous groups but instead might lead to an unstable estimation, similar to multicollinearity in regression. For example, catchment elevation might be correlated with catchment size, as small head catchments usually are located higher up in the mountains of a basin. Similarly, the elevation is correlated with soil, land use and vegetation, as in higher elevation the soils will change to be less fertile, climatic conditions are becoming increasingly harsh and hence less agriculture will take place. Therefore, it is recommended to always perform a pre-selection of attributes before defining groups of hydrologically similar catchments based on these attributes. There exist several possibilities for such a pre-selection, e.g., the variance inflation factor or the consideration of the eigenvectors. The most simple approach to assess correlation between attributes is the cross-correlation matrix, where each attribute is compared to each other attribute regarding correlation. What has to be taken into account is the type of each attribute. As mentioned above, some might be continuous and some nominal. Hence, one cannot consider one correlation measure for all. Fischer and Schumann (2021) recommend the use of three different measures:

- Spearman's correlation ρ_S introduced in Chap. 3 for the comparison of two continuous variables
- Cramer's V for the comparison of two nominal variables with values in $[0, 1]$, 0 indicating no correlation of the variables, 1 indicating perfect correlation (Cramer 1946)
- the Intraclass Correlation Coefficient (ICC) Class 3 for the comparison of a continuous and a nominal variable with values in $[0, 1]$, 0 indicating non-similarity of the variables, 1 indicating high similarity (Shrout and Fleiss 1979).

Cramer's V is a transformation of the $\mathcal{X}^2$ test statistic:

$$V = \sqrt{\frac{\mathcal{X}^2}{n \cdot \min(m, k)}}, \tag{13.6}$$

where n is the sample size and m and k are the number of columns and rows, respectively. In terms of attributes, m and k would be the number of categories each of the two attributes that are compared have. The ICC was originally developed for the agreement of ratings. It compares the variability of ratings of the same raters to the total variation among all raters and ratings (Shrout and Fleiss 1979).

All these correlation measures then can be visualised in a correlation matrix. Using a threshold of correlation or some other decision criterion, attributes are selected for defining hydrologically similar regions that are not correlated with each another.

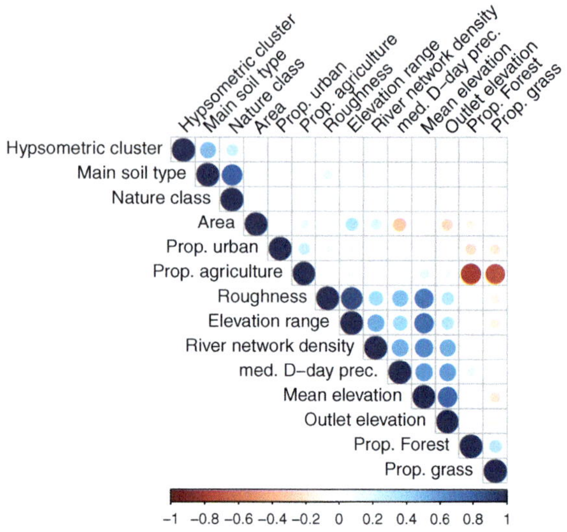

Fig. 13.5 Cross-Correlation matrix of several catchment attributes

For example, in Fig. 13.5, it becomes visible that the different proportions of land use classes are interrelated, i.e., large proportion of forest or grass automatically leads to small proportion of agriculture. These three variables hence should not be considered simultaneously. Similarly, roughness and elevation range, mean elevation, river network density and mean D-day precipitation (at least for flood types R2 and R3) of a catchment are all inter-correlated. However, which of these variables should be used for defining homogeneous groups of catchments depends on the study region and the purpose of regionalisation. To take advantage of the flood types, the selected attributes vary between the flood types.

After the selection of appropriate, non-inter-correlated attributes, groups of catchments that are hydrologically similar can be defined for each flood type. For this purpose, clustering approaches are useful, as they require few assumptions on the underlying relations between the attributes and hence might reveal hidden structures in the data that might not be obviously visible. Fischer and Schumann (2021) propose the use of hierarchical clustering (Sect. 13.2) as this approach delivers interpretable clusters of catchments.

As an example, catchment, precipitation and runoff attributes for 164 catchments in Bavaria, Germany, are considered (Sect. A.1). More precisely, the nature class (a definition of geographical regions according to flora, fauna, soil and other topographic characteristics provided by the Bavarian state office for the environment), the hypsometric curve (in terms of a cluster membership derived by CLD), the range of elevation, the D-day precipitation sum, maximum daily precipitation intensity and retention capacity are used as input data. The last three attributes are type-specific and hence differ between the flood types. All input data are normalised to the range between 0 and 1 by using a min-max-scaling. Then, hierarchical clustering is applied (Fig. 13.6). The resulting clustering tree delivers possible groupings of the

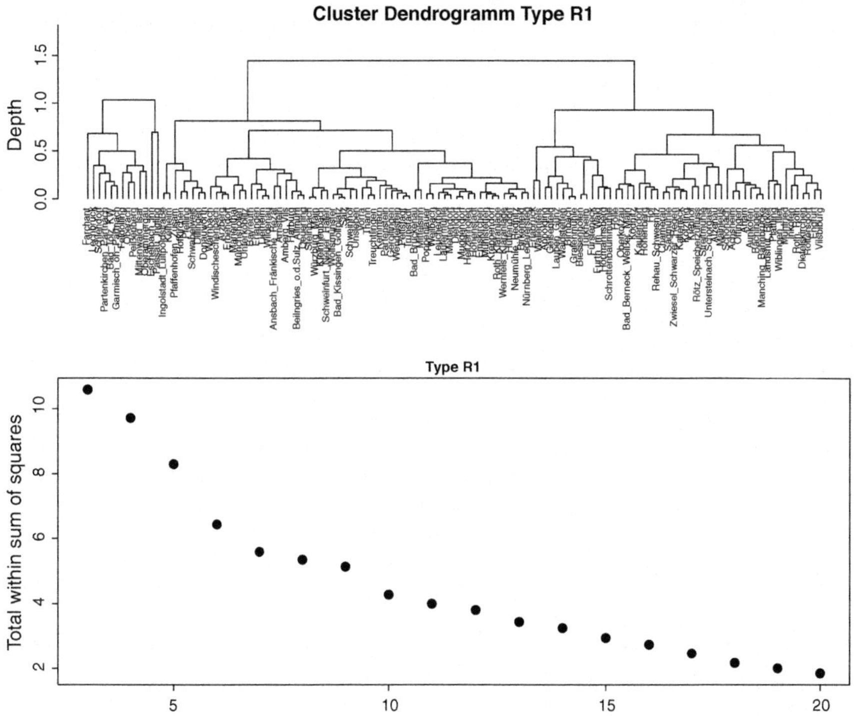

Fig. 13.6 Tree of hierarchical clustering of catchment and runoff attributes (top) and TSS as criterion to define the optimal number of clusters (bottom) for flood type R1 for catchments in Bavaria, Germany

catchments. To define the optimal number of clusters, the TSS criterion is applied (Eq. 13.1). For flood type R1 in Fig. 13.6, it was determined to be seven, as then the decrease of variance starts to converge much slower against zero. The same number of clusters was also determined for all other flood types, though theoretically this number might differ since also the runoff attributes differ. The resulting clusters of hydrologically similar catchments are shown in Fig. 13.7. Despite using identical catchment attributes as input data, the type-specific runoff attributes that were considered lead to type-specifically different clusters. For the snowmelt-induced floods, a dependence on the elevation is visible, while for heavy-rainfall floods almost all catchments north of the Danube are grouped into one cluster. For all flood types, spatial dependency can be detected, leading to spatially connected clusters. The resulting clusters are identical for some cases, especially for the southern catchments for the flood types R1 and R2 and the northern catchments for flood types R3 and S2. Clusters for flood type S1 have only little similarity regarding catchment membership to the remaining flood types.

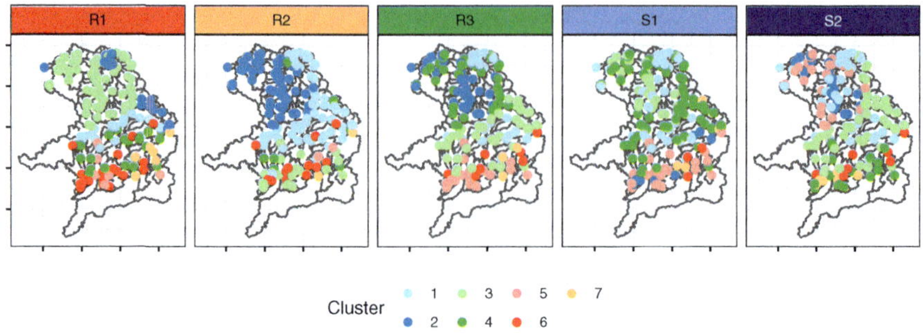

Fig. 13.7 Clusters derived by hierarchical clustering of catchment, precipitation and runoff attributes to obtain hydrologically similar regions in Bavaria, Germany

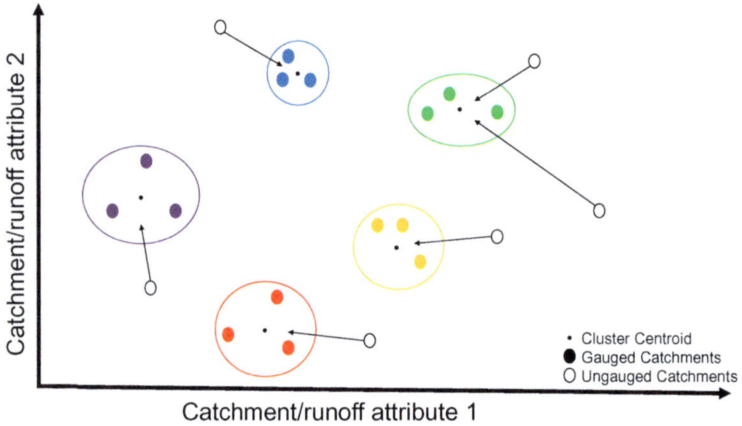

Fig. 13.8 Example of ungauged catchments that are assigned to homogeneous clusters according to the distance of their catchment/precipitation/runoff attributes to the cluster centroids

The input data used here for defining the clusters were not chosen type-specifically, i.e., the same kind of input data were used for all flood types (though the values may be type-specifically, see $Prec_D$). Of course, their selection could be done type-specifically, too. In such a case the input data could be selected according to a goodness criterion for the regionalisation. Indeed, such a type-specific selection of input criteria can improve the results of regionalisation. This will be discussed in more detail in the following chapter.

Once the homogeneous clusters have been defined, ungauged catchments can be assigned to these clusters. This is done by minimizing the distance of the considered catchment, precipitation and runoff attributes to the centroids of the clusters (Fig. 13.8). This approach guarantees a spatially consistent assignment of ungauged catchments.

The definition of hydrologically similar regions is crucial for almost all regionalisation approaches. Usually, such a definition is based on selected catchments, precipitation and runoff attributes. Taking into account the flood type and hence the flood-generating mechanisms improves the definition of such regions and their hydrological interpretation. Classical clustering algorithms like hierarchical clustering deliver a simple yet unsupervised possibility to define similar regions according to runoff and catchment attributes.

References

Blöschl G, Sivapalan M (1995) Scale issues in hydrological modelling: a review. Hydrol Process 9(3–4):251–290

Burn DH (1997) Catchment similarity for regional flood frequency analysis using seasonality measures. J Hydrol 202(1–4):212–230

Calinski T, Harabasz J (1974) A dendrite method for cluster analysis. Commun Stat 3(1):1–27

Cramér, H (1946) Mathematical methods of statistics, vol 282. Princeton University Press, Princeton

Dalrymple T (1960) Flood frequency analyses. USGS, Water Supply Paper 1543A:11–51

De Michele C, Rosso R (2002) A multi-level approach to flood frequency regionalisation. Hydrol Earth Syst Sci 6(2):185–194

Fischer S, Schumann A (2020) Spatio-temporal consideration of the impact of flood event types on flood statistic. Stoch Environ Res Risk Assess 34:1331–1351

Fischer S, Schumann AH (2021) Regionalisation of flood frequencies based on flood type-specific mixture distributions. J Hydrol X 13:100107

Genolini C, Ecochard R, Benghezal M, Driss T, Andrieu S et al (2016) kmlShape: an efficient method to cluster longitudinal data (Time-Series) according to their shapes. PLoS ONE 11(6):e0150738

Merz R, Blöschl G (2005) Flood frequency regionalisation-spatial proximity vs. catchment attributes. J Hydrol 302(1–4):283–306

Nathan RJ, McMahon TA (1990) Identification of homogeneous regions for the purposes of regionalisation. J Hydrol 121(1–4):217–238

Odry J, Arnaud P (2017) Comparison of flood frequency analysis methods for ungauged catchments in France. Geosciences 7(3):88

Salvador S, Chan P (2004) Determining the number of clusters/segments in hierarchical clustering/segmentation algorithms. In: 16th IEEE international conference on tools with artificial intelligence, pp 576–584

Shrout PE, Fleiss JL (1979) Intraclass correlation: uses in assessing rater reliability. Psychol Bull 86:420–28

Sivapalan M, Takeuchi K, Franks SW, Gupta VK, Karambiri H, Lakshmi V, Liang X, McDonnell JJ, Mediondo EM, O'Connell PE, Pomeroy JW, Schertzer D, Ulenbrook S, Zehe E (2003) IAHS decade on Predictions in Ungauged Basins (PUB), 2003–2012: shaping an exciting future for the hydrological sciences. Hydrol Sci J 48(6):857–880

Chapter 14
Regionalisation of Statistical Distribution Parameters: The TMPS Model

Abstract When aiming to regionalise statistical distributions directly, instead of regionalising the quantiles, one has to regionalise the parameters of the model. In the case of the TMPS model, this means that the distribution parameters of each flood type have to be regionalised separately based on hydrologically similar regions as for example defined in Chap. 13. The advantage when considering flood types in the regionalisation lies in the possibility to define type-specific similar regions. The parameters then can be easily obtained by either regression approaches or k-nearest-neighbour method. The regionalisation procedure for the TMPS model is described in this chapter.

The TMPS model introduced in Chap. 5 incorporates the distributions of the different flood types. Hence, it consists of a set of parameters for each flood type. When regionalising these parameters, one has to define type-specific hydrologically similar regions or relations to catchment attributes. Each type-specific distributions consists of six parameters, three of the GPD distribution for the exceedances of the threshold, and three for the GEV distribution of all flood events of the given flood type. In fact, for each flood type j, two shape parameters, κ_j of the GPD and ξ_j of the GEV, two scale parameters, β_j of the GPD and σ_j of the GEV, as well as two location parameters, the threshold u_j of the GPD and μ_j of the GEV, have to be regionalised. However, these parameters cannot be treated alike, as they are quite differently related to catchment, precipitation and runoff attributes. For example, the location parameters which control the mean flood regime are much related to the catchment size. This is a fact that also other regionalisation approaches make use of, e.g., when estimating the index flood. On the other hand, the shape parameters that control the skewness and the tail heaviness might not be that directly related to any catchment attributes but may depend on the statistics of heavy rainfall events. It is therefore proposed to consider each parameter separately and to investigate different regionalisation strategies. The approach proposed in this chapter is step-wise. First, the optimal regionalisation results are obtained for each parameter separately, while considering the remaining parameters as known. Then, a global optimisation over all parameters is performed to obtain the optimal regionalisation result.

S. Fischer and A. H. Schumann, *Type-Based Flood Statistics*, Water Science and Technology Library 124, https://doi.org/10.1007/978-3-031-32711-7_14

When evaluating regionalisation approaches regarding their performance, the predicted variable, in this case the flood quantile, has to be compared with the observed variable. Of course, for ungauged catchments, which are the catchments of interest in regionalisation, this is not possible as no (or not enough) such observed variables are available. Instead, the sample of observed catchments is split, i.e., one sub-sample is used to estimate the regionalised parameters and the other sub-sample is used for the assessment of the prediction (Klemes 1986). More precisely, one or more gauged catchments of the study region are selected and treated as ungauged, such that the parameters of the TMPS model for these catchments are estimated by regionalisation from the remaining catchments. The resulting predicted flood quantiles are then compared to the observed at-site flood quantiles, where observed in this case means that the flood quantiles have been estimated by fitting the TMPS model to the observed floods. This procedure is repeated for all catchments in the sample until each was treated once as ungauged. This method is well-known also under the name of cross-validation, leave-one-out or jackknife approach. The predicted and observed flood quantiles then can be compared by classical performance measures, such as (normalised) root mean squared error or (absolute) bias (see Eq. 5.5). These measures are applied here to evaluate the results of each step of the regionalisation and hence the impact of each step on the total regionalisation performance as well as for the whole regionalisation framework.

When evaluating the performance of the regionalisation, the standard jackknife approach could be applied. However, for highly nested regions or those with a high gauge density, as is the case for the catchments in Bavaria, Germany, it is recommended to treat more than one gauge as ungauged in the jackknife procedure, since this omits that catchments that are nested with the ungauged catchment are used to transfer too much information to the ungauged catchment.

Fischer and Schumann (2021) applied an extension of the classical jackknife approach, the delete-d jackknife approach, for evaluating the performance of regionalisation. Delete-d in this case means that not only one catchment is treated as ungauged at a time, but a number of d catchments. In such a case, the random selection of ungauged catchments is repeated for a given number of runs.

14.1 Type-Specific Regionalisation of the Shape Parameters κ_j and ξ_j

The shape parameter of both GEV and GPD distributions is determined by the skewness of the sample. Therefore, it contains information on the extreme flood events. It is also often used as an indicator for the heaviness of the tail. Positive shape parameters indicate a heavy right tail and hence the occurrence of single extreme floods. Since the estimation of the shape parameters relies heavily on this information on extreme floods, it is also the parameter whose estimation is afflicted with highest uncertainty. Given that in hydrology only comparably short records of time series of flood peaks are available that seldom exceed 100 years, the occurrence of extreme floods in these records is purely coincidental. Indeed, it might occur that extreme flood peaks are included in one sample, while in the sample of the neighbouring gauge they are not, simply due to a different observation period. Fischer and Schumann (2022) illustrate how the information contained in different samples has an impact on the regionalisation. On the other hand, the estimation of the shape parameter is of great importance, as it is highly relevant for the major flood quantiles and thus for the estimation of design floods. Additionally, the shape parameter usually is less linked to the catchment and runoff attributes like catchment size or mean runoff of a catchment. This is why the shape parameter is most difficult to regionalise. Instead, often the shape parameter is assumed to be constant within a region. For example, in Bulletin 17B of the United States, regions in the United States with constant GEV shape parameters are defined England et al. (2019). Such an approach assumes that certain regions, due to similar catchment and climate conditions, will have a similar probability of extreme floods. The approach proposed here is similar to that assumption, though the definition of regions with similar shape parameter is based on catchment and runoff attributes, i.e., on hydrologically similar regions, and type-specific. In Chap. 13, such an approach was introduced. It is now applied to estimate the type-specific shape parameters κ_j and ξ_j. First, homogenous clusters of gauged catchments are estimated for each parameter and flood type as described in Chap. 13. Then, ungauged catchments are assigned to these clusters (see Chap. 13). After that, the type-specific parameters κ_j and ξ_j for ungauged catchments can be simply estimated by the median of the respective shape parameters of the gauged catchments in the same homogeneous cluster. The impact of the choice of the median will be discussed later on. Alternatively, other measures could be considered, like the often-used weighted mean, where the weights are chosen according to the record length of the gauged catchments.

In the previous chapter, it was already pointed out how the type-specific selection of input attributes could improve the definition of homogeneous clusters and hence the regionalisation. However, it is not a-priori clear how to select the type-specific catchment, precipitation and runoff attributes for the definition of homogeneous clusters. There exist various attributes and combinations of these and an appropriate choice leading to the smallest errors in regionalisation depends much on the study region. However, since type-specific clusters are considered, a selection motivated by the flood type characteristics should be applied. For example, for heavy-rain floods, the rainfall intensity and the retention capacity of a catchment are relevant characteristics, as they will lead to similar catchments reactions. The elevation, however, has a lesser impact on these floods. Catchment elevation or more precisely the range, though, is highly relevant for the snow-induced flood types, as the elevation of a catchment is an indicator for the existence and frequency as well as depth of a snow cover. Fischer and Schumann (2021) investigated the optimal selection of catchment and runoff attributes for the flood types proposed in Sect. 6.2 for catchments in Bavaria, Germany (see Appendix A.1). They found out that indeed for the shape parameter κ_j of the distribution of the exceedances, the selected attributes vary between the flood types, as expected. For heavy-rain floods (R1), the runoff attributes rainfall intensity, the 80%-rainfall quantile and retention capacity of the catchment together with the catchment attribute nature class should be used to define homogeneous cluster of catchments. Since heavy-rain floods are generated by rainfall events with high intensities and thus high amounts of rainfall in a short time, knowledge on the similarity in the rainfall events and their temporal distribution improves the regionalisation of the shape parameter, which controls the right tail and thus the impact of extreme events. The nature class incorporates geographical factors like soil as well as information on flora and fauna and serves as a spatial indicator of similarity. It is, together with retention capacity, relevant for defining clusters for the estimation of shape parameters regardless of the flood type. For flood type R2, similar attributes are relevant for defining the clusters, except for the rainfall amount which is no longer considered. When defining clusters for flood type R3 and hence for long-duration rainfall events or a sequence of rainfall events, the rainfall intensity is no longer relevant. Instead, retention capacity and amount of rainfall as well as nature class and the hypsometric clusters are considered. The hypsometric cluster might provide information on the runoff formation in the catchment, a critical aspect for this type of floods. The same holds true for snowmelt-induced floods (S2) and rain-on-snow floods (S1), though for S1 floods again the intensity of rainfall should be considered, too. The intensity of rainfall has an impact on how much energy is set free in the snow cover and hence how much runoff is generated at once. Catchments with similarly high intensities of the extreme events of rainfall tend to generate a similar (relative) amount of runoff from a snow cover and hence have a similar shape parameter. Rainfall intensity is not relevant for snowmelt floods, since here rainfall plays a minor role. However, as mentioned before, for defining clusters for the shape parameter of the distribution of these type of floods, the elevation range is important, as this provides information on the snow cover, its thickness and extent. The selection of attributes does not differ between both shape parameters, κ_j and ξ_j, as the extreme events, which mainly control these parameters, are included in both samples, the one of the exceedances of the threshold only and the one that includes all flood events. Which catchment and runoff attributes have been selected by Fischer and Schumann (2021) for the regionalisation of the two shape parameters are summarised in Table 14.1.

Table 14.1 Selection of catchment, precipitation and runoff attributes to define homogeneous clusters for the regionalisation of the type-specific shape parameters of the TMPS model

	R1	R2	R3	S1	S2
Catchment	Nature class	Nature class Hypsometric curve	Nature class hypsometric curve	Nature class hypsometric curve	Nature class elevation range
Precip./Runoff	Retention capacity intensity 80% quantile rainfall	Retention capacity	Retention capacity mean rainfall	Retention capacity intensity mean rainfall	Retention capacity

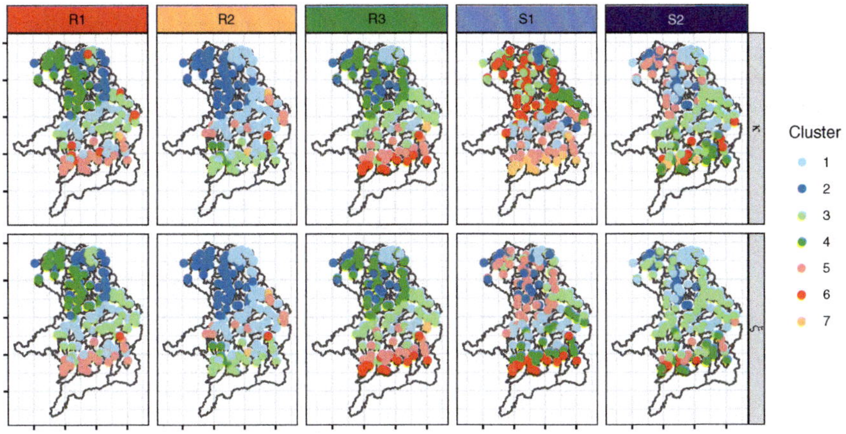

Fig. 14.1 Type-specific clusters of hydrologically similar catchments in Bavaria, Germany, for the regionalisation of the shape parameters of the TMPS model (modified from: Fischer and Schumann 2021)

The resulting clusters for catchments in Bavaria, Germany, are given in Fig. 14.1. Since not all resulting clusters are spatially disjunct, ungauged catchments are assigned to a cluster by estimating the joint distance of all considered attributes to the cluster centroids. The catchment is then assigned to the cluster with minimum distance.

14.1.1 Performance of the Regionalisation of the TMPS Model: Shape Parameters

When evaluating the goodness of the regionalisation of the shape parameters of the TMPS model only, i.e., using the regionalised parameter values for the shape parameters in the jackknife but keeping the scale and location parameters that have been estimated from the observed floods for the respective gauge, it can be seen that the mean relative bias for the TMPS model lies in the range of 8–13% (Table 14.2). Moreover, except for flood type R1, the errors are comparably small for the small return period of 10 years. This is likely due to the fact that the shape parameter does not impact the quantiles for small return periods much, but has an impact on the quantiles for large return periods, where the skewness becomes recognisable. The errors of the relative bias in general are in a reasonable range, while it can be seen for the absolute relative bias that the total errors are higher. Nonetheless, the use of the median for regionalisation does not introduce a higher uncertainty than standard procedures, at least when being used for regionalising the shape parameter. The robustness of the median omits the overestimation of the shape parameter, when

Table 14.2 Mean (absolute) relative bias of the regionalisation procedure of the type-specific distributions and the TMPS model when only considering the shape parameter for regionalisation, with application of the delete-10 jackknife and 100 repetitions. The regionalised quantiles of the type-specific distributions and the TMPS model were compared to those obtained from the fit of the type-specific distributions and the TMPS model to the observed flood events

	Return period	TMPS	R1	R2	R3	S1	S2
rel. bias	10	0.122	–0.518	0.047	0.016	0.051	0.010
	100	0.129	–0.144	0.105	0.092	0.155	–0.123
	500	0.089	0.034	0.138	0.146	0.213	–0.147
rel. abs. bias	10	0.322	0.187	0.096	0.084	0.126	0.096
	100	0.329	0.427	0.308	0.309	0.365	0.314
	500	0.400	0.575	0.436	0.485	0.513	0.511

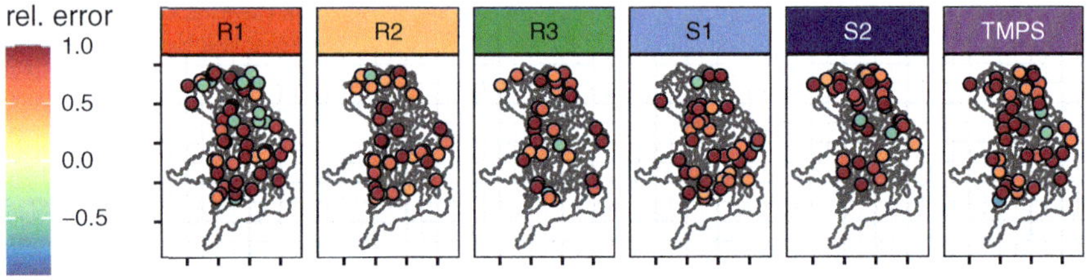

Fig. 14.2 Location of the gauges with relative errors of more than 50% in regionalisation of the flood quantile with return period 100 years when regionalising the shape parameters of the TMPS model only

a large shape parameter occurs for only one gauge in the respective cluster, e.g., due to short record length and one extraordinary flood event.

When having a closer look on where the largest errors occur in the study region, certain patterns become visible (Fig. 14.2). As already indicated by Table 14.2, most of the largest errors occur for flood types R1 and S1. For R1, a general underestimation is visible, while only in the south-west the regionalisation tends to overestimate. For S1, no general spatial patterns occur. The magnitude of the error does neither depend on the catchment area (with the exception that catchments of the upper macroscale are generally not affected by high errors) nor on the record length. However, when having a look at when extraordinary large floods occurred in this region, the patterns of these floods and the errors for flood type R1 seem to coincide (Fig. 14.3). Especially the spatial extension of the 1999 and 2005 floods in the south-west and that of the 1946 flood in the north coincide with the regions of highest errors for flood type R1. These flood events have been identified as floods of type R1 for most catchments. For flood type S1, the flood event in January 2011 plays a major role when comparing the spatial distribution of the highest errors of the regionalisation of the shape parameter. It is not surprising that extraordinary large floods have a crucial impact on the estimation of the shape parameter. The main issue in this case is that often these events are

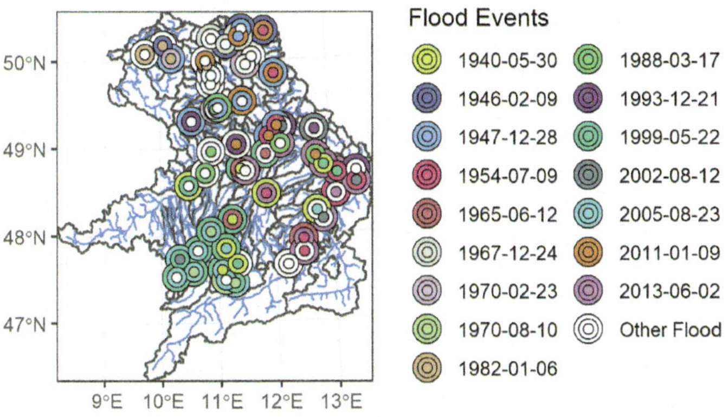

Fig. 14.3 Location of the 16 largest flood events in the study region of Bavaria, Germany

included in the records for some catchments, while they are not included for others, due to a different record period. This is the case for the 1946 event, which is only included in the records of a few gauges. When combining gauges with and without this event in the records in one cluster, the use of the median for regionalisation might lead to an underestimation of the shape. However, it prevents the overestimation of the shape parameters for those gauges that do have the flood event included in their records. The errors of the regionalisation for each flood type then is propagated to the TMPS model. A possibility to overcome this issue was proposed by Fischer and Schumann (2022) by compensating the effects of record length in regionalisation.

14.2 Type-Specific Regionalisation of the Scale Parameters β_j and σ_j

The scale parameter of a distribution is determined by the variance of the data and is thus related to the second moment. The larger the scale parameter, the more scatter the data. In many regionalisation approaches, this parameter is assumed to be scale-dependent, i.e., the parameter value is related to the catchment size. Usually, it is assumed that larger catchments also have a larger scale parameter. This can be explained by the catchment processes and the different possibilities of rainfall to overlap with a catchment. According to the type of rain and the weather conditions, a rain event may hit only a part of the catchment. In such a case, the retention of soil as well as the river network and the distance to the gauge will balance the temporal variability of rainfall such that the rainfall intensity plays a minor role. Only in the case where a convective rain cell hits the catchment close to the outlet gauge, the rainfall intensity has an impact on the generation of a flood wave. Such rainfall events may result in extraordinary high peaks. In mountainous regions, snowmelt and orographic rain depend on the spatial variability of elevation zones. This variability is usually higher in larger catchments. With this in mind, the assumption of increasing variability for increasing catchment size due to a higher variability of possible rain scenarios respectively snowmelt in mountainous regions is meaningful. In such a case, the scale parameter can be estimated simply by the catchment size in a regression model.

However, can this reasoning really be applied to all catchments? In small catchments, where the river network is less branched and the distance to the gauge site is short, the catchment reaction to rainfall is more direct than in large catchments. It is more likely that local rainfall events hit large parts or the whole catchment at quite the same time. The runoff then forms a flood wave which is determined by the drainage conditions of the catchment as a whole. In other cases, the flood wave is strongly influenced by the spatial hotspot of the rain event. The same holds true for snowmelt: in small catchments which are located at a plateau either there is snowmelt or not, but it will occur likely simultaneously in the whole catchment. In such cases, a scale-dependence of the variability of flood peaks is questionable and the direct approach of regression might not be applicable for the scale parameters. Based on this reasoning, in the following different approaches for catchments depending on their size are proposed. Of course, the definition of scale-dependence in hydrology is hard and (following Klemes 1983)

> Formulations appropriate at a given level usually are not applicable at the immediately adjoining levels.

For Central Europe, a common classification of scales exists that tries to take into account such reasoning. Becker (1995) proposes a distinction into four classes: the lower mesoscale (<300 km^2), the upper mesoscale (300–1000 km^2), the lower macroscale (1000–10000 km^2) and the upper macroscale (>10000 km^2). The motivation behind this classification was the modelling perspective. Becker (1995) argued that in the upper macroscale, global circulation models are most relevant, while for the lower macroscale regional climate models are required. For the mesoscale, instead, the catchment attributes are most relevant. For the upper mesoscale, spatial patterns, e.g., the above-mentioned nature classes, are a key information for modelling. For the lower mesoscale, more small-scale information is required, e.g., soil layers or plants. Of course, this is only one of many alternatives to define spatial scales in hydrology. However, since it is based on hydrological attributes, it can be well integrated in the regionalisation. Therefore, in the following the distinction between mesoscale and macroscale catchments is applied, where the differentiating characteristic is a catchment size of 1000 km^2.

14.2.1 Regionalisation for Mesoscale Catchments

When regionalising the scale parameters β_j and σ_j of the TMPS model for mesoscale catchments, i.e., catchments smaller than 1000 km^2, no linear dependence of the parameters on the catchment size becomes visible for any flood type for catchments

Table 14.3 Selection of catchment and runoff attributes to define homogeneous clusters for the regionalisation of the type-specific scale parameters of the TMPS model

	R1	R2	R3	S1	S2
Catchment	Nature class size elevation range	Nature class size elevation range	Nature class size elevation range	Nature class size elevation range	Nature class size elevation range
Precip./Runoff	Retention capacity intensity	Retention capacity mean rainfall	Retention capacity mean rainfall	Retention capacity intensity 80% quantile rainfall	Retention capacity

in Bavaria, Germany. Instead, the variability of the scale parameters is quite high. Such a variability can be expected for many mesoscale catchments in the world, as for these the small-scale processes play a crucial role in the flood generation. Hence, as long as there is a high variability in the spatial and temporal distribution of the rainfall, the same will be expected for the floods in those catchments. Yet, one still has to differentiate between the lower and upper mesoscale (the distinction is made here on the basis of a catchment size of 300 km^2), as there again different processes become relevant. Therefore, (Fischer and Schumann 2021) recommend to perform different regionalisation approaches for the lower and upper mesoscale. The idea is the same as for the shape parameters: while no linear dependence on the catchment attributes can be assumed, at least homogeneous clusters of the catchments can be defined. Therefore, the procedure for the definition of homogenous clusters for the shape parameters (Sect. 14.1) is repeated for the scale parameters, though this time catchments are grouped additionally into either lower or upper mesoscale catchments.

As an example, once again the catchments in Bavaria, Germany, are considered. The attributes selected to define the homogeneous clusters of scale parameters also differ between flood types (Table 14.3). However, these attributes do not differ between the lower and upper mesoscale, implying that the same basic processes might be relevant for both catchment classes but with a different impact on the variability. The selection of attributes reflects the differences of the flood-generating processes between the flood types. Again, the heavy-rain and the rain-on-snow flood clusters profit from considering the rainfall intensity, while the clustering of the remaining flood types makes use of the rainfall amount. For all flood types, catchment size, elevation range and nature class are considered, implying that though there does not exist a linear dependence on any of these attributes, at least an implicit dependence has to be considered. The resulting clusters are displayed in Fig. 14.4.

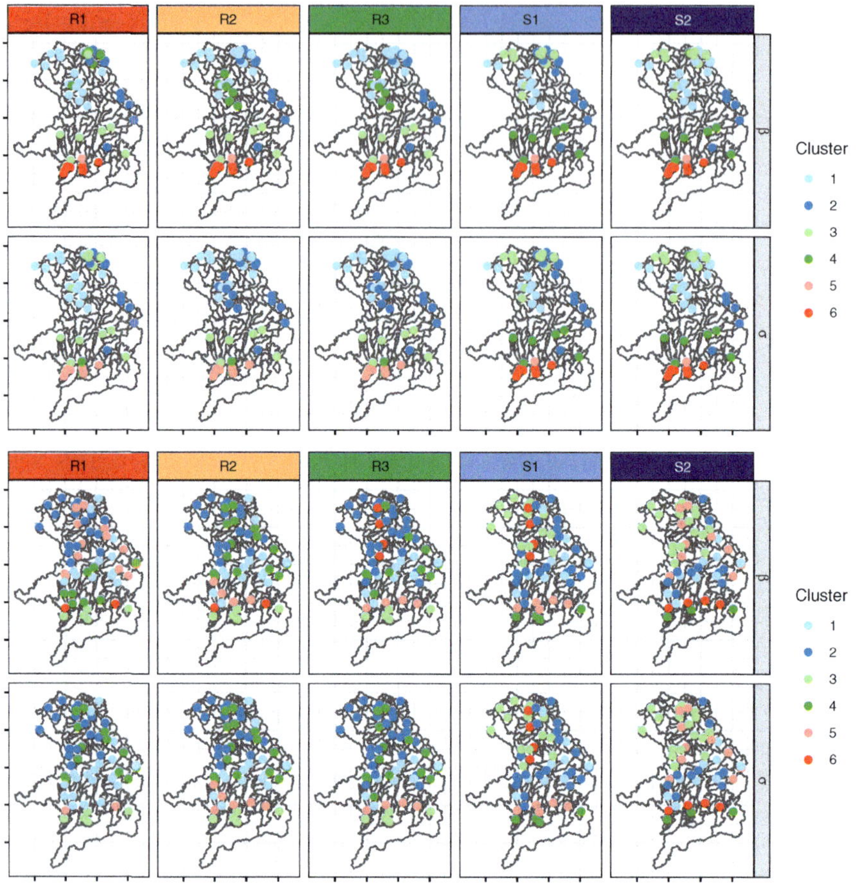

Fig. 14.4 Type-specific clusters of hydrologically similar catchments in Bavaria, Germany, of the lower mesoscale (top) and the upper mesoscale (bottom) for the regionalisation of the scale parameters of the TMPS model (modified from: Fischer and Schumann 2021)

14.2.2 Regionalisation for Macroscale Catchments

When regionalising the scale parameters of the TMPS model for macroscale catchments, i.e., catchments larger than 1000 km^2, the scale-dependence of these parameters has to be considered. As pointed out before, the large size of these catchments leads to a balance of the variability of rainfall events, though at the same time the number of possible flood scenarios increases due to the catchment size and flood superposition. Thus, on the one hand, the catchment reaction is dampened if only parts of the catchment that are located far away from the outlet are affected by heavy rain. On the other hand, spatially extended rainfall or heavy rainfall close to the outlet can lead to quite large flood peaks. Based on these consideration, (Fischer and

Schumann 2021) propose to estimate the scale parameters by a type-specific linear regression on the catchment size. Again, a differentiation is made between the lower and the upper macroscale, as different regional or more global aspects have to be considered. Therefore, the following relationship is assumed:

$$
\begin{aligned}
\beta_{j;LM} &= \alpha_{1;j;LM} + \gamma_{1;j;LM} AE \\
\sigma_{j;LM} &= \alpha_{2;j;LM} + \gamma_{2;j;LM} AE \\
\beta_{j;UM} &= \alpha_{1;j;UM} + \gamma_{1;j;UM} AE \\
\sigma_{j;UM} &= \alpha_{2;j;UM} + \gamma_{2;j;UM} AE,
\end{aligned}
$$

where AE is the catchment size, $j = 1, \ldots, 5$ indicates the flood type and LM respectively UM denotes lower or upper mesoscale.

Therefore, four models of linear regression have to be estimated, one for each parameter and catchment class.

Consider once again the catchments in Bavaria, Germany. The proposed regression was applied to regionalise the scale parameters for the macroscale catchments (Fischer and Schumann 2021). However, an additional consideration was made: the split up of the catchments of each class into two groups, performing a two-fold regression (Fig. 14.5). Such a split can be done by simply optimising the fit of each regression, e.g., by means of the coefficient of determination as already explained in Sect. 6.2 for defining the rainfall-driven flood types. Though this procedure might increase the predictive power of the model (Fischer and Schumann 2021 report coefficients of determination well above 0.7), it has one major drawback: ungauged catchments cannot be simple plugged into the model, as it is not apriori clear to which of the two regressions they belong. Instead, they have to be assigned to one group either according catchment information or due to similarity to other catchments in the group, e.g., regarding distance or catchment attributes. Therefore, such a distinction should be only done when there is clear evidence for more than one group in the regression. This, of course, depends much on the study region, as more homogeneous regions might only show one group while very heterogeneous regions might even show several groups in regression.

14.2.3 Performance of the Regionalisation of the TMPS Model: Scale Parameters

The combined results for the error of the regionalisation of the scale parameters for meso- and macroscale catchments is given in Table 14.4. The regionalisation of the scale parameters results in smaller errors compared to the shape parameters. Nevertheless, the contribution of this error to the overall regionalisation is considerable, especially for flood type R1. In general, the error tends to increase for increasing return periods, as the scale parameters are determined by the variance of the data and hence have an impact on the values which are not close to the mean.

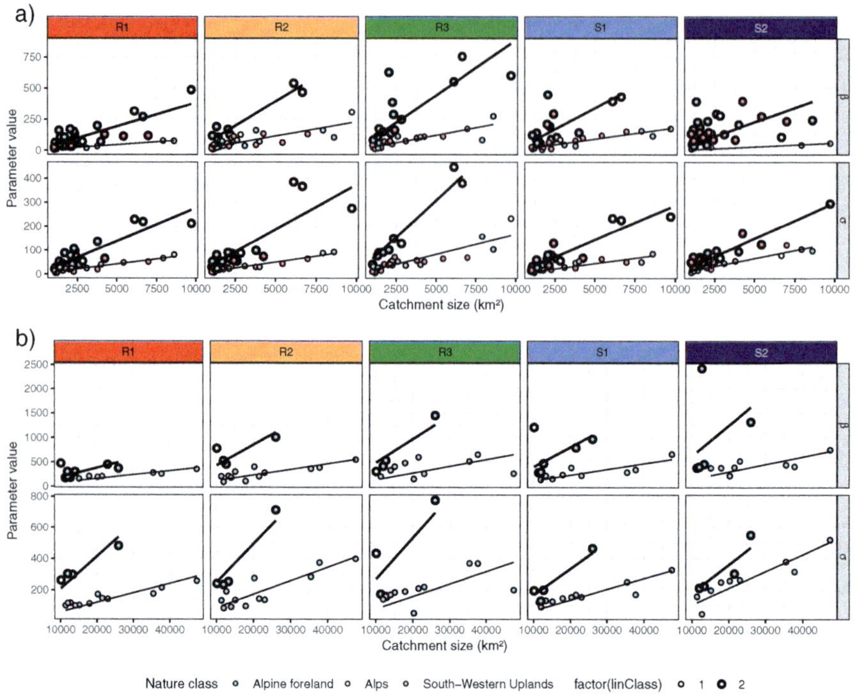

Fig. 14.5 Type-specific regression between catchment size and scale parameters of the TMPS model for catchments in Bavaria, Germany, of **a** the lower macroscale and **b** the upper macroscale. Line thickness displays membership to each regression group

Table 14.4 Mean (absolute) relative bias of the regionalisation procedure of the type-specific distribution and the TMPS model when only considering the scale parameters for regionalisation, with application of the delete-d jackknife ($d = 10$) and 100 repetitions. The regionalised quantiles of the type-specific distributions and the TMPS model were compared to those obtained from the fit of the type-specific distributions and the TMPS model to the observed flood events

	Return period	TMPS	R1	R2	R3	S1	S2
rel. bias	10	0.127	0.198	–0.011	–0.007	–0.011	–0.063
	100	0.198	0.163	0.044	0.107	0.053	–0.090
	500	0.217	0.179	0.053	0.127	0.073	–0.081
rel. abs. bias	10	0.231	0.369	0.262	0.342	0.203	0.0294
	100	0.278	0.377	0.260	0.381	0.288	0.244
	500	0.333	0.400	0.286	0.416	0.473	0.268

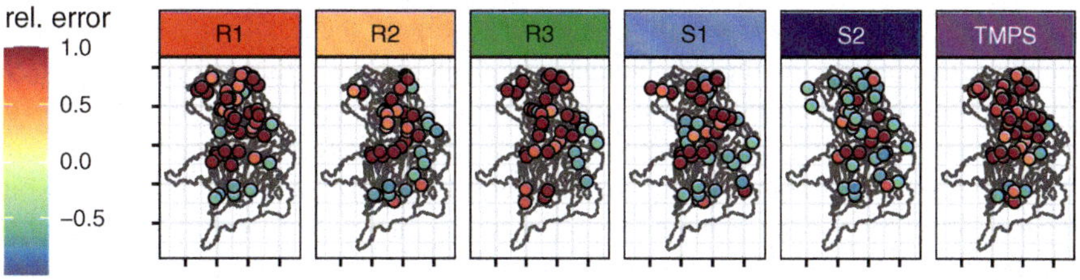

Fig. 14.6 Location of the gauges with relative errors of more than 50% in regionalisation of the flood quantile with return period 100 years when regionalising the scale parameters only

When comparing the catchments where the errors of regionalising the scale parameters are largest, the flood types do not differ much regarding the number of catchments with high errors (Fig. 14.6). However, mostly those catchments have high errors which have a catchment size smaller than 1000 km^2. For these catchments, the clustering approach was applied to regionalise the scale parameters, since a regression approach would result in even higher errors here. This is already an indicator that the regionalisation of the scale parameters for small catchments is complex and depends on many factors, leading likely to high uncertainty in the estimation. What can be seen when having a look at the spatial patterns of the errors is a clear difference between northern and southern catchments for almost all flood types. While in the south, the flood quantiles are underestimated, in the north they are overestimated when regionalising the scale parameter only. A coherence with the record length, however, cannot be observed. The spatial differences cannot be explained by the cluster membership, as different clusters are affected and not all cluster members show high errors. Again, it has to be assumed that the occurrence of extraordinary large events and thus their effect on the estimation of the scale parameters is the reason for the large errors in some catchments. It was shown before in Fig. 14.3 that there are distinct differences in the occurrence of such events for the Main basin in the north and the Danube basin in the south. These events increase the variance much when included in the sample and thus the scale parameters are also much higher in the case where such an extraordinary large event is in the records. The approach of Fischer and Schumann (2022) reduces this uncertainty.

14.3 Type-Specific Regionalisation of the Location Parameter μ_j and the Threshold u_j

The location parameter of the GEV distribution as well as the threshold parameter of the GPD distribution determines the value of the flood peak where most of the observed flood peaks are scattered around. It is closely related to the first moment and represents the average flood peak in a catchment. Hence, it is highly dependent on the catchment size, which mainly determines the mean flood regime in a catchment.

Therefore, many regionalisation approaches estimate the location parameter by a function depending on the catchment size. This can be either a linear or a non-linear relation, depending on the flood regime in the region of interest. Such an approach has the advantage that not much information is required and the procedure is easily applicable to any ungauged catchment. The approach is analogous to the one proposed for the scale parameters of macroscale catchments (Sect. 14.2.2), though for the location parameters, it is applicable to all catchment classes (lower and upper mesocale, lower and upper macroscale), resulting in eight different regressions, one for each parameter and catchment class.

When having a look at the regionalisation error introduced by regionalising the location parameters only, i.e., having values for shape and scale parameters estimated from the observed floods, the error is negligible. Only for the small return periods, where the location parameter mainly defines the flood quantiles, errors above 1% can be observed. Hence, the regionalisation of the location parameter does not have a large impact on the overall regionalisation error (Fig. 14.7 and Table 14.5).

14.4 Type-Specific Regionalisation of the Poisson Parameter

With the approaches for shape, scale and location parameters introduced in the previous section, all parameters required for the TMPS model can be regionalised. However, one also might want to regionalise the annual distributions of the respective flood types. This might provide valuable information on the respective flood types as was demonstrated in Chap. 7. For this purpose, not only the type-specific parameters of the GPD distribution of the exceedance of the threshold u_j have to be known but also the Poisson parameter λ_j (see Eq. 5.3). The Poisson parameter describes the mean annual number of flood events per year for each flood type, i.e., the number of events of each flood type per year, required for the type-specific distributions. Since floods occur spatially and are usually not limited to a single catchment, the regionalised value of this parameter can be estimated by using the information of the surrounding gauged catchments. For this, a weighted kMeans approach can be applied, where the weights are chosen according to similarity in catchment size: catchments with similar catchment size to the ungauged target catchment are assigned with highest weight. This is similar to the approach described in Sect. 13.3.1. The errors of the regionalisation of the Poisson parameter are below 5% for all flood types and thus not reported here in detail.

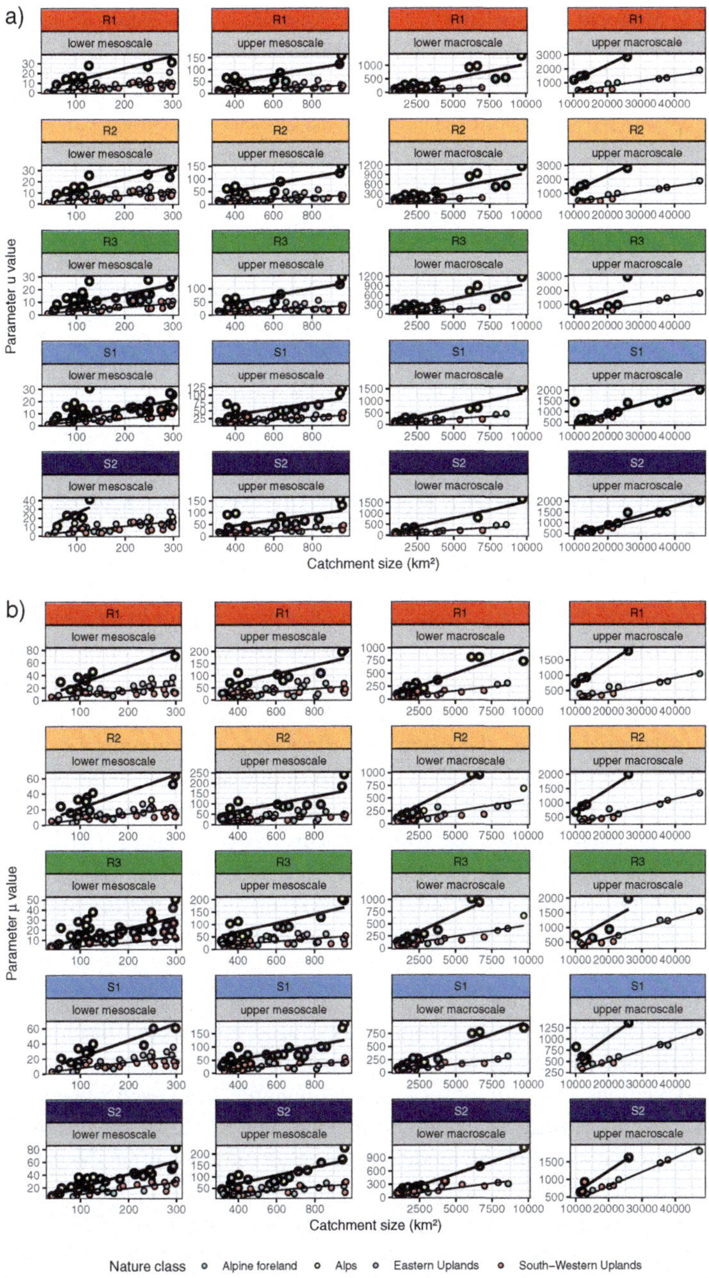

Fig. 14.7 Type-specific regression between catchment size and **a** threshold parameter respectively **b** location parameter of the TMPS model for catchments in Bavaria, Germany, of different catchment classes. Line thickness represents the regression group

Table 14.5 Mean (absolute) relative bias of the regionalisation procedure of the type-specific distribution and the TMPS model when only considering the location parameters for regionalisation, with application of the delete-d jackknife ($d = 10$) and 100 repetitions. The regionalised quantiles of the type-specific distributions and the TMPS model were compared to those obtained from the fit of the type-specific distributions and the TMPS model to the observed flood events

	Return period	TMPS	R1	R2	R3	S1	S2
rel. bias	10	–0.003	–0.002	0.004	–0.012	–0.004	–0.012
	100	–0.008	0.004	0.004	–0.002	–0.006	–0.008
	500	0.002	0.003	0.003	–0.002	–0.008	–0.006
rel. abs. bias	10	0.003	0.003	0.003	0.040	0.004	0.011
	100	0.004	0.004	0.004	0.006	0.003	0.008
	500	0.005	0.005	0.004	0.005	0.004	0.007

14.5 Performance of the Regionalisation of the TMPS Model

In the previous sections, the regionalisation of each parameter together with its corresponding error was introduced. However, for the final regionalisation, all parameters have to be regionalised jointly. The selection of appropriate attributes to define the clusters for regionalising shape and scale parameters therefore is based on the globally best performance, i.e., the overall smallest error in regionalisation. For such a case, it might happen that the individual errors of the estimation of the parameters are higher, but their combination leads to a balance of positive and negative deviations and thus to an overall better performance. This has to be taken into account when comparing the individual errors of the previous sections with those reported here for the whole regionalisation framework. The results for the performance evaluation of the regionalisation approach with delete-d jackknife with $d = 10$ are given in Table 14.6.

The regionalisation error lies well within the typical range of 20–40% bias as reported by the extensive study of regionalisation approaches in Blöschl et al. (2013). However, the bias might lead to falsified conclusions about the error, as negative and positive deviations from the observed flood quantile level each other out, leading to a bias close to zero as for example for flood type S2. This is especially the case when no systematic bias of the model can be observed, as is often the case in hydrologically diverse regions. Instead, the absolute bias should be reported, where the absolute value of the (normalised) difference between predicted and observed quantile is considered. When comparing the results of Table 14.6 to those derived when regionalising one type of parameters only, i.e., shape, scale or location parameters, it becomes obvious that the main error is obtained by the regionalisation of shape and scale parameters. While the error of the regionalisation of the scale parameters has an impact especially on the quantiles of small return periods, the error of the regionalised shape parameters have an impact on the quantiles of large return

Table 14.6 Mean (absolute) relative Bias of the regionalisation procedure of the type-specific distribution and the TMPS model, derived with a delete-10 jack-knife and 100 repetitions for different return periods. The regionalised quantiles of the type-specific distributions and the TMPS model were compared to those obtained from the fit of the type-specific distributions and the TMPS model to the observed flood events. (modified from: Fischer and Schumann 2021)

	Return period	TMPS	R1	R2	R3	S1	S2
rel. bias	10	0.119	1.636	0.108	0.212	0.028	0.088
	100	0.103	0.309	0.159	0.171	0.175	–0.008
	500	0.035	0.319	0.162	0.191	0.229	–0.005
rel. abs. bias	10	0.340	1.803	0.334	0.560	0.387	0.521
	100	0.389	0.556	0.401	0.422	0.451	0.342
	500	0.442	0.635	0.453	0.488	0.537	0.374

periods. Thereby, both parameters also seem to compensate each other, as the combined error of both parameters is smaller compared to the summed up error of the separate regionalisations. This is mainly due to the fact that the shape parameters sometimes are underestimated, while the scale parameters have a general tendency of overestimation.

It has to be noted that the results highly depend on the density of the available gauging network. The more dense the network, the more information from closely located gauged catchments (and thus potentially hydrologically similar catchments) or even from nested catchments can be obtained for regionalisation. However, a dense network is usually rare when applying regionalisation methods. The delete-d jackknife tries to take into account the impact of the density of the network by excluding a number of d gauges at a time. However, also the choice of d depends much on the available gauging network. For the region in Bavaria, Germany, which has a rather high density of gauged catchments, an increasing number of catchments that are assumed to be ungauged and are therefore not included in the regionalisation leads to an only slow increase of the bias. When reducing the number of catchments under consideration by about 50%, the bias increases by about 10% (Fig. 14.8). Fischer and Schumann (2021) report on similar results, though an increase of the variance of the bias is observed for an increasing number of ungauged catchments.

When considering not only the regionalisation error of the regionalised TMPS model but additionally also that of the type-specific distributions, conclusion on the origin of the error can be drawn. For example, the flood type R1 contributes much to the overall error compared to the remaining flood types which have much smaller bias. R1 is also the flood type with the heaviest tail of the distribution of the relative error (Fig. 14.9a), which is then inherited by the TMPS model. All remaining flood types have relative errors scattered around zero with much lighter right tails. This indicates a meaningful performance of the regionalisation approach without any systematic bias. When having a closer look at the distribution of the bias for the regionalised flood peak of return period $T = 100$ years (Table 14.7), this impression

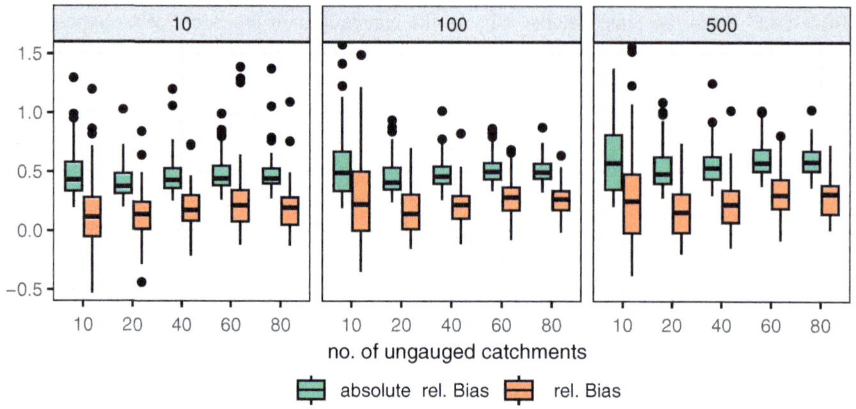

Fig. 14.8 Relative Bias and absolute relative Bias for return periods 10, 100 and 500 years and delete-d jack-knife method with a number of $d = 10, 20, 40, 60, 80$ of catchments that are treated as ungauged (of a total number of 168 catchments). The scale of the y-axis is limited to [–0.5,1.5] for better readability. (modified from: Fischer and Schumann 2021)

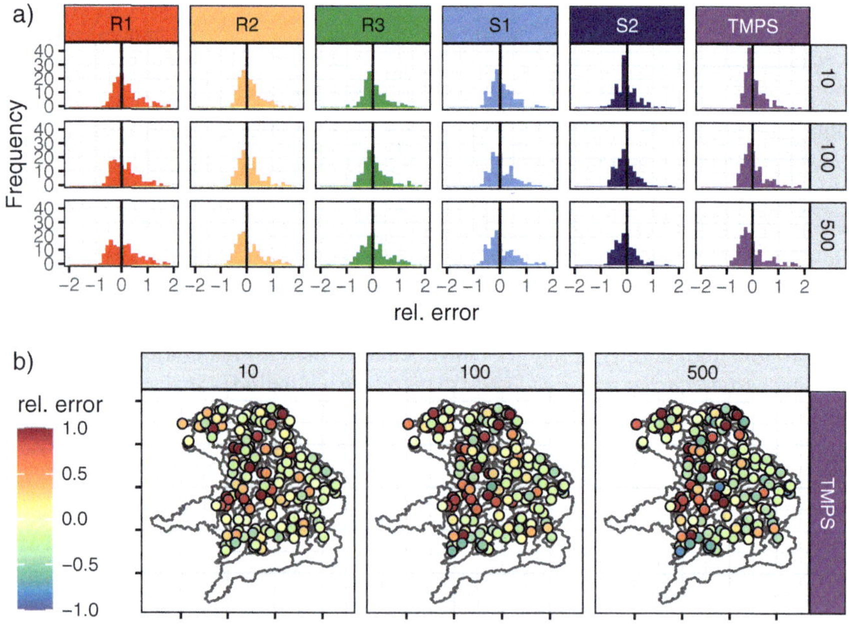

Fig. 14.9 Relative error between predicted and observed flood quantile for the regionalisation procedure **a** as a histogram and **b** spatially distributed for the catchments (where the range of the values is limited to [–1,1] for better readability). (from: Fischer and Schumann 2021)

Table 14.7 Proportion of catchments with an absolute bias in regionalisation (derived with delete-d jackknife, $d = 10$) of the flood peak with return period $T = 100$ in certain quantile ranges, differentiated by flood type and for the TMPS model

Bias range	TMPS	R1	R2	R3	S1	S2
$<-50\%$	0.057	0.050	0.088	0.050	0.082	0.113
-50% to -25%	0.183	0.176	0.189	0.195	0.232	0.245
-25% to 0%	0.258	0.208	0.245	0.270	0.252	0.296
0% to 25%	0.214	0.132	0.170	0.151	0.132	0.164
25% to 50%	0.069	0.088	0.119	0.151	0.113	0.082
$>50\%$	0.220	0.346	0.189	0.182	0.188	0.100

is confirmed: For the TMPS model, almost 50% of the catchments have a relative bias between –25% and 25%. Similarly, for flood types R2, R3, S1 and S2, the majority of catchments has a bias in this range. Only a small proportion of catchments (<20%) has a bias greater than 50%. Large negative biases occur for almost no catchments. Again, R1 floods form a different picture. For this flood type, about 35% of all catchments have a bias greater than 50%. These heavy-rain floods prove to be a challenge in regionalisation. Due to their often spatially limited occurrence, where only few catchments are affected by one such a event, the correct prediction of such floods by using spatial information becomes difficult. Though regions might be hydrologically similar, they often did not have experienced the same heavy-rain floods in the past. For future research, it is recommended to pay special attention to this crucial problem.

The type-specific regionalisation of the parameters of the TMPS model offers a possibility to take into account the flood-generating processes in a catchment as basis for the information transfer to ungauged catchments. It has to be considered how the different distribution parameters are related to catchment, precipitation and runoff attributes and how these relations change with the flood type and catchment size. Despite the high number of parameters in the TMPS model, the regionalisation still results in comparably small errors. A challenge in regionalisation proved to be the flood type R1, which is hard to capture for all catchments in a hydrologically similar region, due to its spatially limited extent. This leads to larger errors in the regionalisation which are inherited by the TMPS model. Yet, the type-specific regionalisation approach offers the opportunity to reveal which flood-generating processes are challenging in regionalisation such that future research can focus on the se specific tasks.

References

Becker A (1995) Problems and progress in macroscale hydrological modelling. In: Feddes, RA (ed) Space and time variability and interdependencies in hydrological processes. International Hydrology Series. Cambridge University Press, pp 134–143
Blöschl G, Sivapalan M, Wagener T, Savenije H, Viglione A (eds) (2013) Runoff prediction in ungauged basins: synthesis across processes, places and scales. Cambridge University Press
England JF Jr, Cohn TA, Faber BA, Stedinger JR, Thomas WO Jr, Veilleux AG, Kiang JE, Mason RR Jr (2019) Guidelines for determining flood flow frequency–Bulletin 17C. U.S. Geological Survey Techniques and Methods, no. 4-B5, p 148
Fischer S, Schumann AH (2021) Regionalisation of flood frequencies based on flood type-specific mixture distributions. J Hydrol X 13:100107
Fischer S, Schumann A (2022) Handling of the stochastic uncertainty of flood statistics in regionalisation approaches. Hydrol Sci J 67(10):1449–1465
Klemes V (1983) Conceptualization and scale in hydrology. J Hydrol 65(1–3):1–23
Klemes V (1986) Operational testing of hydrological simulation models. Hydrol Sci J 31(1):13–24

Chapter 15
Regionalisation of Hydrographs and Flood Scenarios

Abstract While in the previous chapters a framework was introduced that allows for the regionalisation of results of flood frequency analyses, it also of importance for ungauged catchments to obtain information on the design flood hydrographs. For this purpose, information on the flood volume as well as hydrograph-specific parameters like rise and recession time is required. With this knowledge, synthetic hydrographs can be obtained analogously to gauged catchments. This section proposes regionalisation approaches for these characteristics and thus extends the model chain introduced in Chap. 9 to the case of ungauged catchments.

Since design floods are often required for catchments where no or only short records of observed floods are available, regionalisation is a key challenge when dealing with design floods (Viviroli et al. 2009). In the previous chapters, the basics of type-specific regionalisation were introduced. In Chap. 14, it was shown how to regionalise the type-specific distribution to obtain regionalised flood quantiles for given return periods. Based on this type-specific regionalisation model, the proposed framework for synthetic flood scenarios introduced in Chap. 9 is extended to the case of ungauged catchments. Machine-learning-based algorithms are used to transfer the information on type-specific characteristics such as the timescale or rise and recession time of the flood events from gauged to ungauged catchments. By this approach, type-specific synthetic design flood hydrographs can be derived for ungauged catchments. The three main steps to derive such regionalised DFH are

1. Derivation of a regionalised flood quantile for a given return period T: see Chap. 14
2. Derivation of a regionalised type-specific flood volume
3. Generation of a regionalised synthetic type-specific design flood hydrograph.

These steps are the regionalised versions of the framework already introduced in Fig. 9.3. Methods to obtain steps 2 and 3 will be presented in this chapter.

S. Fischer and A. H. Schumann, *Type-Based Flood Statistics*, Water Science and Technology Library 124, https://doi.org/10.1007/978-3-031-32711-7_15

15.1 Regionalisation of Type-Specific Flood Volume

To obtain the type-specific flood volume for a given flood peak, in this case the flood quantile of return period T, the type-specific flood timescale as described in Sect. 6.2 is required, i.e., the slope of the regression between (direct) peak and volume, to describe the general relation between flood peak and volume. Of course, for ungauged catchments such a timescale is not available and hence has to be regionalised. But, regionalisation of this variable is not straight-forward. Gaal et al. (2012) report that

> the catchment area is not the most important control on the flood timescales.

A simple regression model based on the catchment area therefore is not suitable to regionalise the flood timescale. Fischer et al. (2019), too, report that only for flood type R1 a slight dependence of the flood timescale on the catchment area becomes visible while this is not the case for all other flood types. There, it is proposed to use the median of all flood timescales, instead. While this was reasonable for a small study region with only few catchments as the Mulde basin, for larger study regions such an estimation might be afflicted with too high uncertainty. Additionally, the flood timescale mirrors somehow the type-specific catchment reaction and hence there must be a relation to several catchment attributes.

When assuming a dependency or a pattern in the data without being able to specify this precisely, machine-learning algorithms might help to reveal these hidden structures. However, machine-learning algorithms often are hard to understand and to verify as they operate as black-box models. An alternative to these black-box models forms, e.g., the CART (Classification And Regression Tree) algorithm for continuous variables, which provides a classification based on decision trees. It solves a regression problem with the catchment characteristics as input data (independent variables) and the type-specific mean flood timescale as output (dependent variable). The regression is evaluated according to an error function, usually the sum of squares which is minimised. One starts with one knot where all input data are considered. Then, this knot is split into to child knots by defining the input variable that is able to split the data into two optimal subsets. Then, these child knots are split again, each into two new child knots. This procedure is repeated until some stop criterion is reached, e.g., the minimum number of data in a sub-sample. The algorithm was first published by Breiman et al. (1984). It is quite popular in hydrological sciences as the resulting classification tree makes the obtained classification hydrologically interpretable. Moreover, the relevance of each input variable for the outcome of the CART can be considered.

In theory, any input variables can be considered and the selection should be made dependent on the study region. To explain the utilisation of this methodology, the example of catchments from Bavaria is used again (See Sect. A.1). For these catchments in Bavaria, Germany, one could consider mean elevation, catchment size, hypsometric curve (clustered, see Chap. 13), nature class, and river network density as well as the D-day median precipitation sum (Eq. 13.5) as input variables for the CART. Then, the CART algorithm is applied type-specifically to all catchments with a training set consisting of 80% of the data and a test set consisting of the remaining 20% of the data. The estimation is repeated 100 times each to consider randomness in the classification into training and test set and the mean of all runs is taken. An example of one such resulting CART for flood type R2 is given in Fig. 15.1. Here, catchments of the lower and upper mesoscale located in the Alps or Alpine foreland with steep convex hypsometric curve have a mean flood time scale of 44 hours for flood type R2. Smaller flood timescales are only derived for catchments with high mean elevation above 730 m. The longest flood timescales of this type are obtained for very large catchments. Each of the resulting classes consist of at least 10 events, ensuring a meaningful classification.

Performance of the CART can be estimated by classical jackknife approaches (see Sect. 14.5) and application of, e.g., the RMSE or relative error. Exemplary results for the 164 catchments in Bavaria, Germany, are provided in Table 15.1.

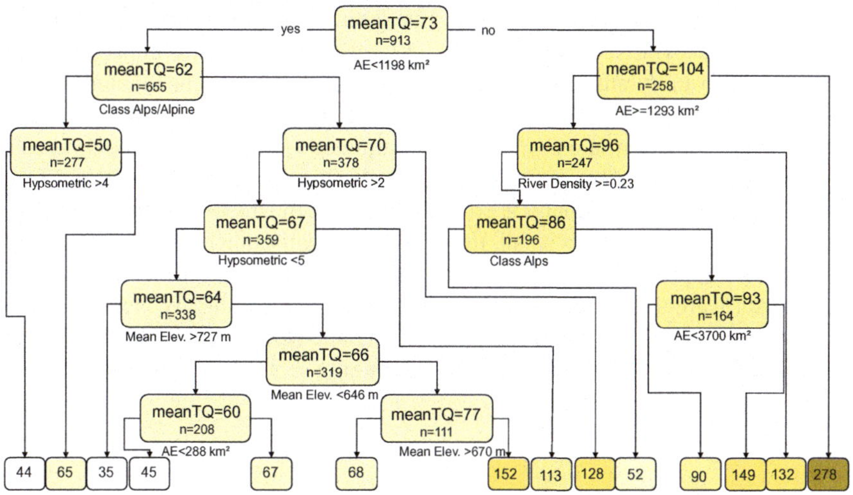

Fig. 15.1 CART of the mean flood timescale for flood type R2 for catchments in Bavaria, Germany

Table 15.1 RMSE, mean relative error and relevance of the catchment attributes for the decision of the CART in the regionalisation of the flood timescale of each flood type

	RMSE	Rel. error	Hypsometric curve (%)	Mean elevation (%)	Catchment area (%)	$Prec_D$ (%)	Nature class (%)	River density (%)
R1	10.31	–0.029	8	24	24	17	7	20
R2	28.09	–0.069	9	10	43	11	12	14
R3	66.83	–0.154	5	10	49	12	15	9
S1	21.41	–0.034	8	19	17	26	11	17
S2	82.72	–0.239	5	33	12	19	2.2	29

The results emphasise that the CART algorithm in general performs well for predicting the flood timescale for ungauged catchments. The relative error is around 5% for flood types R1, R2 and S1, which are usually characterised by single-peaked flood hydrographs of short to moderate duration. For flood types R3 and S2, the error is higher with larger variance. The flood events of these types usually are more complex due to several parallel flood-generating processes in the catchment. When evaluating the relevance of the catchment attributes, it is reasoned that the CART algorithm is hydrologically meaningful. While the timescale of type R1 is mainly defined by catchment elevation and size as well as river density, for the other two rainfall-induced flood types, R2 and R3, mainly the catchment size is relevant. For the short and steep hydrographs of flood type R1, associated with heavy rainfall, the catchment structure plays a crucial role in the flood generation: steep and small catchments with a high density of the river network and large differences in elevation generate faster and shorter hydrographs than flat catchments. For flood types R2 and R3, associated with long-duration rainfall respectively several successive rainfall events, however, mainly the catchment size defines the duration and therefore the flood timescale. The snow-impacted flood timescales are highly impacted by the catchment elevation range, which is of course correlated with the amount of snow cover. Flood timescales of type S1, rain-on-snow floods, additionally are impacted by the D-day precipitation sum $Prec_D$ (D was estimated to 4 days). For flood timescales of floods mainly caused by snowmelt, type S2, also the river density is relevant. Thus, the relevant catchment attributes are directly linked to the flood-generating processes and since the overall error is small, the CART-regionalisation of the flood timescale is a suitable approach for ungauged catchments. Different partitions into training and test set were tested, too, where the relative error only slightly increased when reducing the training sample to up to 50% of all data. However, the RMSE increased much and hence, at least for this region, the training sample size should be 70–80% of the data.

15.2 Regionalisation of the Synthetic Design Flood Hydrographs

When having regionalised the design flood peak and the flood timescale (and hence the flood volume) for each flood type, the next step is to obtain regionalised type-specific design flood hydrographs. To derive flood type-specific hydrographs for ungauged basins, the required flood event characteristics used in the framework introduced in Chap. 9 have to be regionalised, in particular the 5%- and 95%-quantiles of rise and recession time for each flood type to derive a parameter space for the fitting of the probability density functions. These attributes can be assumed to be similar for a hydrologically similar region and similar catchments size, since they are scale-dependent. Therefore, the k-nearest neighbour approach with catchment-size-based weights is used to derive these attributes for ungauged catchments, similar to the approach used in Chap. 13. More precisely, for each flood type, all flood events of the respective flood type of the ten nearest catchments (according to the catchment centroid) within the catchment class (i.e., lower or upper mesoscale respectively macroscale) are selected. For these events, the 5%- and 95%-quantiles of rise and recession times respectively are taken. Next, the integer-rounded weighted mean of all catchments is used to estimate a regionalised value, where the weights are chosen according to the similarity of the area of the respective catchment to the ungauged target catchment. This way, hydrological and catchment-size similarity are taken into account. The k-nearest neighbour approach is sufficient here, as only a rough span of the parameters is required for the optimisation of the distribution function. Therefore, small errors are acceptable.

For the estimation of the regionalised baseflow and the probability of each flood type to result in the given flood peak, additionally to the similarity of the regions also the flood peak magnitude has to be taken into account. The baseflow may vary much depending on the flood peak magnitude and so does the probability of each flood type, as was shown in Chap. 7. Therefore, for the k-nearest neighbour approach, only those events are considered which have a peak in the probability range of the flood quantile of the target return period T. For example, when aiming to obtain a type-specific flood hydrograph for the design flood of return period $T = 100$, then the ten nearest catchments are considered and for each catchment the observed flood events with a peak in the range of the largest 1% $(1/T)$ are considered, when having catchments with a record length close to the target return period. If the record length is not as long as required, the highest observed flood events in these time series should be used. To consider the scale-dependence of the baseflow, not the baseflow itself is taken for regionalisation but the relation between baseflow and catchment area. This means that the baseflow of the observed flood events of the gauged catchments is divided by the catchment area and the mean of those events in the specified peak range of the ten nearest catchments is taken. This relative baseflow is then rescaled to be considered for the flood event of the ungauged catchment by the catchment area of the ungauged catchment to obtain the regionalised baseflow.

The mean relative error, with a trimming of 5% to reduce the impact of single extraordinary large or small errors, for the regionalisation of each characteristic is summarised in Table 15.2 and the relative error for all catchments is given in Fig. 15.2. The relative errors for all three characteristics are comparably small except for flood types R3 and S2. These flood types were identified as challenging before due to the complex rainfall and snowmelt processes in different elevation zones in the catchment that lead to varying rise and recession times and multiple peaks and hence the higher errors are not surprising. The error for the 5% rise and recession times of flood type R1 and S1 is often close to zero, as these events are rather short and hence the lower threshold for the rise and recession time is one day. However, at the same time, an overestimation of these recession times of one day automatically leads to a relative error of 1. At this stage, it can be concluded that the k-nearest neighbour approach is suitable for regionalising flood event duration and timing of peak for each flood type except R3 and S2 (which will be discussed later on), when considering the catchment size and events of similar (relative) magnitude. The regionalisation of baseflow is

Table 15.2 Mean (absolute) relative error of the regionalisation of rise and recession time quantiles as well as baseflow, differentiated by the flood type

	Flood type	5% rise	95% rise	5% recession	95% recession	Baseflow
Rel. error	R1	0.001	0.037	0.062	0.094	0.113
	R2	0.019	0.048	0.086	0.088	0.102
	R3	0.143	0.173	0.105	0.077	0.003
	S1	0.047	0.062	0.094	0.127	0.123
	S2	0.243	0.221	0.383	0.220	0.093
Rel. abs. error	R1	0.156	0.209	0.181	0.291	0.397
	R2	0.260	0.289	0.276	0.300	0.401
	R3	0.866	0.762	0.712	0.664	0.311
	S1	0.208	0.270	0.305	0.327	0.414
	S2	0.483	0.514	0.630	0.438	0.422

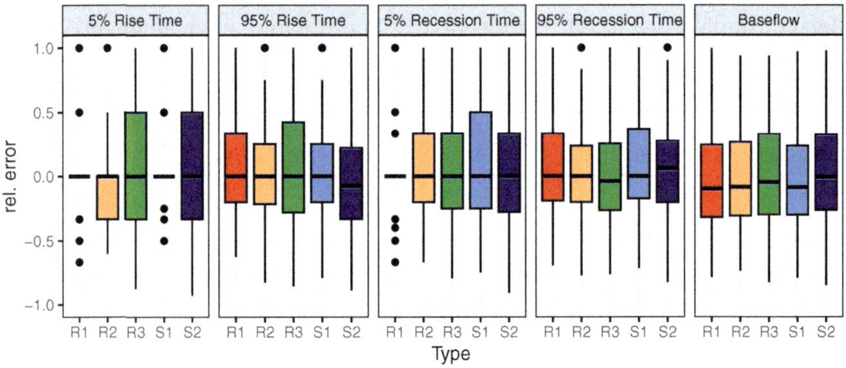

Fig. 15.2 Relative error of the regionalisation of the hydrograph characteristics differentiated by flood type

affected by higher errors but baseflow is only a proxy to obtain a meaningful level of the hydrograph. Thus, it does not impact the generation of the hydrograph itself.

To finally obtain regionalised synthetic design flood hydrographs with the framework proposed in Chap. 9, one last parameter has to be regionalised: the proportion of multiple-peaked flood events for flood types R3 and S2. Multiple-peaked flood events occur due to two phenomena: either successive rainfall events (flood type R3) respectively successive snowmelt periods in different elevation ranges (S2) or superposition of floods from different tributaries in a catchment. While the first phenomenon is clearly related to the meteorological conditions of the floods and (for events with multiple peaks) is affected strongly by high soil moistures, the latter is related to the catchment structure. While for meteorological conditions it can be expected that they are similar for neighbouring catchments, flood superposition depends mainly on the size of the catchment. The larger the catchment, the more differences are to be expected in the temporal distribution of precipitation and thus between the resulting flood discharges of the subbasins. Based on these assumptions, a type-specific regionalisation approach for the proportion of multiple-peaked events is chosen. For flood type R3, the meteorological conditions clearly dominate the origin of multiple-peaked events. Since precipitation is available also for ungauged catchments, it can be used for regionalisation. More precisely, for the ungauged catchment, all precipitation events are separated from the daily areal precipitation series. This is done by considering all days with a precipitation larger than 0.1 mm, a typical threshold for identifying precipitation (Reiser and Kutiel 2009). Then, only those precipitation events are considered that are assumed to be related to an R3 flood event. These are events which exceed a certain duration, typical for R3 floods. This duration can be obtained by from neighbouring gauged catchments of the same catchment class (i.e., lower or upper meso- or macroscale) by taking the 10% quantile of all durations of the precipitation corresponding to R3 flood events. The use of the 10% quantile ensures that extraordinary events are excluded from the estimation. Finally, the proportion of precipitation events that probably lead to multiple-peaked floods is estimated. Such events have to have a clear multiple-peaked temporal distribution of rainfall. Hence, the same criteria as for the identification of multiple-peaked flood events is applied: the second-largest precipitation peak has to be at least 50% of the maximum precipitation peak and has to be more than 30% larger than the minimum between both precipitation peaks. By this approach, the assumption for the identification of multiple-peaked flood events is transferred to their meteorological causes.

For the snowmelt floods (S2), precipitation only plays a minor role in the generation of multiple-peaked events. Instead, the topographical structure of the catchment is crucial. Depending on the different elevation ranges in a catchment, different phases of snowmelt occur that then lead to multiple peaks of flood events. Therefore, it can be assumed that catchments with similar hypsometric curve have a similar proportion of multiple-peaked flood events. To take this into account, the hypsometric clusters introduced in Sect. 13.3.2 are considered and the regionalised proportion of multiple-peaked flood events for flood type S2 is obtained by taking the mean proportion of all gauged catchments that are in the same hypsometric cluster as the ungauged catchment.

The mean relative error with a trimming of 5% of the regionalisation of the proportion of multiple-peaked flood events then lies by −11.4% for flood type R3 and 9.9% for flood type S2, respectively 36.0% and 25.7% when considering the absolute relative error. It has to be considered that only two-peaked flood events are modelled by the framework, as described in Chap. 9.

With the regionalised type-specific hydrograph characteristics rise and recession time, baseflow and the proportion of multiple-peaked flood events together with the type-specific regionalised peak from Chap. 14 and the regionalised flood volume, synthetic design hydrograph for ungauged basins can be derived analogously to Chap. 9. The methodology to obtain the required regionalised variables to generate synthetic type-specific design hydrographs is summarised in Table 15.3. The

Table 15.3 Summary of the methodology to obtain regionalised variables for the generation of type-specific synthetic design flood hydrographs

Variable	Input data	Methodology
Flood peak (TMPS-model):		
· Shape parameters	Type-specific: nature class, hypsometric curve, elevation range, retention capacity, intensity, $Prec_D$	Hierarchical clustering and median
· Scale parameters	· Mesoscale (type-specific): nature class, catchment size, elevation range, retention capacity, intensity, $Prec_D$	Hierarchical clustering and median
	· Macroscale: catchment size	Two-fold type-specific regression for each catchment class
· Location parameters	Catchment size	Two-fold type-specific regression for each catchment class
Flood volume (flood timescale)	Hypsometric curve, mean elevation, catchment size, $Prec_D$, Nature class, river density	Type-specific CART
Hydrograph characteristics:		
· 5%/95% rise time	–	Weight. type-sp. k-nearest neighbour
· 5%/95% recession time	–	Weight. type-sp. k-nearest neighbour
· Baseflow	–	Type-specific k-nearest neighbour of events in same peak range
· Proportion of multiple- peaked events	R3: areal precipitation	Mean
	S2: hypsometric curve	

final step of the framework, the generation of hydrographs, does not depend on any catchment-specific assumptions, as probability density functions are used. The resulting hydrographs are given in Fig. 15.3, where the catchments Achsheim, Garmisch and Oberach (the same as in Fig. 9.4) were treated as ungauged and all parameters were estimated based on regionalisation. The resulting hydrographs are quite similar to those obtained when using observed input characteristics (Fig. 9.4). Again, it has to be taken into account that the results might profit much from the dense gauging network in the region under consideration. In data sparse regions, the error of regionalisation might increase or certain input data like the nature class might not be available. In such a case it is recommended to revise the framework by explicitly taking into account the region-specific flood-generating processes.

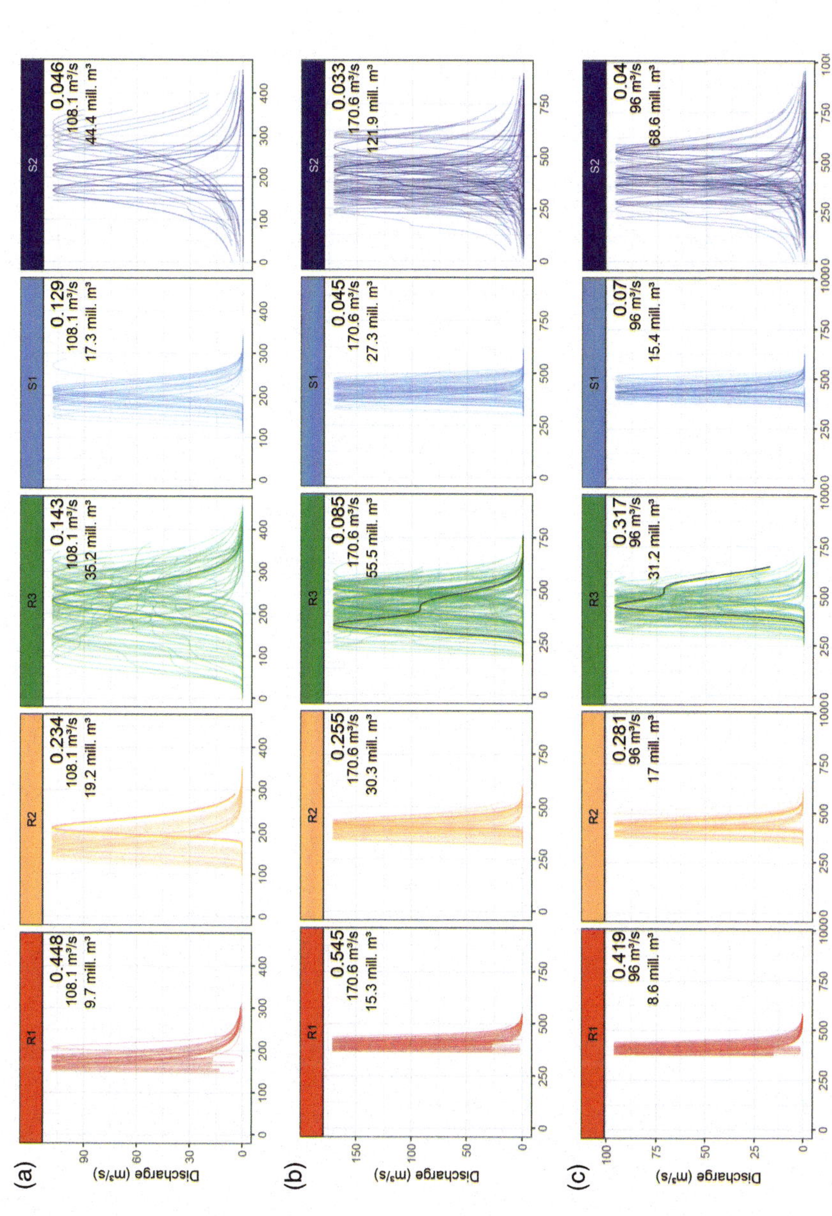

Fig. 15.3 Regionalised synthetic design flood hydrographs for catchments **a** Achsheim (359 km^2), **b** Garmisch (249 km^2) and **c** Oberach (96 km^2)

References

Breiman L, Friedman JH, Olshen RA, Stone CJ (1984) CART: classification and regression trees. Wadsworth, Belmont, CA

Fischer S, Schumann A, Bühler P (2019) Timescale-based flood typing to estimate temporal changes in flood frequencies. Hydrol Sci J 64:457–470

Gaál L, Szolgay J, Kohnová S, Parajka J, Merz R, Viglione A, Blöschl G (2012) Flood timescales: understanding the interplay of climate and catchment processes through comparative hydrology. Water Resour Res 48(4)

Reiser H, Kutiel H (2009) Rainfall uncertainty in the Mediterranean: definitions of the daily rainfall threshold (DRT) and the rainy season length (RSL). Theoret Appl Climatol 97:151–162

Viviroli D, Mittelbach H, Gurtz J, Weingartner R (2009) Continuous simulation for flood estimation in ungauged mesoscale catchments of Switzerland-Part II: parameter regionalisation and flood estimation results. J Hydrol 377(1–2):208–225

Chapter 16
A Deterministic-Based Model to Apply Type-Based Statistics for Ungauged Catchments

Abstract In the previous Chap. 15, a methodology was proposed to directly regionalise the statistical model for consistent stochastic simulations of flood peaks, volumes and shapes of hydrographs under consideration of flood types. However, in practise often a different approach is required: the regionalisation of a deterministic model to generate flood series for which then flood frequency analyses can be performed. By using a deterministic model, it is possible to determine the complete flood hydrograph using long precipitation series, which are usually available more frequently than runoff series. However, the demands on deterministic models are high if they are to be applied to unobserved watersheds and the diversity of possible flood events. The parameters should have a physical meaning and the model should be applicable for different hydrological processes alike (e.g., flow from saturated areas, runoff processes driven by rainfall intensities, snow retention etc.). There is some contradiction in using a model with a small number of meaningful parameters and representing the complex processes of flood formation. A deterministic model is suitable for all situations in which the processes taking place are completely determined by known initial and boundary conditions. This is not the case for flood generation, where only the meteorological drivers are known. However, deterministic models can also incorporate stochastic aspects by incorporating randomness into certain parameters or inputs. Here, a simple deterministic model is proposed that takes into account the differences between flood types by considering different input distributions for the initial conditions, the main parameters and the meteorological drivers. More precisely, a model chain is proposed that can be used to obtain not only the peak for a given return period, but the flood-type specific flood volume corresponding to the flood peak. The generation of these pairs, which are identical in their statistical distribution to the observed ones, is obtained solely by combining the distribution of the precipitation sum, the runoff coefficient and a relation between flood peak and volume. This allows for a regionalisation of the stochastic event generator without considering additional catchment, precipitation or runoff attributes. Based on these generated samples, flood frequency analyses can be performed for the ungauged basins, confidence bands can be derived as well as a generation of synthetic design hydrographs.

S. Fischer and A. H. Schumann, *Type-Based Flood Statistics*, Water Science and Technology Library 124, https://doi.org/10.1007/978-3-031-32711-7_16

Classical regionalisation approaches like those introduced in the previous chapters aim to transfer runoff information from hydrologically similar gauged catchments to an ungauged catchment. An alternative offer deterministic models, where a given design precipitation is used and the model delivers the corresponding flood event. The latter approach has the advantage that it is easily applicable and, once calibrated, can be flexibly used to obtain several rainfall-runoff events. However, it has the drawback that the model parameters have to be calibrated, which requires data from gauged catchments and a regional transfer of the estimated model parameters to ungauged catchments. Moreover, most deterministic models cannot be extrapolated beyond the range of observed discharge, as their calibration and validation is based on the observed floods. At the same time, often the assumption is made that the probability of the input precipitation event is equal to the probability of the outcome flood event. Of course, this assumption cannot be justified physically and is highly questionable, having in mind the high impact of the catchment state on the runoff generation.

Instead, it should be the aim to generate a sample of flood events which is representative for the variety of floods which has to be expected based on information from flood analyses of gauged neighbouring catchments. This can be achieved by a stochastic event generator. Instead of fixed input parameters, probability distributions of the input parameters are used in a deterministic model. This increases the flexibility of the output and takes into account the uncertainty of the input variables. Examples are the Stochastic Event Flood Model (Schaefer and Barker 2002) or the seasonal approach by Filipova et al. (2019). Still, in these approaches the deterministic component remains and flood-generating processes have to be modelled. One could go also one step further: Instead of deterministic models, the flood characteristics peak and volume are simulated statistically. Instead of each event, the individual distribution is simulated here. From the statistical perspective, not the exact simulation of each flood event is important but the general long-term statistical behaviour. If this behaviour is known, the distribution function can be derived and quantile estimation can be performed. This can be derived by regionalising the statistical model directly, as shown in Chap. 14, or by applying a stochastic generator, as shown in this chapter.

A drawback of most deterministic models is that they are often not able to distinguish between flood-generating mechanisms without being hopelessly overparametrised. Because the observed events are caused by different meteorological factors, the deterministic model cannot account for the specific aspects of runoff formation without randomness. An example is the problem of distinguishing between surface runoff resulting from waterlogged soils and surface runoff resulting from high precipitation intensities. Often, both processes exist synchronously in a catchment, but in different spatial manifestations. Again, the two flood events in 2002 and 2013 in the Mulde basin, already used in Chap. 4, provide an excellent example on how different the generating mechanisms of floods can be (Fig. 16.1). In 2002, clearly the high intensity of rainfall led to the extraordinary flood peak. In 2013, instead, the rainfall first led to wetted soils which then quickly generated runoff when the next rainfall event hit the catchment. Deterministic models should be able to distinguish between these mechanisms but often lack the ability to do so. Another unsolved

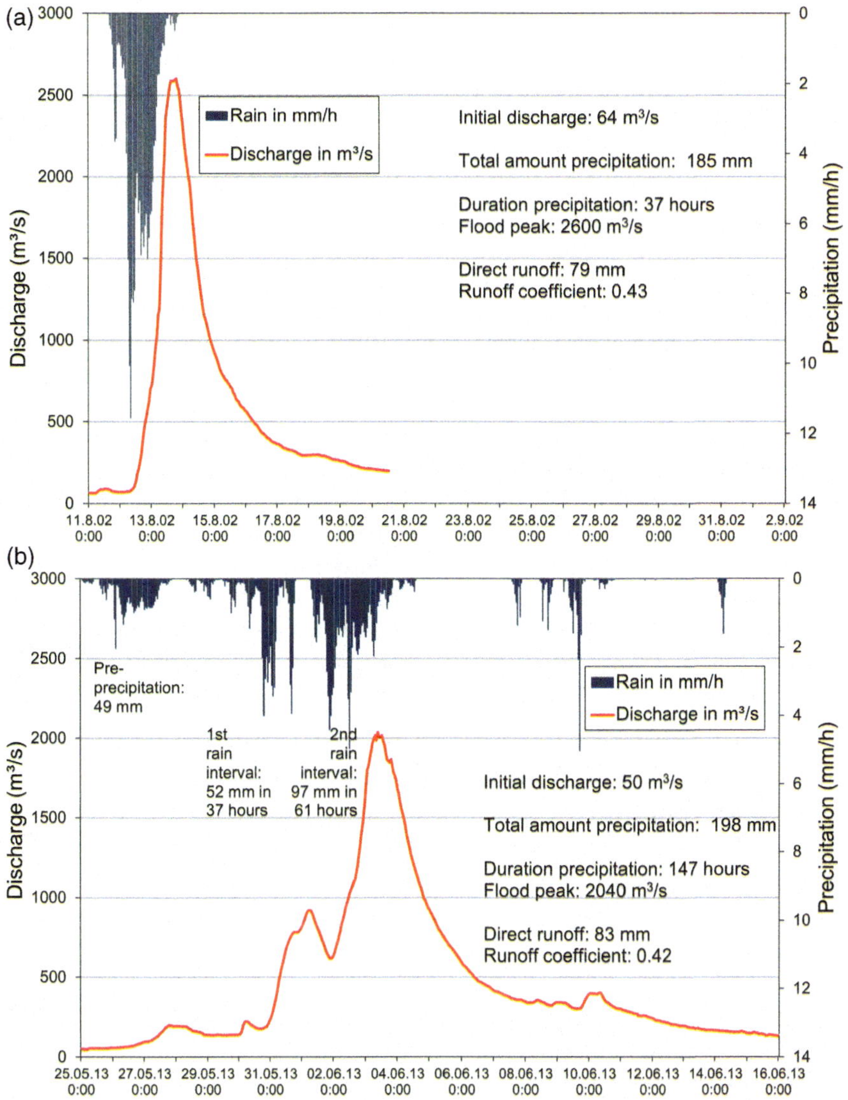

Fig. 16.1 **a** 2002 and **b** 2013 flood events in the Mulde river basin, eastern Germany, and the corresponding precipitation events

problem is the change of the hydraulic regime in flood wave propagation models if the bank-full runoff is exceeded.

To overcome this issue, it is attempted to compensate the stochastic character of the input and state variables by mean or effective parameters, but this does not allow sufficient consideration of the inherent uncertainty of the model applications. In the framework proposed here, the stochastic characteristics of different meteorological drivers, initial moisture conditions and runoff coefficients are considered explicitly. By application of a flood typology (Sect. 6.2), the large range of the meteorological drivers is subdivided into homogeneous classes. By using type-wise simulations, the different geneses of the flood events can be considered. Such a stochastic event generator for different flood types was proposed, e.g., by Fischer et al. (2019). Here, an extension of this approach, considering the hydrological components more directly, will be introduced in detail in this chapter. The simulation is based on the main factors of flood-generation: rainfall, snowmelt and the runoff coefficient, describing the retention capacity of a catchment. However, such a simplification is not applicable to any catchment. At the macroscale, catchments are affected by more complex spatial interactions, such as heterogeneous distributions of precipitation, flood superpositions from tributaries (Chap. 11) or different multiple processes at various locations in the catchment. Hence, such a statistic generation should be limited to catchments of the mesoscale with catchment size smaller than 1000 km^2. Due to the simplified structure of the generator, it is much easier to regionalise it compared to classical deterministic models. If observed time-series of precipitation are available for ungauged catchments, only the runoff coefficients as well as the relation between flood peak and volume have to be regionalised. This will be demonstrated in this chapter. The resulting flood events can be used for flood frequency analyses and compared to those derived in the previous Chap. 14 by regionalisation of the statistical model.

16.1 Generation of Type-Specific Pairs of Flood Peak and Volume

Flood types differ notably in their flood-generating processes. In the most simplified way, these flood-generating processes can be described by estimating the volume of direct runoff from the event precipitation sum P_{ev} with a runoff coefficient Ψ:

$$Vol_{dir} = P_{ev} \cdot \Psi. \tag{16.1}$$

The parameter "runoff coefficient" steers how much of the available precipitation is transferred into runoff, thus describing basically the available soil storage capacity and the percolation into deeper soil zones. For rainfall-induced flood types R1, R2 and R3, the duration and available amount of precipitation as well as the runoff coefficient differ significantly, such that a joint consideration is not meaningful. For example, R3 flood events associated with long-duration rainfall and several successive rainfall events are usually related to wet soils such that the runoff coefficient is high.

For snowmelt-induced floods, additionally the amount of water obtained from snowmelt corresponding to the event, S_{ev}, has to be considered:

$$Vol_{dir} = (P_{ev} + S_{ev}) \cdot \Psi. \tag{16.2}$$

In the stochastic event generator, the observed precipitation, snowmelt and runoff coefficient are considered by their statistical distributions only, and a suitable statistical model has to be found for these. Moreover, all three variables (P_{ev}, S_{ev}, Ψ) have to be simulated for ungauged catchments, i.e., based solely on statistical information from gauged catchments in the region. Depending on the required spatial and temporal resolution, precipitation modelling can be quite complex. Hence, there exist various statistical models that aim to describe this behaviour, such as the Neyman-Scott or the Bartlett-Lewis processes, where simple distributions like the Poisson or Exponential distributions are combined in a nested way (Cowpertwai et al. 2002). By a distinction into flood types, which implies a categorisation of flood inducing precipitation, the complexity of the precipitation distribution is reduced. Implicitly, precipitation is differentiated by intensity and duration, e.g., by differentiation into heavy rainfall events or long-duration events. This simplifies the statistical model that is required to describe the statistical distribution of the amount of rainfall, as single distributions for each type have to be found instead of a mixture model. Hence, for the deterministic-stochastic event generator, the following variables have to be simulated by a suitable type-specific statistical distribution:

- rainfall sums related to floods for five different flood types: R1, R2, R3, S1, S2
- snowmelt sums related to floods for two different flood types: S1, S2
- runoff coefficients for five different flood types: R1, R2, R3, S1, S2.

In theory, any statistical distribution may be used to model the three initial characteristics. When considering catchments in Central Europe, distinct distributions proved to be superior: By application of the Anderson-Darling test and the information criteria AIC and BIC, the PearsonIII distribution is selected as best-fitting for P_{ev} from a sample of ten standard distribution functions for each flood type separately. The same holds true for the distribution of the amount of snowmelt S_{ev}, which is required for flood types S1 and S2. Of course, the selection of a fitting distribution may differ when the generator is applied to other regions in the world, where the flood-generating processes differ.

In theory, also the runoff coefficient can be modelled by a suitable type-specific distribution function. For example, (Fischer et al. 2019) propose to consider the Beta distribution for the runoff coefficient. Then, the flood volume is estimated by taking all possible combinations of precipitation sum and runoff coefficient to obtain a broad spectrum of flood volumes for a given flood type. However, this is contra-intuitive to hydrological process knowledge. In fact, the runoff coefficient depends to a certain extent on the sum of precipitation. Another factor which determines the reaction of a catchment is the initial soil moisture at the beginning of the flood event. In a wet catchment, even a relatively small amount of rain can cause a flood. At the same time, a large amount of rain will lead to a flood event, even if the soil storage is almost empty when the rain starts. The relation between precipitation and runoff coefficient depends much on the soil storage capacity in the catchment but also on the temporal distribution of rainfall, especially on the occurrence of high rainfall intensities. The importance of both factors becomes visible when comparing the relation between antecedent soil moisture, runoff coefficient and the empirical

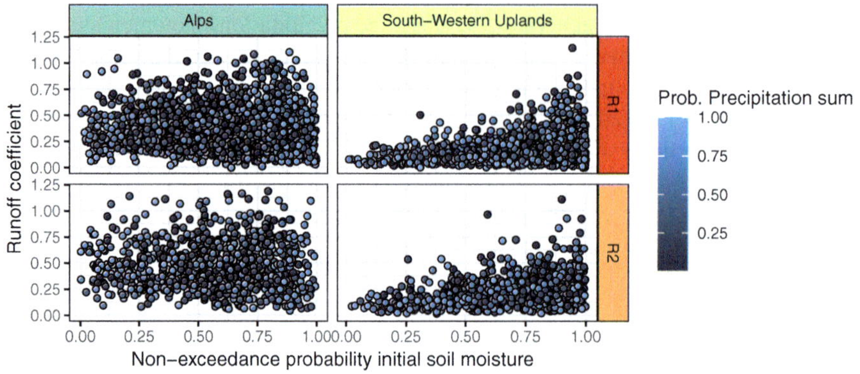

Fig. 16.2 Relationship between initial soil moisture (specified by the proxy initial discharge), runoff coefficient and precipitation sum for catchments in Bavaria, Germany and selected flood types, differentiated by the nature class

non-exceedance probability of the rainfall sum of the respective flood type for the catchments in Bavaria, Germany (Fig. 16.2). As a proxy for the initial wetness state of soils in a catchment, the empirical non-exceedance probability of the discharge at the beginning of each flood event estimated from all discharges at the same calendar day which do not belong to a flood event is used. While in the Alps the runoff coefficients are uniformly distributed over all degrees of antecedent soil moisture, a clear increase of the runoff coefficient becomes visible for the South-western Uplands for increasing initial soil moisture. In the Alps, the soil layer is thinner than in the uplands, leading to a smaller storage capacity and hence flood retention even when the soils are dry. The soil moisture in these regions thus plays a minor role. In the uplands, where the upper soil layer is thick, the initial soil moisture has a crucial impact on the storage capacity of the soils and hence the precipitation that is transformed into runoff. Moreover, for a small antecedent soil moisture, high precipitation sums are required to result in a flood at all. For wet soils, even small a amount of precipitation may result in a flood event with small to moderate runoff coefficients. If high soil moisture occurs together with large precipitation sums, the runoff coefficients are highest, leading to large amounts of runoff. The relation of antecedent soil moisture, precipitation and runoff also depends much on the intensity of the rainfall events. For flood type R1—the heavy-rain floods with high intensity—these dependencies become more tight with higher runoff coefficients than for floods caused by long-duration rainfall (R2), where the intensity is usually lower.

The detected dependencies between the antecedent soil moisture, the precipitation and the runoff coefficient can be used in the stochastic event generator to derive meaningful combinations of rainfall (for flood types S1 and S2, in combination with snowmelt), initial soil moisture and runoff coefficient and hence a meaningful range for the resulting flood volume. This is hydrologically more reasonable than simply taking the full range of all possible runoff coefficients into account and combining these with all possible precipitation sums, as would be the case when using the Beta distribution. For this purpose, a CART approach is used to determine the runoff coefficients, taking into account these influencing factors (see Sect. 15.1 for more details on CART). More precisely, for each flood type, the CART is trained type-specifically on the sample of the type-specific precipitation sum and the discharge at the beginning (again as a proxy for antecedent soil moisture) of each observed flood event in the study region. To consider scale-dependence, not the precipitation itself is considered but its empirical probability, derived by application of the empirical distribution function of all rainfall events (even if those did not result in a flood event). It was shown in the previous investigation (Fig. 16.2) that the runoff coefficient does not only depend on the flood type but also on the region and the catchment size. Therefore, a further distinction is made for the CART: the catchments are grouped according to their nature class (in this case Alps, Alpine foreland, South-western Uplands and Eastern Uplands) and according to their catchment size (<300 km^2, 300–500 km^2, 500–1000 km^2). One could also think about a further classification, e.g., according to the land use. Catchments with a large proportion of forest will probably have a smaller runoff coefficient than catchments with large proportion of agricultural areas. However, there was no evidence in the data for such coherences and hence a further reduction of the sample size for training the CART is omitted. However, it has to be noted that extreme cases, e.g., large urban areas in a catchment, cannot be captured by this approach. For other study regions, different classifications might be used. Hence, before application of the CART, always an analyses of the coherence between initial soil moisture, precipitation and the runoff coefficient should be done. After training with observed events, in the synthesis the CART is applied to the simulated samples of rainfall (and snowmelt, for S1 and S2) using the distributions selected above and the probability of initial soil moisture to specify a runoff coefficient. To specify the exceedance probability of the initial wetness conditions, a uniform distribution in the interval [0, 1] is used. As input for the CART, all possible combinations of the input data are considered. This then results in a sample of runoff coefficients that can be assigned to a certain input of rainfall and soil moisture conditions. From the trained CART, it also known how often each class (i.e., the leaves of the CART decision tree) was addressed. These relative frequencies of elements in the leaves are transferred to the sample of simulated pairs of rainfall and runoff coefficient such that a representative frequency distribution of the random combinations is obtained. This way, combinations of rainfall and initial soil moisture that do not lead to a flood event are automatically excluded from the sample, as they cannot be assigned to any class of the trained CART (the CART is trained based on flood events only).

To test the applicability of the CART, it is first applied to the catchments (grouped by catchment size and nature class) in Bavaria in Germany with different sizes for training and test sets. It turned out that even 60% of training data lead to comparably small errors in prediction. The predicted values for the runoff coefficient in mean are close to the observed ones with a mean error of less than 10% (Fig. 16.3). Moreover, the relevance of each input variable for prediction clearly represents the flood-generating mechanisms of each flood type. For R1 and S1 floods, which depend strongly on rainfall intensities, precipitation is much more important than the initial

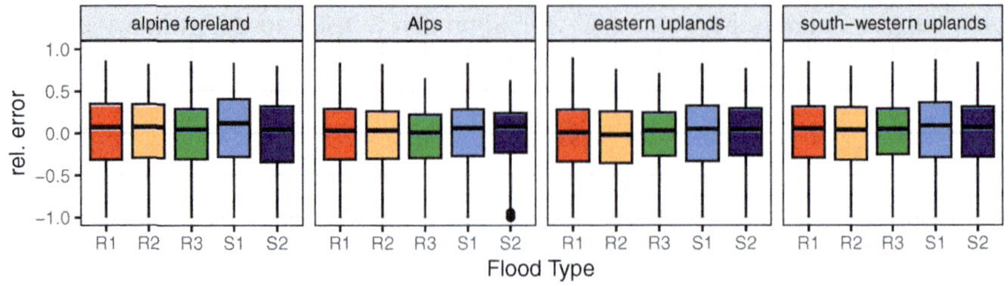

Fig. 16.3 Relative error between observed and predicted runoff coefficient (via CART, 60% training data) differentiated by nature class and flood type

wetness state, while it is the opposite for R3 and S2 floods. For R2 floods, both characteristics are equally important for the prediction.

However, there are also some quite large deviations, especially for flood types R1 and S1 in the Alpine Foreland and the South-western Uplands where high runoff coefficients are underestimated. There are two reasons for these differences:

1. The data base to train the CART is small as only few flood events have such high runoff coefficients.
2. As the unknown rainfall intensities may have a great impact on runoff coefficients for these flood types, only a small or no causal relationship between runoff coefficients, initial soil moisture and precipitation sums could be determined by the CART.

The second aspect is a general challenge for other flood types, too. As already shown in Fig. 16.2, large runoff coefficients may occur mainly when high initial soil moisture is combined with high precipitation. However, the same combinations can also result in small runoff coefficients, leading to a broad variance of runoff coefficients in this spectrum. Thus, the CART algorithm cannot make out any relationship between the input variables and the runoff coefficient. From the hydrological perspective, this phenomenon can be explained by two factors: Using daily precipitation data, it is impossible to consider the temporal variability of rainfall intensities sufficiently. In the CART algorithm, only precipitation sums over several days were considered. However, especially for floods of type R1 and S1 which are affected by convective cells and other short rain events with high intensities, the runoff formation depends on rainfall intensities that exceed the infiltration intensities at the upper soil zone. If the same amount of rainfall falls down in a short time span, this will increase the runoff coefficient as less water can be infiltrated into the soil compared to rainfall events with a uniform intensity over a longer period. Second, also the error in rainfall records has to be considered. Short and intense rainfall events often occur on a small spatial scale. Under unfavourable conditions, the measured rainfall data are not representative for a catchment. Any underestimation of the event rain may cause an overestimation of the runoff coefficient. Again, flood types R1 and S1 are affected most by this issue. In view of these problems with large runoff coefficients for flood types R1 and S1, CART cannot be used for the full spectrum of runoff coefficients.

Instead, large values of runoff coefficients are obtained by simulating these values by application of a statistical distribution, similar to precipitation and snowmelt. More precisely, runoff coefficients up to the 75% quantile are derived type-specifically by the CART algorithm, while the upper 25% are derived from the upper 25% of the sample generated with a type-specific statistical distribution. For this purpose, the Beta distribution is selected. The Beta distribution, denoted with $Beta(\alpha, \beta)$, has the distribution function

$$F(x) = \begin{cases} 0, & \text{if } x \leq 0 \\ \frac{1}{B(\alpha,\beta)} \int_0^x u^{\alpha-1}(1-u)^{\beta-1}\mathrm{d}u, & \text{if } 0 < x \leq 1 \\ 1, & \text{if } x > 1, \end{cases} \tag{16.3}$$

where $B(\alpha, \beta)$ is the Beta-function defined as

$$B(\alpha, \beta) = \frac{\Gamma(\alpha)\Gamma(\beta)}{\Gamma(\alpha+\beta)} = \int_0^1 u^{\alpha-1}(1-u)^{\beta-1}\mathrm{d}u$$

and serves as norming factor. $\alpha, \beta > 0$ are the parameters of the distribution and define the shape of the distribution function. The Beta distribution can exhibit quite different density shapes and is flexible when aiming to model different frequency distributions (Fig. 16.4). The upper limit of one corresponds to the upper limit of the runoff coefficient, since no more than 100% of precipitation can be transformed into runoff. However, for some events runoff coefficients greater than 1 might be observed. Such artefacts result from the precipitation observations, which might be underestimated due to interpolation from point observations to the area. When

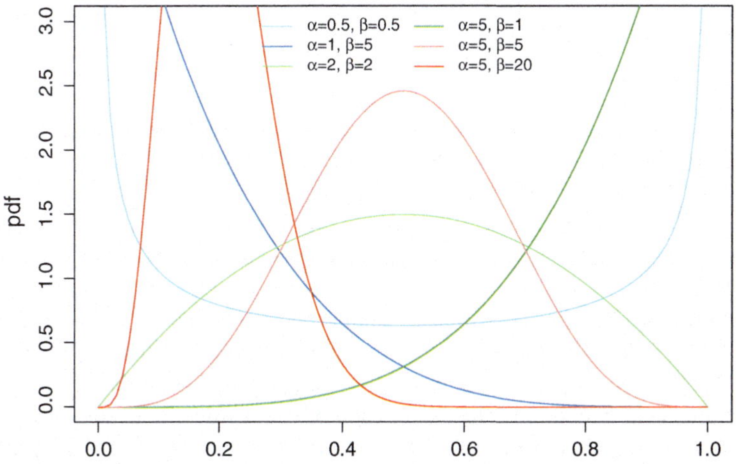

Fig. 16.4 Probability density function (pdf) of the Beta distribution for different parameter combinations

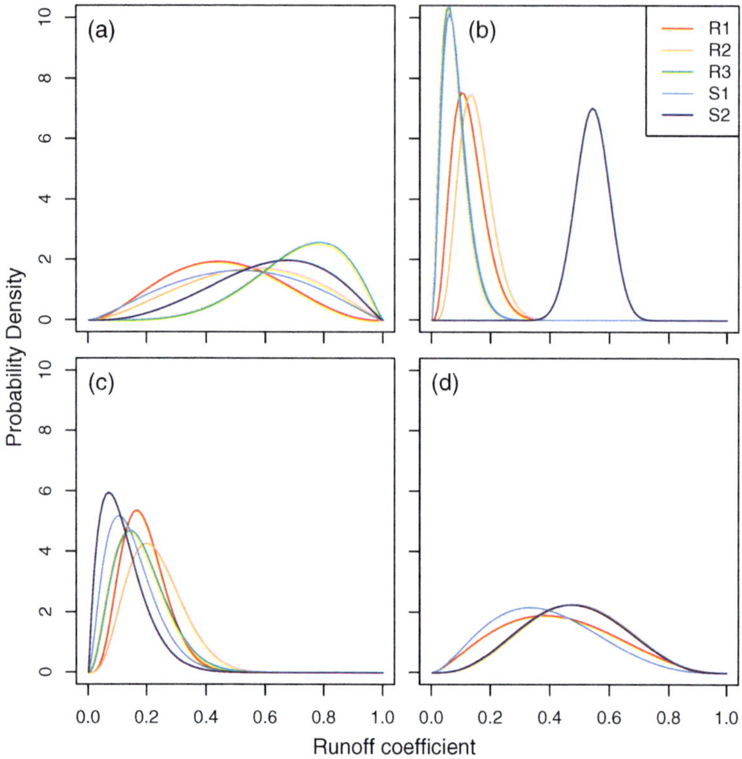

Fig. 16.5 Examples of the probability density function of the Beta distribution fit to the sample of runoff coefficients of flood events of different flood types for **a** a catchment in the Alps (Oberach/Weissach, 96 km^2), **b** a catchment in the Alpine foreland (Achsheim/Schmutter, 359 km^2), **c** a catchment in the South-western uplands (Bayreuth/Roter Main, 330 km^2) and **d** a catchment in the Eastern uplands (Wallenfels/Wilde Rodach, 96 km^2)

applying the Beta distribution to the runoff coefficients of flood events in catchments in Bavaria, Germany, indeed quite different shapes of the probability density can be observed (Fig. 16.5). Depending on the region of the catchment and the flood type, symmetric or skewed densities occur with varying location and variance. A distinction into different flood types and the consideration of the region hence is required.

In summary, for the deterministic-stochastic event generator, first a CART is trained with observed samples of precipitation, snowmelt, initial soil moisture and runoff coefficients, based on clusters of catchments grouped by their catchment size and nature class. Second, for the catchment under consideration, simulated samples of precipitation (and snowmelt) are obtained with the PearsonIII distribution and used as input data for the trained CART to estimate the corresponding runoff coefficient. The so derived samples of simulated precipitation sum and the corresponding runoff coefficient are then combined to obtain the type-specific flood volume using Eqs. 16.1

or 16.2. Knowing the flood volume, the corresponding flood peak for rainfall-driven floods is obtained by application of the type-specific mean flood timescale (see Eq. 6.3). For the snow-induced floods, however, the application of the flood timescale is not meaningful (see Sect. 6.2). Instead, a non-linear model of the form

$$Q_P = \frac{\alpha}{Vol^{\beta}} + \gamma_1 Hyps + \gamma_2 Elev,$$

is fit to all catchments with Q_p and Vol being the normalised flood peak and volume (i.e., divided by the catchment area) of flood type S1 respectively S2, $Elev$ being the mean elevation of each catchment and $Hyps$ being the hypsometric cluster (see Chap. 13), a categorical dummy variable. The parameters $\alpha > 0$ and $\beta \in (-1, 0)$ result in a concave shape of the resulting model. This way, a non-linear relation between flood peak and volume for each snow-induced flood type can be derived and the flood peak can be obtained from the simulated flood volume.

Hence, type-specific samples of flood peak and volume of arbitrary length but with the same statistical distribution as the observed samples can be generated.

The methodology for the stochastic event generator is condensed in Fig. 16.6, where also exemplary probability densities of each component and flood type are given. It can be seen how the runoff coefficients are derived from these probabilities of the total amount of precipitation, snowmelt and initial moisture state of the catchment in a type-specific way. For example, the sum of precipitation for flood type R1 is more skewed than for type R3. A joint consideration of all events hence would falsify the distribution and the simulation results. A flood typology can help to derive the differences between the events and to correctly implement them in the generator.

So far, some of the assumptions made above for the application of the stochastic event generator are still based on observations and records in a catchment that are not available for ungauged catchments. This applies to the distribution of precipitation and snowmelt which is fit only to those characteristics that belong to a flood event. However, not every precipitation (and snowmelt) leads to a flood. For ungauged catchments, only precipitation data are available but no information whether these lead to a flood event or not. In the CART, this is considered as only those precipitation and snowmelt events above a certain probability are considered. This threshold is derived from the probabilities of the observed rainfall and snowmelt sums which lead to floods in gauged catchments and are used for training the CART. Moreover, only those combinations are considered that may lead to a flood, as only those classes exist for the trained CART. For regionalisation, a similar threshold has to be considered, too. Second, the distribution of the runoff coefficients used for flood types R1 and

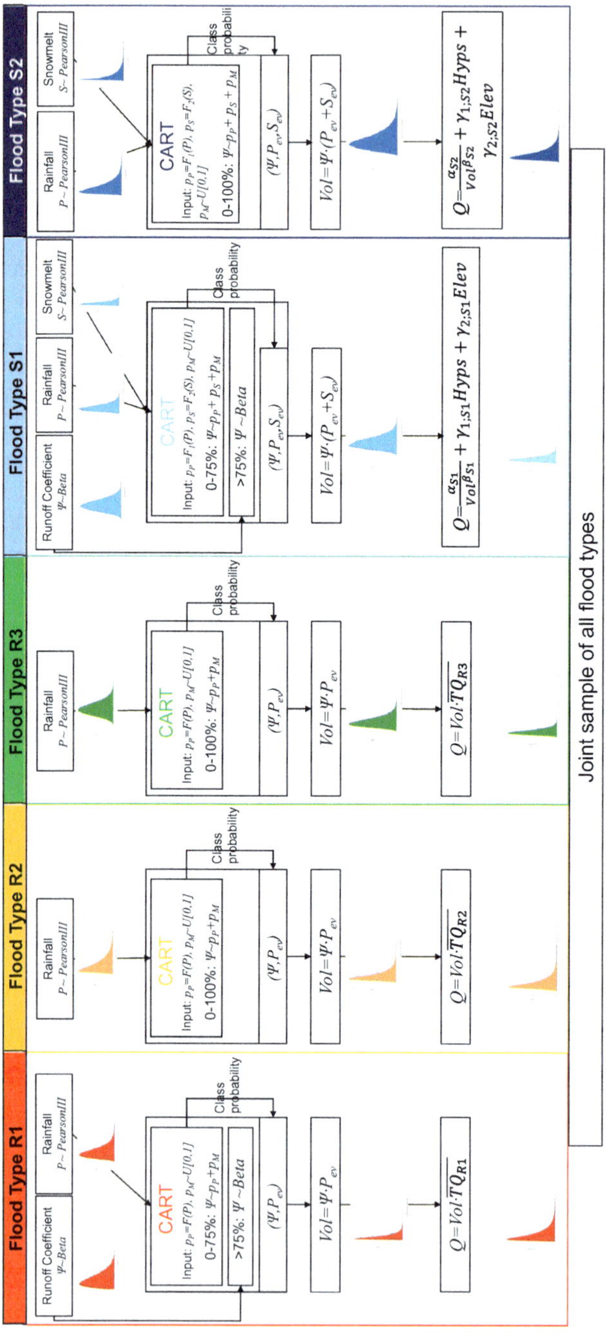

Fig. 16.6 Schematic overview of the methodology for the stochastic event generator and exemplary densities of each component and flood type

S1 is unknown for ungauged catchments. Finally, also the flood timescale for the rain-driven floods has to be regionalised. How all this can be done is described in the next section.

16.2 Application of the Deterministic-Stochastic Event Generator to Estimate the Sampling Effect in Flood Statistics

The sampling effect is a major challenge in flood frequency analysis and leads to a large uncertainty in the estimation of flood quantiles. Often, only short records are available and one extraordinary large flood modifies the statistical parameters and the extrapolation of the distribution function. To take into account stochastic uncertainty resulting from the sampling effect, often confidence bands are used to describe the range of possible flood quantiles for a given return period depending on the size of the sample. Standardly, two approaches are used: the parametric and the non-parametric bootstrap. The first approach assumes a certain distribution of the sample of flood peaks and uses this distribution to generate samples of the same sample size as the observed records. Then, the assumed model is fit to these samples and the resulting quantiles are compared. This way, confidence bands of the estimation of flood quantiles are obtained. For the second approach, the non-parametric bootstrap, no distribution function is assumed but new samples are drawn from the observed sample and the quantiles are estimated based on these drawn samples. While both are approaches are relatively easy to apply, they have considerable drawbacks. The parametric bootstrap relies heavily on the distribution that is assumed for the flood peaks. Hence, the results are basically what is already assumed to be the true distribution, including possible heavy tails. The non-parametric bootstrap, however, is very much limited to the observed sample. When aiming to have samples of the same size as the observed sample, these have to be drawn with repetition, i.e., flood events will occur twice or more in the sample. When this applies to extreme floods, the bias of the sample might be very high. An alternative to these standard approaches offers the deterministic-stochastic event generator. With the generator, samples of arbitrary size of flood peak and volume can be generated, while only assuming the distribution of precipitation and using knowledge on flood generation. The type-specific framework also allows for a wide application of possible models, such as the TMPS model (Chap. 5). As an example, the TMPS model is applied for the Achsheim/Schmutter gauge in Bavaria, Germany, which has been considered before. The catchment has a size of 359 km^2 and mean elevation of 541 m a.s.l. The observation period has a length of 53 years.

Large samples of flood peak and volume are generated for each flood type, using 1000 simulated realisations of each precipitation (for S1 and S2 plus snowmelt) and the soil moisture for each flood type. In a next step, samples with an assumed observation period of n years are drawn from all samples of flood types, while

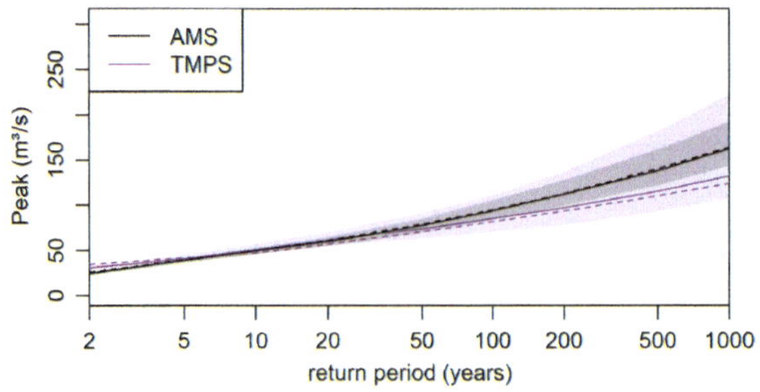

Fig. 16.7 Confidence bands for the TMPS and AMS models obtained with the deterministic-stochastic event generator for Achsheim gauge in Bavaria, Germany. Solid lines represent the median of all runs, dashed lines are the distribution functions obtained from the observed sample

considering the relative frequency of each flood type and the mean number of events per year (see Chap. 14). These samples are used to fit the TMPS model as well as the AMS model for comparison. The drawing of n-year samples of mixed types of floods is repeated 1000 times and the 5 and 95% quantiles of all quantile estimates for each return period are derived. This way, a repetitive drawing of single events is omitted and flood peaks (and volumes) are generated according to their geneses. The results are given in Fig. 16.7.

It can be seen that the AMS model and the TMPS model both have large confidence bands, especially for large return periods. The TMPS model has larger confidence bands in this range compared to the AMS model, which is due to the higher number of parameters of the model. The upper confidence range is larger than the lower range, indicating that single large events have an impact on the confidence bands. This is due to the small sample size of only 53 years that is considered here. The AMS and TMPS models based on the observed sample lie well within the confidence bands, indicating that the deterministic-stochastic event generator delivers reliable samples of flood peaks. With the derived confidence bands, the sample uncertainty can be evaluated. For example, for a flood quantile of return period 100 years, one would assume a range of 50 to 90 m^3/s for the TMPS model. Depending on the task at hand, one can then choose the design flood peak from this range. Similarly, confidence bands can also be derived for the type-specific distributions (Fig. 16.8). There it can be seen that the main uncertainty in the TMPS model is inherited from the flood type R2, which has the widest uncertainty bands. The R1-floods, which result in the largest flood quantiles for all return periods, do not have such a large uncertainty, even for the comparably small sample size of 53 years. The smallest confidence bands are derived for flood type S2.

The deterministic-stochastic event generator can not only be used to derive confidence bands corresponding to the observed samples, but it can also investigate the uncertainty dependent on the sample size. For this purpose, samples of varying size n are drawn from the large simulated sample and the TMPS model as well as the AMS model are fit. The confidence bands then give notice on how robust the models

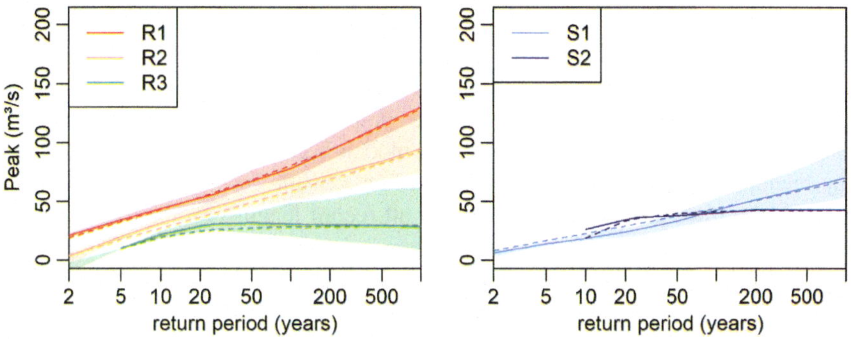

Fig. 16.8 Confidence bands for the type-specific distributions obtained with the deterministic-stochastic event generator for Achsheim gauge in Bavaria, Germany. Left: rainfall-driven floods, right: snowmelt-induced floods. Solid lines represent the median of all runs, dashed lines are the distribution functions obtained from the observed sample

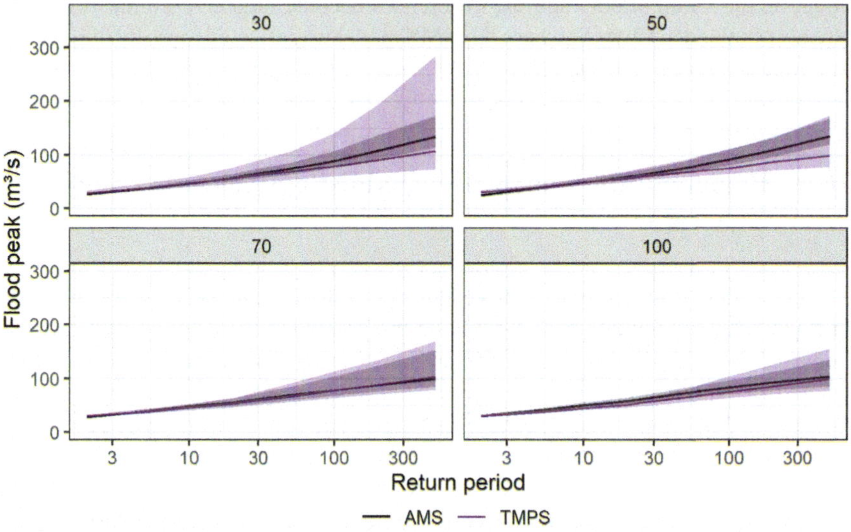

Fig. 16.9 Confidence bands of the TMPS and AMS models obtained with the deterministic-stochastic event generator for Achsheim gauge in Bavaria, Germany, with different sample sizes n (in years). Solid lines represent the median of all runs

are and how large the effect, e.g., of single extraordinary large events is. An example is given in Fig. 16.9, where the sample size is varied between $n = 30, 50, 70,$ and 100 years. These are typical sample sizes for runoff series.

It can be seen that the confidence bands vary much between the different sample sizes, especially for the TMPS model. For a sample size of 30 years, the confidence bands are wide, leading, e.g., to a spectrum from 50 to 150 m^3/s for the flood quantile of return period 100 years of the TMPS model. The large confidence bands occur due to the large number of parameters of the TMPS model. These are more sensitive to the sampling effect than the three parameters of the AMS model. From a record length of 70 years on, however, the confidence bands for AMS and TMPS model become similar and hence both models are comparable regarding the uncertainty in estimation. A major difference between the models becomes visible when comparing the median quantiles of the models (solid lines): while for the TMPS model, the median quantiles are almost identical for all record lengths, the median quantiles of the AMS model vary much until they become identical to the TMPS model for a record length of 70 years on. This convergence to the TMPS model shows that, despite the higher number of parameters, the TMPS model delivers more robust estimates that do not vary much for different record lengths, at least in median. This can be explained by the consideration of the different flood types. Though single extraordinary flood events have an effect on the estimation of single flood type distributions, the overall distribution of each flood type is captured well even for small sample sizes. For the AMS model, however, single large events have an effect on the whole model and the differences become larger. A similar result was already obtained in Chap. 8, where the AMS model was affected more by the changes in frequency of single flood types.

16.3 Regionalisation of the Deterministic-Stochastic Event Generator

The deterministic-stochastic event generator is able to generate type-specific pairs of flood peak and volume based solely on the precipitation and snowmelt distributions and the initial soil moisture. So far, this generator is applicable to catchments where discharge is observed and the flood type-specific runoff coefficient as well as the distribution of flood-generating precipitation and snowmelt can be derived. As was demonstrated in the previous chapters, this is only the case for a small part of a basin, since the density of all gauging networks is limited. Even if a station was installed in the last decades, the available records might be too short to estimate reliable distributions required for the event generator. Hence, an application to ungauged basins is of high interest. If deterministic models are applied instead of the stochastic approach, the parameters of the model have to be regionalised. In this case, the parameters are usually transferred from gauged catchments to ungauged but hydrologically similar catchments. A similar approach is used in the case of this deterministic-stochastic event generator. There exist three input variables, whose distributions have to be known: precipitation, snowmelt and runoff coefficient. Additionally, the type-specific flood timescales for rainfall-driven floods are required. The non-linear relation between flood peak and volume for the snow-induced floods is already generalised for all catchments and can be easily applied to ungauged catchments. The availability of precipitation and snow data usually does not depend on whether or not a catchment is equipped with a discharge measuring station, but it must be known which meteorological events are likely to lead to a flood. The other required input variable for the stochastic event generator, the runoff coefficient, is much easier to regionalise due to the framework introduced in the previous section. The CART algorithm already provides a regionalised estimation for the runoff coefficient when using several gauged catchments of the same catchment class for training. Yet, still the Beta distribution has to be regionalised to obtain the largest runoff coefficients for flood types R1 and S1. Similarly, the mean flood timescale for the rainfall-driven floods has to be derived from catchment attributes. All three approaches will be described in more detail in the following.

16.3.1 Regionalisation of the Probability Distribution of Rainfall and Snowmelt

Precipitation is available in many gridded data sets basically for the whole globe. For example, the E-OBS data sets (Cornes et al. 2018) provide daily gridded precipitation in 0.1 degree resolution in Europe and the CPC data set from NOAA (CPC 2020) provides a similar product for the United States. Another possibility would be to use stochastic weather generators such as the one by Haberlandt et al. (2015), since only the statistical distribution of the events and not the exact timing is required for the generator. Based on these data sets, type-specific rainfall events can be defined using the type-specific thresholds defined in Sect. 13.3.1 for precipitation events. For these rainfall events, their distribution can be estimated analogously to the case of gauged catchments. The same holds true for snowmelt. If no gridded data sets exist, hydrological models can be applied to derive snowmelt for ungauged catchments.

Yet, there is one challenge to solve: in the deterministic-stochastic event generator, only those precipitation events should be considered that are related to a flood event. Indeed, not each precipitation event will generate a flood event. To consider this, the results of the CART algorithm can be used. As described before, the CART algorithm was trained for gauged catchments to derive the runoff coefficient from the empirical probability of all rainfall events by type. Hence, when considering an ungauged catchment and having identified all rainfall events, the probability that a certain amount of rainfall results in a flood event is implicated in the CART algorithm trained for the same catchment class (i.e., in the same range of catchment size and in this case belonging to the same nature class) by the minimum empirical probability of rainfall events used for classification. For example, if the precipitation events associated with R1-floods have an estimated duration of one to three days, then all one- to three-day precipitation events from the daily precipitation series are identified. This is done by considering all precipitation events which have a precipitation of more than 0.1 mm per day for one or at most three consecutive days. For the precipitation sum of these events, then the empirical probability is estimated. This probability is compared to the empirical probabilities of the precipitation sums related to R1-floods from gauged catchments that have been used for training the CART and only those precipitation sums are considered for the event generator that have a probability equal to or larger than the minimum probability of the CART-precipitation-sums.

This way, a lower threshold is considered to identify rainfall events that probably generate floods. Similarly, such a threshold can be obtained for the snowmelt events used for flood types S1 and S2.

16.3.2 Regionalisation of the Runoff Coefficient

The runoff coefficient for flood types R2, R3 and S2 for ungauged catchments can be derived simply from the trained CART algorithm introduced in the previous section using a simulated sample of precipitation and snowmelt events derived from the PearsonIII distribution. Since the CART is trained for a group of gauged catchments with a similar catchment size and nature class, it can be directly used for an ungauged catchment of the same class by using the simulated precipitation and snowmelt samples as input. However, only those precipitation and snowmelt sums are inserted that are above the lower probability threshold of events resulting in a flood used for training the CART. For flood types R1 and S1, the CART algorithm can be used in the same way for the lower 75% of the runoff coefficients. However, it was shown that the upper 25% runoff coefficients for these flood types cannot be predicted based on soil moisture and precipitation and instead the Beta distribution has to be used to generate runoff coefficients in this range (Eq. 16.3). Instead of regionalising the runoff coefficient itself, thus, the parameters of the type-specific distributions of the coefficient are regionalised. Fischer et al. (2019) reported that the relation between α and β resulted in clearly disjoint clusters for the three rainfall flood types for their case study of mesoscale catchments in eastern Germany. Such a clear disjunction may not always be present, depending on the complexity of the catchments under consideration. Yet, it can be expected that clear differences exist between the parameters of the type-specific distributions. For example, disjoint clusters cannot be reported for the large data set of mesoscale catchments (<1000 km^2) in Bavaria, though still clusters are visible for the different flood types (Fig. 16.10). Therefore, the following methodology by Fischer et al. (2019) is applied to obtain the two distribution parameters for ungauged catchments. Though only the Beta distribution for flood types R1 and S1 will be needed for the deterministic-stochastic event generator, results for all flood types are given here to demonstrate the differences between the flood types.

The first step to regionalise the Beta distribution is the generalisation of the relation between both parameters α_j and β_j for each flood type j. There exist several approaches based on regression or clustering for this purpose. The selection of a suitable approach depends on the complexity of the relation and thus on the complexity of the catchments under consideration. In Central Europe, as shown in Fischer et al. (2019), the relation is less complex and simple linear regression is sufficient to estimate the mean relation between both parameters. For a larger study region, such as the United States, it is recommended to additionally make use of the catchment attributes such as topography or soil characteristics. By knowing the relation between α_j and β_j, it is possible to calculate the mean μ_j by $\mu_j = \frac{1}{1-\frac{\beta_j}{\alpha_j}}$ of the Beta

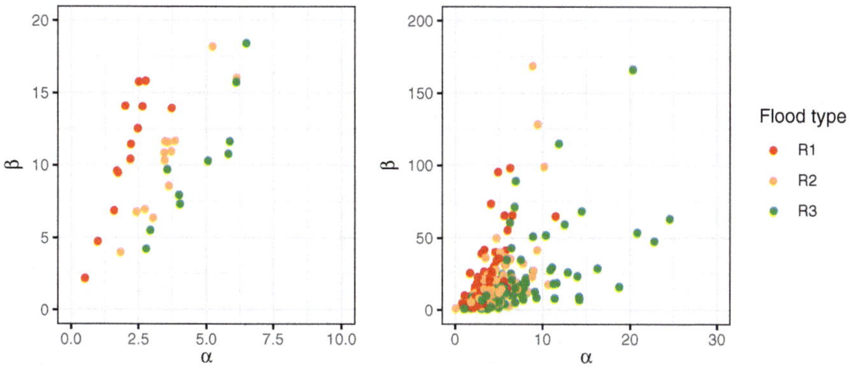

Fig. 16.10 α and β parameters of the Beta distribution of runoff coefficients for all rainfall-induced flood types: for catchments with similar runoff behaviour (Mulde basin in eastern Germany, modified from Fischer et al. 2019) and catchments with diverse runoff behaviour (Bavaria, Germany)

distribution for each flood type $j = R1, R2, R3, S1, S2$. The knowledge on the mean is an alternative to know the exact distribution parameters. For complete knowledge of the distribution, however, the variance has to be known, too. Based on the catchment relations for the parameters, which are derived by (non)-linear regression, (Branscum et al. 2005) proposes a Bayesian approach that includes the estimate of the mean as so-called expert guess and derives the parameters of the distribution by taking into account the uncertainty bands of the linear regression (Branscum et al. 2005). Usually, the confidence level is chosen as 90%. Here, the uncertainty bands are obtained by a weighted k-nearest-neighbour approach where the weights are chosen according to catchment size similarity, as described in the previous chapters. Hence, as long as some catchments are gauged in a basin, their relation between the parameters of the Beta distribution can be generalised for ungauged catchments.

The regionalisation of the runoff coefficient is most difficult for flood type R3, the floods associated with long-duration rainfall or several successive rainfall events (Fig. 16.11). The complex flood-generating processes do not allow for a precise estimation of the runoff coefficient and thus also the determination of the distribution parameters is more difficult. This flood type usually is assigned to the highest estimation error. The estimation of the distribution parameters for flood type R1, though, is much easier since the flood-generating processes for a heavy-rainfall event are less complex: the soil wetness condition is less important compared to the precipitation intensity, which is often temporarily high during these events and leads to high runoff coefficients. The same holds true for the snow-induced flood types, where S1 has a much smaller error than S2. The results emphasise that the intensity-driven flood types R1 and S1, for which the simulation of runoff coefficients by making use of the Beta distribution is required, can be regionalised well. For the remaining flood types, such an approach would not be meaningful, as for these cases the runoff coefficient depends mainly on the interplay between soil moisture and rainfall and a prediction based on these variables leads to hydrologically more meaningful and better results.

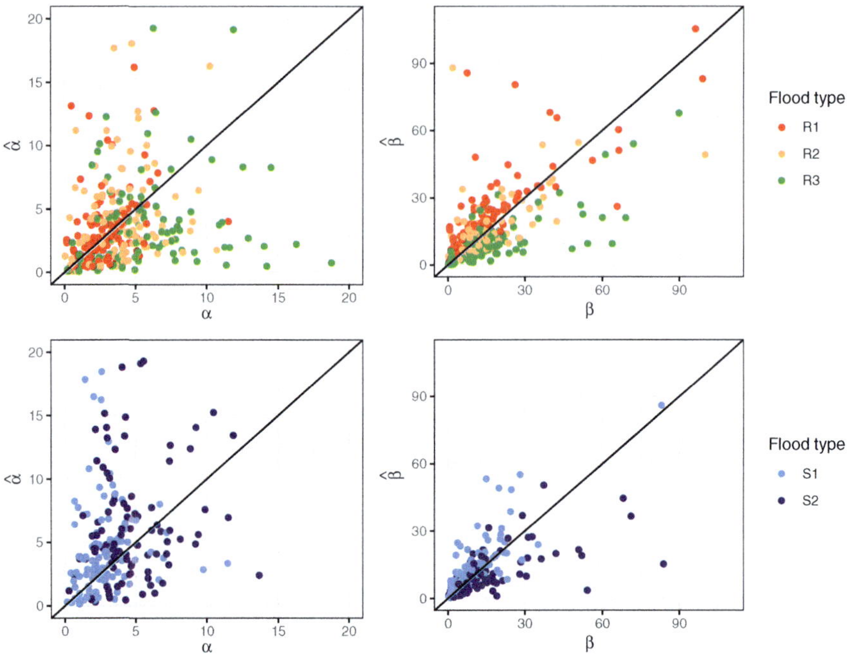

Fig. 16.11 Regionalised (denoted with a $\hat{\alpha}$ resp. $\hat{\beta}$) and observed α and β parameters of the Beta distribution of runoff coefficients for all flood types

Therefore, as proposed before, the CART algorithm should be used for these flood types foe the complete range of runoff coefficients.

16.3.3 Regionalisation of the Mean Flood Timescale

A key role in the deterministic-stochastic event generator, as well as in many other flood-type-related analyses and models, plays the type-specific relation between flood peak and volume, the so-called flood timescale or TQ-value (Eq. 6.3). While in the previous chapter the mean flood timescale was used to estimate the type-specific volume for a given flood peak, in the deterministic-stochastic generator, the type-specific flood peak is estimated for a given flood volume derived from precipitation and runoff coefficients for the rainfall-driven floods. Still, the procedure to derive the regionalised type-specific mean flood timescale is analogous to Sect. 15.1 and not repeated here. As stated before, the non-linear relation between flood peak and volume for the snow-induced flood types does not have to be regionalised as the model is already fit to all catchments in the study region with mean elevation and hypsometric curve as dummy variables. Hence, it can be easily applied to ungauged catchments, using the simulated flood volume as input.

The framework of the regionalised deterministic-stochastic event generator is summarised in the following. For each flood type:

1. Training of the CART to predict the runoff coefficient for gauged catchments grouped by catchment size and nature class. Input data: empirical probabilities of precipitation (for types S1 and S2: plus snowmelt) and initial soil moisture.
2. Separation of type-specific rain and snowmelt events for the ungauged target catchment from long series of precipitation. Snowmelt can be specified by hydrological models, e.g., from the Degree-Day Method.
3. Estimation of the distribution of the rainfall (and snowmelt) event sums.
4. Stochastic generation of samples of rainfall (and snowmelt) sums.
5. Estimation of the probabilities of each rain (and snowmelt) sum and generation of uniformly in [0, 1] distributed values for the probability of the initial soil moisture. The probabilities are then used as input for the trained CART (trained for the respective flood-type, catchment size and nature class), where only those combinations of probabilities are considered that are above the threshold of minimal required rainfall probabilities determined from the trained CART. This results in a sample of precipitation (and snowmelt) sum and the corresponding derived runoff coefficients. For flood types R1 and S1, additionally the Beta distribution has to be regionalised and the upper 25% of the runoff coefficients are derived by this distribution.
6. Selection of tuples of precipitation (snowmelt) sum and runoff coefficient according to the leaf-probabilities of the CART.
7. Estimation of the type-specific direct flood volume by multiplication of rainfall (plus snowmelt) sum with the corresponding runoff coefficient.
8. Regionalisation of the mean type-specific flood timescale for the rain-induced flood types.
9. Estimation of the type-specific non-linear relationship between flood peak and volume for the snow-induced flood types using mean catchment elevation and the hypsometric curve as dummy variables.
10. Estimation of the direct flood peak by multiplication of the direct flood volume and the mean flood timescale for the rainfall-induced floods and by application of the non-linear model for the snow-induced floods.

Finally, all type-specific pairs of flood peak and volume are combined in one sample.

16.3.4 Performance of the Deterministic-Stochastic Event Generator

Similar to the spatial multivariate flood event generator (Chap. 11), QQ-plots can be used to compare the distribution of the simulated and observed flood peaks. Again, it is made use of the advantage of the stochastic event generator to be able to generate arbitrary large samples. It is applied for each flood type separately. Later on, it is shown how to derive combined samples of all flood types. For example,

$n = 1000$ values for the event precipitation (and snowmelt) of each flood type can be generated as well as $n = 1000$ values of empirical probability of soil moisture (by using a uniform distribution on the interval [0, 1]). Then, only those values are taken which have a probability above the lower probability threshold of precipitation (or snowmelt) derived from the CART and are used as input values for the type-specific CART trained to those gauged catchments which belong to the same nature and catchment classes. From the resulting triples of precipitation, soil moisture and corresponding runoff coefficient then a new sample is drawn which has the same relative proportion of each CART class as the trained CART. For flood types R1 and S1, runoff coefficients are additionally generated by the regionalised Beta-distribution and replaced in the sample for probabilities larger than 0.75. Based on these samples, the type-specific flood volume is obtained and the regionalised mean flood timescale respectively the non-linear model is used to obtain the corresponding flood peak. This leads to type-specific samples of flood peak and volume. From these samples, sub-samples are drawn with the same sample size as the observed flood peaks of the same flood type. This is repeated 1000 times and the sub-samples are compared to the observed sample in a QQ-plot (Fig. 16.12). The dots in the plots are the mean quantiles derived from 1000 repetitions and the grey shades are the confidence intervals. More precisely, each subsample is compared to the observed quantiles regarding its quantiles and displayed in the QQ-plot as the grey-shaded area. The mean quantiles of the 1000 sub-samples are used to derive the dots in the QQ-plot.

The graphical comparison emphasises that in general the stochastic event generator provides reasonable samples mirroring the distribution of the observed flood events. A challenge are the extreme flood events, which are few in the observed samples but have a large impact on the distribution. Nevertheless, most of the sample lies well in the uncertainty range, indicating a reasonable performance of the generator. The most difficult flood type to simulate is S1, the rain-on-snow floods. For this flood type, two different random variables (rainfall and snowmelt) have to be combined, while only few observed events of this type are available for the study region. This increases uncertainty for this flood type.

Additionally, also the error measure Trimmed Normalised Root Mean Squared Error (TNRMSE, Eq. 11.1) with a 5% trimming is applied to obtain an objective goodness measure. The TNRMSE gives a quantitative evaluation of the deviation of the simulated to the observed flood quantiles. For example, in the study basin in Bavaria, Germany, the stochastic event generator resulted in a mean deviation between observed and estimated quantiles of 21.9% for flood type R1, 23.4% for flood type R2, 22.5% for R3, 20.9% for S1 and 23.7% for S2. The magnitude of the deviation does not depend on the catchment size and indicates that the event generator is suitable for all mesoscale catchments (Fig. 16.13). There appear to be different numbers of catchments with high errors for each flood type. The largest numbers of catchments with TNRMSE greater than 0.5 occur for flood type R2, followed by flood type R1. This applies to less than 5% (R1) respectively 7% (R2) of all catchments and only for catchments in the Alpine foreland and the South-western Uplands. For these flood types, there exist catchments where the runoff behaviour is not captured well by the generator. This may be due to factors in the catchment

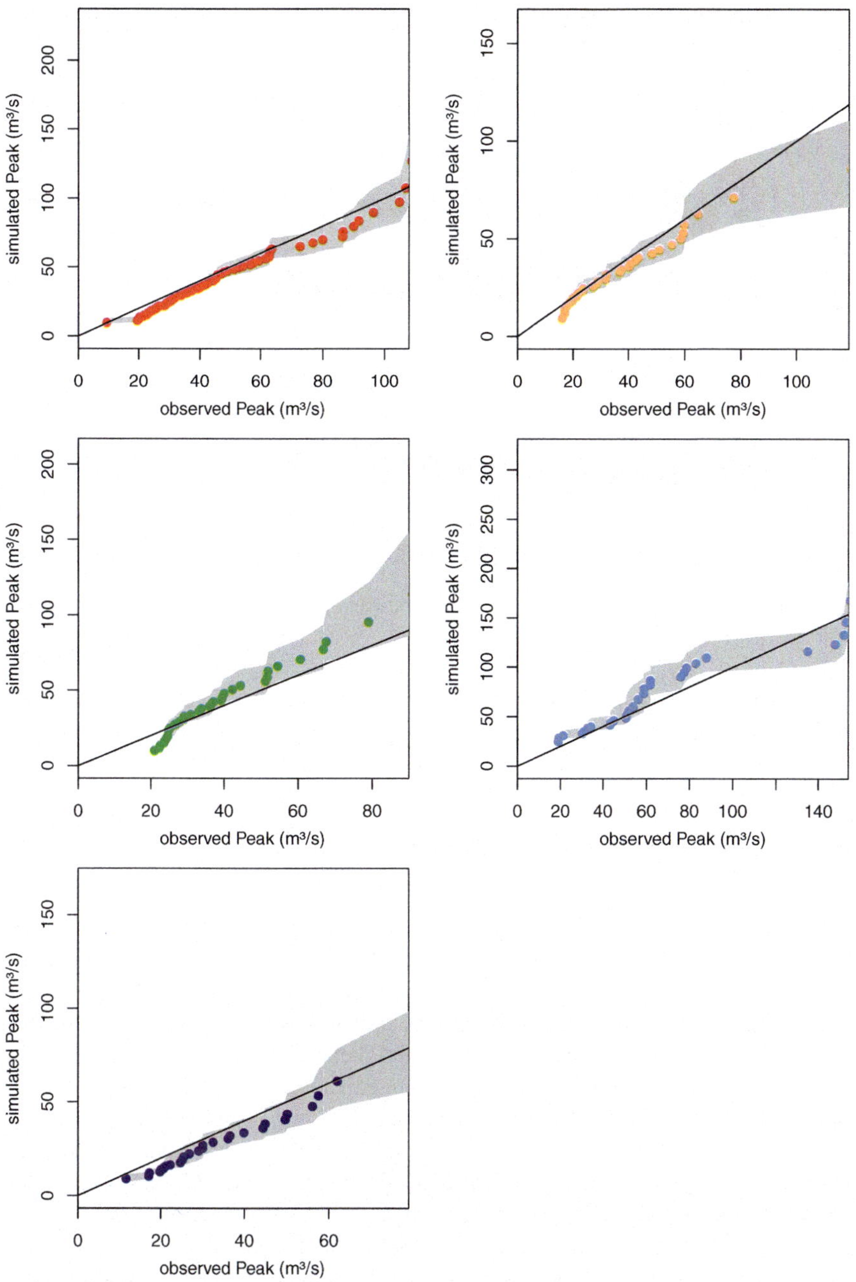

Fig. 16.12 QQ-plots with uncertainty bands considering sampling effect for the observed and generated flood peaks of different flood types: R1 (top left), R2 (top right), R3 (centre left), S1 (centre right), S2 (bottom left)

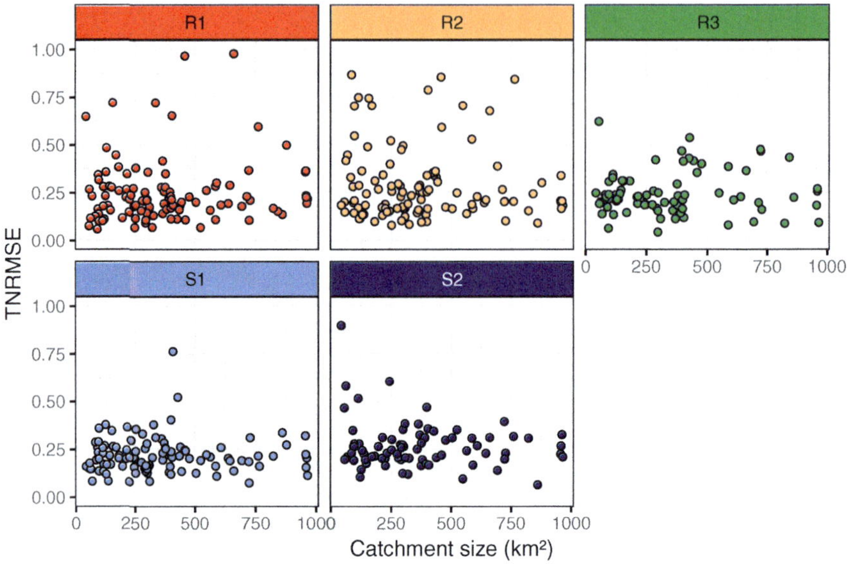

Fig. 16.13 TNRMSE of the deterministic-stochastic event generator for different flood types

that remain unconsidered in the generator, e.g., the land use or soil type. Catchments with a low retention of the soil might not be modelled well by the generator.

16.3.5 The Use of the Regionalised Stochastic Event Generator for Analysing Flood Events

In the previous sections, it was shown how a deterministic-stochastic event generator can be applied to ungauged basins to derive type-specific pairs of flood peak and the corresponding flood volume. Now, it is shown how these samples can be used for flood frequency analyses. More precisely, the samples are used to derive flood quantiles by the TMPS model for an ungauged catchment. Since arbitrarily large samples can be drawn, also the confidence bands of the estimation can be derived. This is a great advantage compared to classical flood frequency analyses, since the sampling effect, i.e., the uncertainty introduced by the record length, can be estimated. Therefore, uncertainty for typical record lengths, i.e., between 30 and 100 years, can be considered in the estimation. The only challenge that remains is the derivation of the relative frequencies of each flood type on the whole sample. Since type-specific samples are generated statistically, the relative proportion of each flood type, which is required for the TMPS model, is not considered. However, one can assume that the proportion of flood types is similar for catchments in the same catchment size class as well as the in same region. Thus, using once again the gauged catchments that have been used for training the CART in the previous sections, the relative proportion

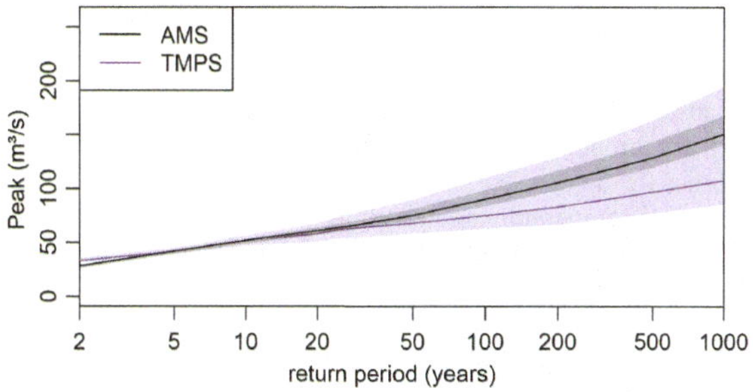

Fig. 16.14 Distribution function of the TMPS model and the AMS model and corresponding 95%-confidence intervals for the Achsheim/Schmutter gauge in Bavaria, Germany, derived with the regionalised deterministic-stochastic event generator

of each flood type can be easily transferred from these. Alternatively, one could also derive the relative frequencies directly from the precipitation and snowmelt data, as each rainfall and snowmelt event has already been associated with a certain flood type.

To demonstrate the applicability, once again the Achsheim gauge from Fig. 16.7 is taken and treated as ungauged. Large samples of flood peak and volume are generated for each flood type, using 1000 simulated values of each precipitation (and snowmelt) and the soil moisture. Then, a sample of n years is drawn from all flood types, while considering the relative proportion of each flood type and the mean number of events per year (see Chap. 14). Based on this sample, the TMPS model as well as the AMS model are fit. This is repeated 1000 times. The resulting mean flood quantiles together with the confidence bands are shown in Fig. 16.14.

It can be seen that the confidence bands for the TMPS model are again wider compared to those of the AMS-model. This can be explained by the larger number of parameters that have to be estimated for this model. The generation of these confidence intervals helps to better estimate the uncertainty related to the estimation of flood quantiles, even for ungauged catchments.

The stochastic event generator offers a simple approach to generate pairs of flood peak and volume based on hydrological process-understanding even for ungauged catchments. By consideration of the type-specific distribution of precipitation and a CART algorithm to predict the runoff coefficient based on precipitation and soil moisture, random samples of type-specific flood volumes can be derived that coincide with the observed flood volumes in their statistical distribution. By making use of the type-specific flood timescale and a non-linear model, based on these sample of flood volume, the corresponding flood peaks can be derived. When applying the generator to ungauged catchments, this requires the regionalisation of the distribution of the runoff coefficient and the flood timescale. By Bayesian and machine-learning techniques, this can be obtained directly. The resulting samples of flood peak and volume then can be used for regional flood frequency analyses.

References

Branscum AJ, Gardner IA, Johnson WO (2005) Estimation of diagnostic-test sensitivity and specificity through Bayesian modelling. Prev Vet Med 68:145–163

Cornes R, van der Schrier G, van den Besselaar EJM, Jones PD (2018) An ensemble version of the E-OBS temperature and precipitation datasets. J Geophys Res Atmos 123

Climate Prediction Center-Monitoring and Data. www.cpc.ncep.noaa.gov

Cowpertwait PSP, Kilsby CG, O'Connell PE (2002) A space-time Neyman-Scott model of rainfall: empirical analysis of extremes. Water Resour Res 38(8)

Filipova V, Lawrence D, Skaugen T (2019) A stochastic event-based approach for flood estimation in catchments with mixed rainfall and snowmelt flood regimes. Nat Hazards Earth Syst Sci 19:1–18

Fischer S, Schumann A, Bühler P (2019) Timescale-based flood typing to estimate temporal changes in flood frequencies. Hydrol Sci J 64:457–470

Haberlandt U, Belli A, Bárdossy A (2015) Statistical downscaling of precipitation using a stochastic rainfall model conditioned on circulation patterns-an evaluation of assumptions. Int J Climatol 35(3):417–432

Schaefer MG, Barker, BL (2002) Stochastic Event Flood Model (SEFM). In: Singh VP, Frevert D (eds) Mathematical models of small watershed hydrology and applications. Water Resources Publications, Colorado, USA, pp 707–748

Chapter 17
Type-Based Flood Statistics: A Short Summary

Abstract In this book, statistical models for flood statistics are introduced that explicitly take into account the flood geneses by considering flood types. Compared to classical approaches, like the annual maximum series or the peak-over-threshold approach, these models are more complex and require more preliminary work. Yet, they also provide more information, e.g., by extending the considered flood event characteristics from peak only to the flood volume or even the whole hydrograph. Additionally, several further approaches (DFH generation, spatial flood scenarios, regionalisation) are presented that all profit from the consideration of flood types but only under specific circumstances and often under the cost of more complex models. In this final chapter, the requirements, advantages and drawbacks of each approach are summarised and a recommendation is given when to use which approach.

In 1936, the USGS published the report "Floods in the United States. Magnitude and frequency" (Jarvis 1936). In this report, (Slade 1936) stated:

> In the writer's opinion the statistical method, in whatever form employed (graphic or analytical), is an entirely inadequate tool in the determination of flood frequencies. When used in conjunction with nonstatistically inferred data, however, it may attain a high order of precision.

A similar, albeit harsher, criticism of flood statistics can be found in the publications of (Klemes 2000a, b). He concluded (Klemes 1987)

> Thus the hydrological problem of a proper FFA is to predict the as yet unobserved outliers by analysing the plausible dynamics of extreme-flood formation …

Both authors agreed that in FFA more hydrological aspects should be integrated. One approach to do so is discussed in this book where statistical models for flood statistics are introduced that explicitly consider the flood geneses by considering flood types. The developed core element is the Type-based Mixture Model of Partial Duration Series (TMPS model).

S. Fischer and A. H. Schumann, *Type-Based Flood Statistics*, Water Science and Technology Library 124, https://doi.org/10.1007/978-3-031-32711-7_17

With this approach, it was tried to overcome two main problems of flood statistics: (1) Floods do not belong to a single category of identically distributed random variables, and (2) extreme floods may belong to a particular flood type that may be under-represented in annual flood series. To address both problems, it is necessary to expand the data base, which was done by applying the POT approach, and to investigate the main factors of flood generation in more detail, which was done by flood event analyses and classification into flood types. As a result, the TMPS model was developed that represents a certain imposition for the users:

- It requires a more complicated procedure for providing the data base compared to using the classical AMS model,
- More parameters have to be estimated to set-up a rather complex statistical model,
- The results, a section-wise fitted mixed distribution function with no strictly monotone form, differs from the classical shape of the usual statistical distribution functions.

However, a hydrologist familiar with flood generation processes from experience or (better yet) from deterministic evaluation of many observed events in a river basin will find some aspects in the TMPS approach that should meet his expectations:

- Close to reality, in type-based flood statistics, it can be considered that different physical forces dominate the given phenomenon "flood" in different ranges of his state. An example in the presented studies is the type- and regional-specific impact of initial moisture conditions on the runoff coefficients that became evident which was useful for the deterministic-stochastic flood generator.
- The differences between the occurrence of small and large flood events are explained by changing dominances of the type-based distributions in the different flood size classes.
- The meteorological drivers and the impact of initial and boundary conditions differ between flood types which is highly relevant for regionalisation.

The opportunities of applications of the type-based flood statistics for typical engineering tasks, such as the estimation of flood probabilities or the estimation of design flood peaks, were discussed in Chap. 7. The amount of required data is higher and the processing of these data is more labour-intensive than when using the usual flood statistics based on the AMS and the POT series (Chap. 6). Long records (more than 10 events per flood type) of daily discharges and monthly peaks should be available as well as long series of daily precipitation and snowmelt data (or an adequate model for snowmelt). However, the application of the TMPS model and hence the additional effort in data preparation is not always required: if the available flood peak records do not show differences in flood origins, the more readily applicable statistics based on the AMS or POT could be applied instead. If the flood types in a sample are well mixed across the range of all quantiles, the statistical results of the TMPS application would be similar to these traditional approaches. Such differences in origin can be investigated, e.g., by weighted circular statistics. For these statistics, the calendar day of each event (of AMS or POT series)

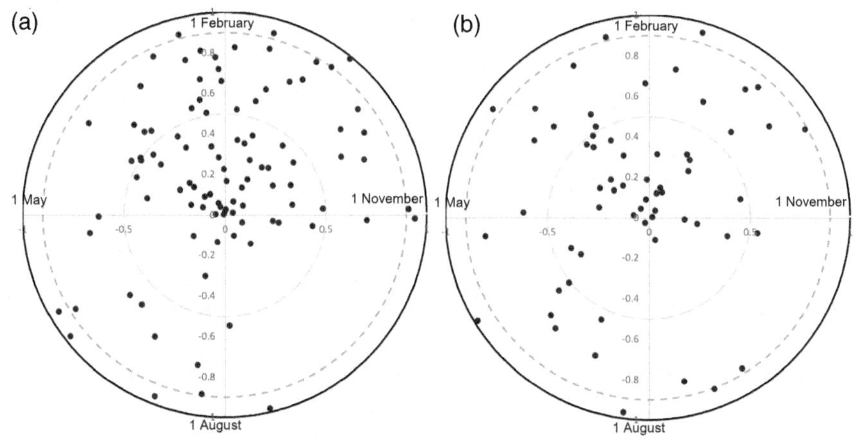

Fig. 17.1 Weighted circular statistics of the AMS series of Marienthal/Regen (**a**) and Kautendorf/Southern Regnitz (**b**) gauges. The grey dashed lines are the 90% quantiles and the green dashed lines are the medians of the empirical probabilities

is transformed to the radian measure by $\theta_i = (\text{calendar day}) \cdot \left(\frac{2\pi}{365}\right)$ such that the radian coordinates of the calendar day of event i become $x_i = \cos\theta_i$ and $y_i = \sin\theta_i$. To take into account the magnitude of the flood peak, these coordinates can be weighted by the empirical probability $p_i = \frac{m}{n+1}$, where m is the rank of the peak of event i and n is the sample size. By this weighting, the largest flood peak lies directly on the unit circle, while smaller flood events lie closer to the centre of the circle. Two examples are given in Fig. 17.1. For the Marienthal gauge, it can be seen that most of the annual maximum floods occur in winter while only few, often large flood peaks occur in summer. Hence, the winter floods dominate the AMS and the flood types are not uniformly distributed in the sample. This is why the Marienthal gauge profited much from the application of the TMPS model (Chap. 7). For the Kautendorf gauge, instead, the floods are well mixed within the year in the AMS sample. Therefore, one can assume that in this case all flood types are represented well. This is line with the results from Chap. 7, where no distinct differences between AMS and TMPS models were observed.

The advantages of the TMPS model become apparent when flood events exhibit type-specific differences in volume and hydrograph shape. In this case, a consideration of these differences leads to more realistic estimates in flood design. Differentiation between flood types provides more homogeneous samples, type-specific probabilities can be estimated and temporal changes in the relevance of flood-generating mechanisms become apparent and can be evaluated statistically. To estimates design flood hydrographs, a statistical generator can be applied to consider the differences of hydrographs between flood types and to reflect the uncertainty of such an assessment.

A basic problem of flood hydrology are spatial analyses of flood events. Often, extreme floods give reason to review flood protection systems in entire river basins, whose performance depend on the spatial distribution of flood discharges among tributaries. Single observed events or model-based assumptions cannot provide a complete picture of the range of flood scenarios which should be considered in such analyses. Again, the spatial interactions between tributaries depend on the flood types. A flood event triggered by snowmelt may differ in its spatial characteristics from rainfall events, which in turn may differ among themselves by the causative weather pattern. One approach to consider these aspects is described in Part V. However, the presented approach to estimate spatially consistent flood scenarios in large river basins requires again long records and a spatial typology of observed flood events in the past to set-up a multivariate stochastic event generator for a basin.

Further advantages of type-based statistics become visible in the regionalisation (Part VI). The coherence between catchment characteristics and flood statistics is all the stronger, the more the flood type considered in each case is influenced by these characteristics (and by the respective statistical properties of the relevant precipitation type). A simple example: a flood caused by a short and intense rain event depends less on the storage capacity of the soil but on the infiltration intensity. For a flood caused by a long-lasting rain with low intensity, these dependencies are reversed. Resulting from these and other mechanisms of flood formation, the relationships between catchment characteristics and flood peaks will differ between flood types R1 and R2. Also, the elevation of a catchment may have different impacts on snow-induced floods or floods which originated from orographic rain. By utilisation of flood types, such differences can be considered. If only short or no records of discharge exist in a catchment, the knowledge of differences in flood geneses in a region, information on catchments characteristics and flood statistics from neighbouring catchments with long records provide opportunities to regionalise TMPS models and to provide not only statistical assessments of flood peaks but also of flood type-specific pairs of peaks and volumes and even of synthetic design flood hydrographs.

The classical problem of hydrology, to estimate the amount of runoff from a pre-given amount of rain, was handled in Chap. 16 from a new perspective. Based on flood types, relationships between precipitation statistics and flood statistics were developed. Here, the type-related statistics of runoff coefficients are applied to transfer precipitation amounts to flood amounts and then derive flood peaks and hydrographs from these. As a result, large samples of flood peaks and corresponding volumes can be generated by a type-specific deterministic consideration of flood generation.

The developed approaches can be applied to solve a number of typical tasks of engineering hydrology on a new basis. An overview of these tasks is given in Table 17.1. In contrast to conventional approaches, it is proposed to include flood scenarios in the design and to consider that reality is more diverse than many hydrological assumptions ("the only certainty is uncertainty").

The results of the presented analyses and also several of the developed methods depend strongly on the available data base. In other climatic regions, a different classification of floods could become necessary and also other methodological approaches could be more appropriate than the methods which were presented here. However, the the main target of all the research presented in this book would be reached if other

Table 17.1 Summary of several classical tasks of engineering hydrology and how AMS and TMPS models handle these tasks

Typical application problems	AMS-based traditional flood statistics	TMPS-based approaches
How large is the flood peak of a flood with a return period of 100 years?	One distribution function	Mixture of distribution functions, the probabilities that a flood with a return period of 100 years belongs to a certain flood type can be estimated
How to specify the hydrograph and volume of such a flood wave?	Specification of a "typical" hydrograph, e.g., by analysis of previous flood events	Type-based volumes and hydrographs can be specified which demonstrates the variabilities of shapes and volumes of floods with similar peaks
How to estimate temporal changes of the flood regime?	Estimation of breaking points and trends in AMS (based on flood peaks)	Separate analyses of temporal changes in flood frequencies and other statistical parameters for each individual flood type
How to specify typical spatial scenarios of floods in large basins?	Analyses of observed flood events, definition of one or several "typical" events	Generation of a large number of spatially distributed flood scenarios with assessments of their plausibility and probabilities
How to interlink flood statistics and catchment characteristics in regionalisation?	Formation of hydrologically similar regions according to the flood statistical results and analyses of catchment characteristics, set-up of statistical relationships between catchment characteristics and flood statistics	Methodological approach similar to AMS, but differentiated by flood type to account for the type-specific hydrological significance of watershed characteristics
How to use precipitation data to provide a data base for flood statistics?	Simulation runs with deterministic models to transfer observed or stochastic generated series of precipitation into discharge series	Statistical analyses of runoff coefficients, differentiated by flood types and regions to transfer precipitation into flood volumes and to derive flood peaks and hydrographs from these volumes

scientists become encouraged to think about flood types, mixed statistical models and other options to integrate deterministic aspects into flood statistics as well as vice versa more statistics into deterministic flood models.

References

Jarvis CS (ed) (1936) Floods in the United States: magnitude and frequency. USGS Water Supply Paper (U.S. G.P.O). Washington, DC

Klemes V (1987) Hydrological and engineering relevance of flood frequency analysis. In: Singh VP (ed) Hydrologic frequency modeling. Springer Netherlands, pp 1–18

Klemes V (2000a) Tall tales about tails of hydrological distributions. I J Hydrol Eng 5(3):227–231

Klemes V (2000b) Tall tales about tails of hydrological distributions. II J Hydrol Eng 5(3):232–239

Slade JJ (1936) The reliability of statistical methods in the determination of flood frequencies. In: Geological Survey (ed) Floods in the United States. Magnitude and frequency. Washington, D.C. (Water Supply Paper, 771)

Appendix A
Information on the Catchments

This chapter introduces the regions and basins that are used in this book to demonstrate the application of type-based statistics. The region of Bavaria in southern Germany with its many small, medium and large catchments and its diverse topography and flood regimes serves as the benchmark example for all applications of type-based statistics. The large river basins of Rhine and Danube, instead, are used to demonstrate the application of spatial type-based statistics. The examples introduced in this chapter are taken from Fischer and Schumann (2021a, b).

A.1 Catchments in Bavaria, Germany

The region of Bavaria in southern Germany is well-known for its diverse hydrological conditions. Located in Central Europe north of the Alps (Fig. A.1a), three main river basins drain the region: the upper part of the Danube, the Main (a tributary of the Rhine) and the upper part of the Saale (a tributary of the Elbe). The Danube serves as a kind of natural and hydrogeological border. South of the Danube the natural area "Alpine foreland" dominates with partially quite extensive and thick gravel and gravel bodies with rich groundwater deposits (Fig. A.1c). This region is characterised by a diverse range of landforms and ecosystems, including mountains, hills, valleys, wetlands, and floodplains. It is subject to a range of flooding events, from seasonal floods associated with snowmelt and heavy rainfall to rare but severe flash floods caused by intense precipitation events. The region's climate is characterised by warm summers and cold winters, with precipitation distributed throughout the year. Snowmelt is an important contributor to the region's seasonal flooding, with snow accumulating in the mountains during the winter months and melting in the spring and summer. Even further south, the Alps begin which are characterised by diverse physiography, ranging from permanent snow cover to hilly landscape with grass and forest area. In the Alps, the highest mountain in Germany, the Zugspitze, with

S. Fischer and A. H. Schumann, *Type-Based Flood Statistics*, Water Science and Technology Library 124, https://doi.org/10.1007/978-3-031-32711-7

2962 m a.s.l. can be found. The Bavarian Alps are a high mountain range with rugged terrain and diverse soil types. The soils here range from shallow and rocky on the mountain peaks to deep and fertile in the valleys. The higher elevations tend to have acidic soils with low nutrient levels, while the lower elevations have more alkaline soils with higher nutrient levels. The area receives heavy precipitation in the form of rain and snow, which contributes to the formation of numerous rivers and streams that flow down from the mountains into the valleys. During the spring and summer months, the melting snow from the peaks adds to the already high levels of precipitation, leading to increased river and stream flows and a heightened risk of flooding. In the north-east of Bavaria at the Czech border, the Eastern Uplands, including the Bavarian Forest (Bayrischer Wald in German), are located, a region characterised by a hilly to mountainous landscape, dense forests, and numerous streams and rivers that drain into the Danube and the Main rivers. The soils here are generally acidic and low in nutrients due to the leaching of minerals caused by the heavy rainfall. The soil in the upper elevation ranges is also relatively thin due to the steep slopes, making it vulnerable to erosion. The flood regime in the Bavarian Forest is influenced by several factors, including precipitation patterns, snowmelt, soil moisture, and the topography of the region. The region receives significant rainfall throughout the year, with the highest amounts occurring in the summer months. This rainfall, combined with snowmelt in the spring, can lead to flash floods and high water levels in the rivers and streams. In addition, the topography of the region plays a significant role in the flood regime. The Bavarian Forest is characterized by steep hills and narrow valleys, which can cause rapid runoff of water and exacerbate flood conditions. The region also has numerous wetlands and bogs, which can act as natural flood control systems by retaining water and releasing it slowly over time. In the north-west, instead, the South-Western Uplands are characterised by a variety of water bodies, including rivers, lakes, and wetlands. The region experiences a temperate climate with significant precipitation throughout the year. As a result, the flood regime in the area is an important natural phenomenon that affects the landscape, ecosystems, and human settlements. The landscape in Bavaria in general is rural, with few large cities like Augsburg, Nuremberg or Munich. Systematic hydrological records in this area date back to the beginning of the 20th century, which makes this region particularly suitable for statistical analyses.

For this study region, 164 catchments have been selected. These catchments all have validated records of discharge and were checked for errors, anthropogenic impacts like reservoirs and/or water transfer to neighbouring catchments. The observation periods cover at least the most recent 30 years of observation, the years 1989–2018. The longest observation period covers 119 years (Fig. A.1d). Mean catchment elevation ranges from 2000 m a.s.l. in the south to 600–1000 m a.s.l. in the uplands in the north and east and to 300 m a.s.l. in the lowlands in the centre of the study region. The selected catchments include all classes of catchment size, with small catchments of 42 km^2 in the lower mesoscale up to more than 10,000 km^2 in the upper macroscale (Fig. A.1b). Daily mean discharges as well as monthly maximum discharges (quasi-instantaneous observations) are available for the analyses. Additionally, several catchment attributes and climate variables such as precipitation and

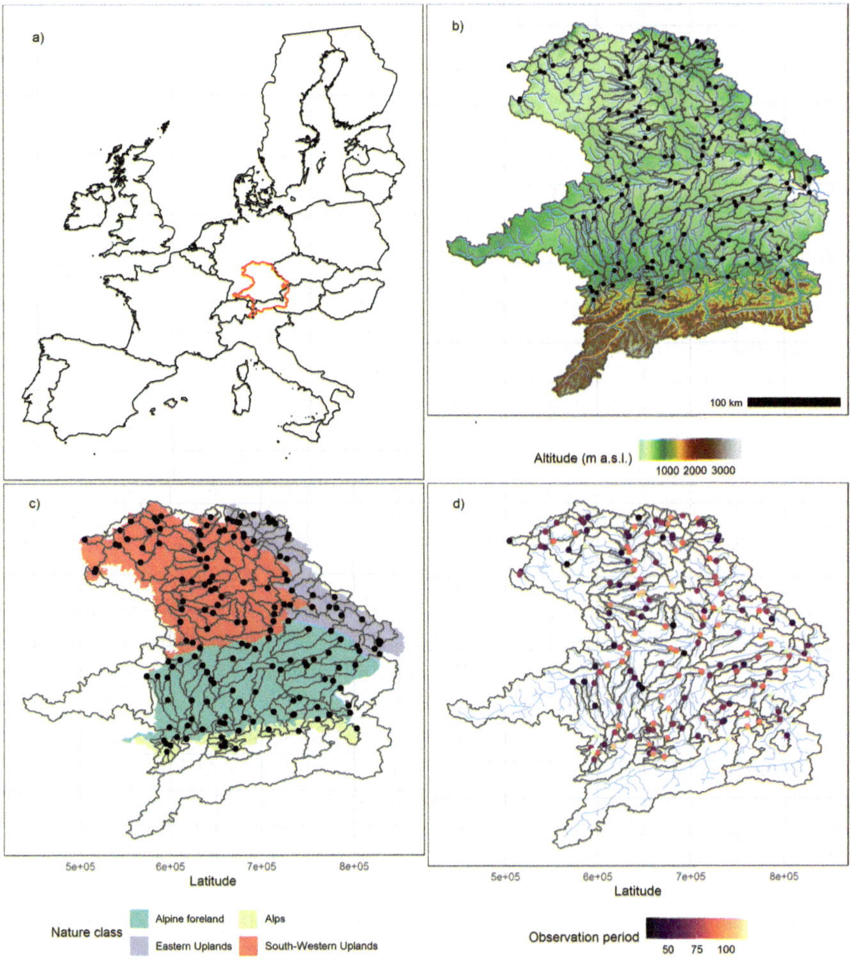

Fig. A.1 **a** Location of the study region Bavaria, Germany, **b** DEM, **c** natural areas and **d** length of observation period of each gauge

snowmelt on a daily basis are considered. An important attribute available for this regions proved to be the nature class. These four classes (South-western uplands, Eastern uplands, Alpine foreland and Alps) summarise different characteristics like geology, morphology, climate, hydrology and land use/soil as well as flora and fauna characteristics.

A.2 The Danube River Basin

The Danube river is the second-longest river in Europe. From its spring in southern Germany, it flows through ten countries until it enters the Black Sea, thus connecting several countries and their economies in Europe since several centuries. Being one of the main traffic routes in Europe, many anthropogenic changes have affected the river course to ensure navigability and to protect the people living in the many large cities along the river. In Germany, the Danube river has a catchment area of approximately 33,000 km^2, and its flood regime is characterised by both high and low flow periods. During the winter months, the Danube river experiences low flows due to reduced precipitation and snow accumulation in the Alps. However, in spring and early summer, the river typically experiences high flows due to snowmelt and increased precipitation. In addition, the river may experience flash floods during the summer months as a result of intense precipitation events. Human activities such as the construction of dams and reservoirs have also influenced the flood regime of the Danube River in Germany. For example, the construction of the Danube Power Plant has resulted in more controlled water releases, which have reduced the frequency and severity of floods in the lower Danube basin. In the recent years, two extreme flood events hit the German part of the Danube in the years 2002 and 2013. The flood event in the year 2013 was a disastrous event that caused incredible damages. A special role in the flood generation plays the Inn tributary, which enters the Danube at Passau, close to the German-Austrian border. In 2013, the superposition of flood events from main river and Inn river resulted in an extraordinary large flood peak and this was not the only time when the Inn river had an impact on the flood generation. Despite of the much smaller catchment of the alpine Inn river, its proportion of the mean runoff in Achleiten is with 52% slightly higher than the runoff originating from the Danube (only 45% at Hofkirchen gauge). Details on the three gauges that have been selected to represent this region and their catchment characteristics are given in Table A.1. Only floods in the joint observation period of all three gauges from 1936 to 2013 were considered (Fig. A.2).

Table A.1 Characteristics of the considered catchments for Danube basin

Gauge	River	Catchment size (km^2)	Mean Elev. (m a.s.l.)	MQ (m^3/s)	Mean gradient (%)	Mean ann. precip. (mm)	River kilometre
Hofkirchen	Danube	47518	627	638	1.77	875	2256.86
Passau-Ingling	Inn	26040	1257	741	7.67	1142	2225.20
Achleiten	Danube	76653	836	1420	3.76	963	2223.05

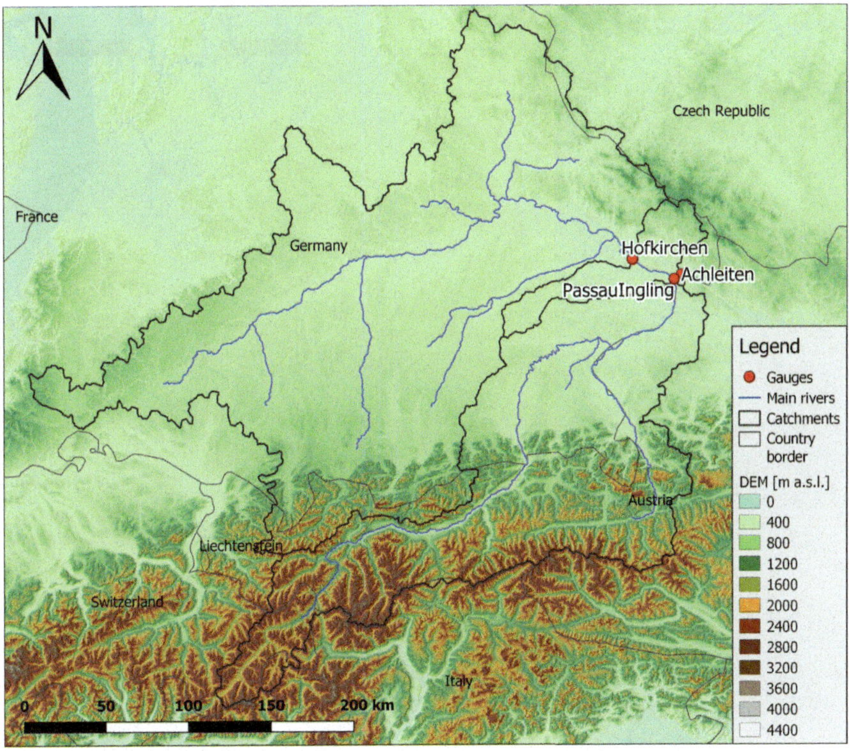

Fig. A.2 The Danube river basin until outlet gauge Achleiten

A.3 The Rhine River Basin

The Rhine river has its spring in the Swiss Alps, from where it flows through Germany to the Netherlands where it finally enters the North Sea. The largest parts of the basin are located in Switzerland, Germany, France, Austria and the Netherlands. The Rhine is one of the most important navigable rivers in Europe, being especially important for the chemical industry. Moreover, several hydroelectric power plants along the Rhine produce large amounts of electricity. The flood regime of the German part of the Rhine river is highly variable, as it is influenced by several factors, including precipitation, snowmelt, and the operation of upstream reservoirs. The Rhine river experiences both regular and extreme floods, with the highest frequency of floods occurring during the winter and spring seasons. Floods in the Rhine river can be caused by prolonged periods of heavy rainfall or by rapid snowmelt, and they can have significant impacts on communities, infrastructure, and the environment. The basin was affected by several large floods in the previous century. Two of the most disastrous floods took place in the winters of the years 1993 and 1995, where the old part of Cologne city was flooded almost completely two times within less than

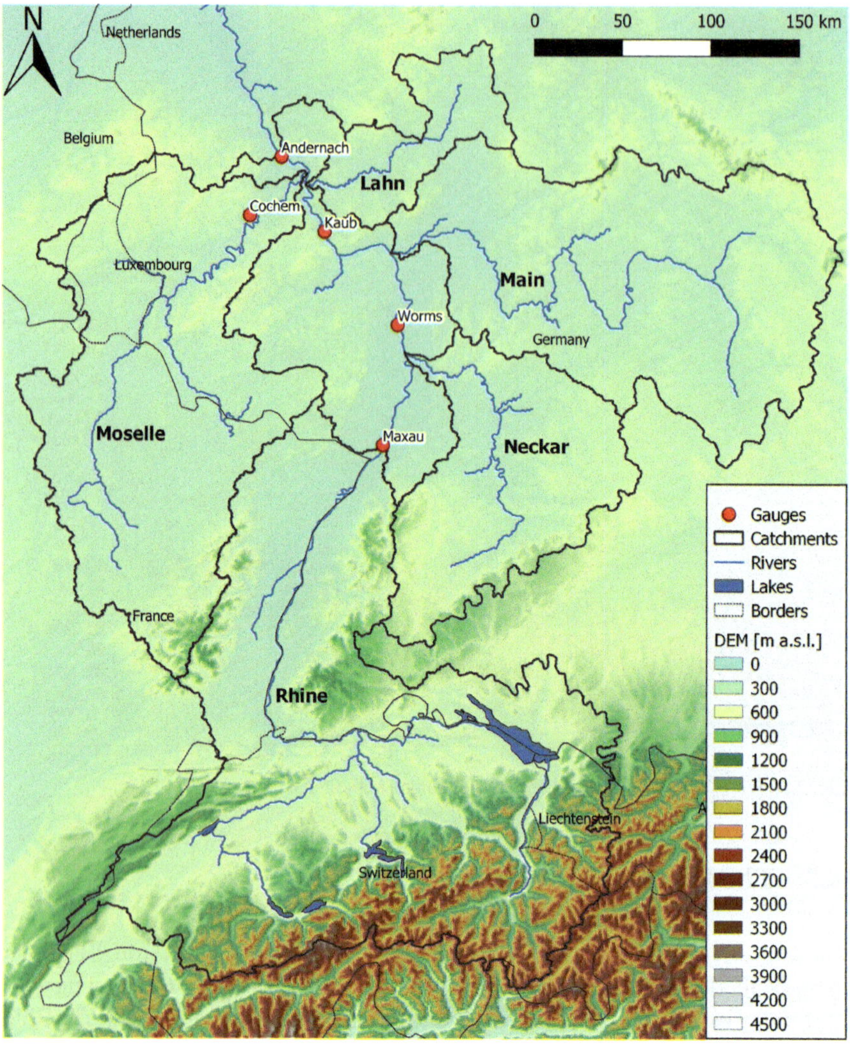

Fig. A.3 The Rhine river basin until outlet gauge Andernach

14 months. To protect people and industry, several flood protection systems have been installed along the river in the previous century. When aiming to understand the flood generation in the Rhine river basin, its complex structure has to be considered. With its several tributaries, some of them being more than 28,000 km^2 large, the role of flood-generation processes is changing seasonally as one travels downstream: in summer and spring, snowmelt and precipitation from the Alps is the main driver, while in winter precipitation from the uplands dominates the flow regime. Travelling downstream, the impact of the tributaries on the flood generation increases and the

Table A.2 Characteristics of the considered catchments for Rhine basin

Gauge	River	Catchment size (km^2)	Mean Elev. (m a.s.l.)	MQ (m^3/s)	Mean gradient (%)	Mean ann. precip. (mm)	River kilometre
Maxau	Rhine	50196	852	1250	5.15	1136	362.3
Worms	Rhine	68827	726	1410	4.12	1045	443.4
Kaub	Rhine	103729	594	1650	3.16	939	546.2
Cochem	Moselle	27088	343	313	1.25	865	51.6
Andernach	Rhine	139549	527	2030	2.69	913	613

impact of the Alps decreases. Tributaries which are know to affect the flood generation most are the Neckar river and the Moselle river, the latter being responsible for extremity of the 1993 and 1995 events. An especially critical region for floods therefore is the part of the Rhine located below the confluence of Moselle and Rhine. The gauges selected to represent the flood regime in the Rhine river therefore have been chosen according to their representativeness of the tributaries and the region downstream of the Moselle confluence, with the Andernach gauge as the gauge after confluence (Fig. A.3). The main characteristics of the selected gauges are listed in Table A.2. The joint observation period for discharge covers the years 1931–2013.

References

Fischer S, Schumann AH (2021a) Regionalisation of flood frequencies based on flood type-specific mixture distributions. J Hydrol X 13:100107

Fischer S, Schumann AH (2021b) Multivariate flood frequency analysis in large river basins considering tributary impacts and flood types. Water Resour Res 57:e2020WR029029

The manufacturer's authorised representative in the EU is Springer Nature Customer Service Centre GmbH, Europaplatz 3, 69115 Heidelberg, Germany. If you have any concerns regarding our products, please contact ProductSafety@springernature.com

Printed and bound by CPI Group (UK) Ltd, Croydon, CR0 4YY
01/10/2026
02225902-0007